AF891553

RICHARD LYDEKKER

CATALOGUE
OF THE
FOSSIL REPTILIA
AND
AMPHIBIA
IN THE
BRITISH MUSEUM
(NATURAL HISTORY)

PART IV

CONTAINING
THE ORDERS ANOMODONTIA, ECAUDATA,
CAUDATA, AND LABYRINTHODONTIA;
AND SUPPLEMENT

Elibron Classics
www.elibron.com

Elibron Classics series.

© 2006 Adamant Media Corporation.

ISBN 0-543-87019-7 (paperback)
ISBN 0-543-87018-9 (hardcover)

This Elibron Classics Replica Edition is an unabridged facsimile of the edition published in 1890, London.

Elibron and Elibron Classics are trademarks of Adamant Media Corporation. All rights reserved.

This book is an accurate reproduction of the original. Any marks, names, colophons, imprints, logos or other symbols or identifiers that appear on or in this book, except for those of Adamant Media Corporation and BookSurge, LLC, are used only for historical reference and accuracy and are not meant to designate origin or imply any sponsorship by or license from any third party.

CATALOGUE

OF

FOSSIL REPTILIA

AND

AMPHIBIA.

PART IV.

CATALOGUE

OF THE

FOSSIL REPTILIA

AND

AMPHIBIA

IN THE

BRITISH MUSEUM

(NATURAL HISTORY),

CROMWELL ROAD, S.W.

PART IV.

CONTAINING

THE ORDERS ANOMODONTIA, ECAUDATA, CAUDATA, AND LABYRINTHODONTIA; AND SUPPLEMENT.

BY

RICHARD LYDEKKER, B.A., F.G.S., ETC.

LONDON:
PRINTED BY ORDER OF THE TRUSTEES.

SOLD BY
LONGMANS & Co., 39 PATERNOSTER ROW;
B. QUARITCH, 15 PICCADILLY; ASHER & Co., 13 BEDFORD STREET, COVENT GARDEN;
KEGAN PAUL, TRENCH, TRÜBNER & Co., 57 LUDGATE HILL;
AND AT THE
BRITISH MUSEUM (NATURAL HISTORY), CROMWELL ROAD, S.W.

1890.

PRINTED BY TAYLOR AND FRANCIS,
RED LION COURT, FLEET STREET.

PREFACE.

This concluding part of the Catalogue of the Fossil Reptilia becomes of the highest importance to students of Vertebrate Palæontology as affording, in a collected form, the results of those recent researches which have been so successfully carried on, by Professors Credner, Fraas, Fritsch, Gaudry, Seeley, and others, concerning the structure and affinities of the Anomodontia and the Amphibia.

Splendid material for working out the anatomy of these oldest and most generalized types of Reptiles has of late years been obtained from the Carboniferous and Permian formations of France, Germany, Bohemia, South Africa, and also from Russia, India, and North America.

The series of South-African Anomodont Reptiles forms one of the most important features in this section of the Museum; indeed, it may be justly claimed to be unrivalled; many additions to the *Pariasauria* having been made by Prof. Seeley while this Catalogue was passing through the press.

The finest Labyrinthodonts discovered are preserved in the Museums of Leipsic, Stuttgart, and Prague; but the series of specimens of *Archegosaurus* from the Permian of Saarbrücken, and of

Loxomma from the Coal-Measures of England and Scotland, in this Museum are unsurpassed by any other. The remarkable genera *Placodus* and *Cyamodus* are represented by several valuable original specimens and by reproductions of von Meyer's types. Lastly, fine examples are to be seen in the cases of *Cryptobranchus scheuchzeri* and *C. tschudii* from the Miocene of Oeningen and Bonn.

HENRY WOODWARD.

Geological Department,
British Museum (N. H.),
25th April, 1890.

INTRODUCTION.

With the present Part the Catalogue of Fossil Reptilia and Amphibia in the Museum is completed.

The Supplement contains such Reptilian remains as have been acquired by the Museum since the previous Parts of the work were published, as well as those which had not been put in their proper serial position when the work was commenced, and consequently did not come under the author's notice in time to be recorded in the earlier parts. Certain modifications and amendments in classification are also recorded in the Supplement. I am again indebted to Mr. Davies for revising the proof-sheets; and my thanks are likewise due to Mr. J. C. Mansel-Pleydell for permission to use the woodcuts numbered 62–64, before their appearance in the 'Proc. Dorset Nat. Hist. Soc.,' for which they were drawn.

I may state that in the diagnosis of the Synaptosaurian branch given in Part II. the statement that sclerotic plates are absent applies only to the Sauropterygia.

The *Baber Collection*, mentioned for the first time in this Part, was purchased in 1889 from the executors of the late Mr. J. Baber, of Knightsbridge.

RICHARD LYDEKKER.

Harpenden,
April 3rd, 1890.

SYSTEMATIC INDEX.

[Existing species are denoted by an asterisk ; and doubtful species by two asterisks.]

	Page
Class **REPTILIA**	1
Ordinal Position Uncertain	1
Group PLACODONTIA	1
Family PLACODONTIDÆ	2
Placodus gigas	2
—— andriani	5
—— quinimolaris	6
**—— impressus	6
Specifically undetermined specimens	6
Cyamodus rostratus	7
—— muensteri	7
—— laticeps	8
—— tarnowitzensis	9
Pleurodus bicolor	9
Theromorous Branch	10
Order ANOMODONTIA	10
Suborder *PROCOLOPHONIA*	12
Family PROCOLOPHONIDÆ	13
Procolophon trigoniceps	13
—— minor	14
—— laticeps	15

Page

Suborder *DICYNODONTIA* 15

Family DICYNODONTIDÆ 16

Dicynodon lacerticeps 18
—— leoniceps 19
—— feliceps 20
—— pardiceps 21
**—— rectidens 23
**—— curvatus 23
—— testudiceps 24
—— tigriceps 24
**—— parvidens 26
Specifically undetermined specimens . . 26
Udenodon baini 29
—— strigiceps 30
—— megalops 30
—— prognathus 31
—— greyi 32
—— sp. 32
Ptychosiagum declive 34
—— latirostris 35
—— microtrema 36
—— murrayi 37
Specifically undetermined specimens . 40
—— orientale 41
Cistecephalus microrhinus 46
—— leptorhinus 47
—— planiceps 47
(?) —— arctatus 48
Specimens of uncertain Generic Position . 49
(*Ptychognathus* [?] *pusillus*) 49
(*Cirognathus cordylus*) 49
(*Eurycarpus oweni*) 50

Family uncertain 62

**Platypodosaurus robustus 62

Page

Family ENDOTHIODONTIDÆ 64

Endothiodon bathystoma 65

—— microps 65

Suborder *THERIODONTIA* 66

Family GALESAURIDÆ 67

Galesaurus planiceps 68

Scaloposaurus constrictus 71

Cynosuchus suppostus 71

Cynochampsa laniaria 73

Cynodraco serridens 74

Ælurosaurus felinus 75

—— sp. 76

—— curvimola 77

Lycosaurus pardalis 78

Hyorhynchus platyceps 80

Tigrisuchus simus 80

Generically undetermined specimens . . 81

Family TAPINOCEPHALIDÆ 81

Tapinocephalus atherstonei 83

Titanosuchus ferox 90

Generically undetermined specimens . . 92

(*Glaridodon*) 92

Family uncertain 95

Genus *non det.* 96

Genus *non det.* 96

Brithopus priscus 98

Deuterosaurus biarmicus 99

Genus *non det.* 100

Family DIADECTIDÆ 100

Empedias molaris 102

—— phaseolinus 104

Diadectes sideropelicus 104

Page

Family CLEPSYDROPIDÆ 104

Naosaurus cruciger 106
—— claviger 107
Embolophorus sp. 108
—— dolloverianus 109
Generically undetermined specimens . . 110

Incertæ sedis 111

Family GORGONOPIDÆ 111

Gorgonops torvus 111

Suborder *PARIASAURIA* 112

Family PARIASAURIDÆ 112

Anthodon serrarius 113
—— sp. 114
Pariasaurus serridens 114
—— bombidens 115
Specifically undetermined specimens . . 117
Propappus omocratus 119

Class **AMPHIBIA** 121

Order ECAUDATA 121

Section Firmisternia 121

Family RANIDÆ 121

Oxyglossus pusillus 122
Rana meriani 122
—— noeggerathi 123

Section Arcifera 124

Family LEPTODACTYLIDÆ 124

*Leptodactylus pentadactylus 124
*—— ocellatus 125
—— sp. 126
*Ceratophrys cornuta 127

Page

Family BUFONIDÆ 128

*Bufo melanostictus 128

Family DISCOGLOSSIDÆ 128

Latonia seyfriedi 129
—— gigantea 129

Family PALÆOBATRACHIDÆ 130

Palæobatrachus diluvianus 130
—— gigas 131
—— meyeri 131
—— bohemicus 132
—— luedeckei 132
—— sp. 134

Order CAUDATA 135

Family SALAMANDRIDÆ 135

Megalotriton filholi 135
Heliarchon furcillatus 136
Molge noachica 136
Genus *non det.* 137

Family AMPHIUMIDÆ 137

Cryptobranchus scheuchzeri 137
—— tschudii 139

Order LABYRINTHODONTIA 139

Suborder *LABYRINTHODONTIA VERA* 140

Family MASTODONSAURIDÆ 141

Mastodonsaurus giganteus. 142
—— granulosus 145
—— keuperinus 145
—— indicus 145
Capitosaurus nasutus 147
—— fronto 149
—— robustus 149

Family MASTODONSAURIDÆ (*continued*).

Page

Metoposaurus diagnosticus 153
Genus *non det.* 153
Trematosaurus brauni 155
Genus *non det.* 157
Genus *non det.* 157

Family ANTHRACOSAURIDÆ 157
Anthracosaurus russelli 158
Macromerium schwarzenbergi 160
—— bayeri 161
—— (?) simplex 161
—— scoticum 162
Loxomma allmani 164
Eosaurus acadianus 166

Family NYRANIIDÆ 167
Nyrania trachystoma 166
Ichthyerpetum bradleyæ 169

Family DENDRERPETIDÆ 170, 223
Dendrerpetum pyriticum 170
—— deprivatum 170
—— acadianum 223
—— oweni 223
Brachyops laticeps 171
Bothriceps australis. 172
—— huxleyi 172
—— (?) major 174
Micropholis granulata 174

Family DIPLOSPONDYLIDÆ 175
Cricotus heteroclitus 176

Family ARCHEGOSAURIDÆ 177
Archegosaurus decheni 178
Platyoposaurus stuckenbergi 183
Actinodon latirostris 184

Family ARCHEGOSAURIDE (*continued*).

	Page
Cochleosaurus bohemicus	186
**Gaudrya latistoma	186
Chelyosaurus vranyi	187
Sphenosaurus sternbergi	188
Sparagmites lacertinus	189
Trimerorhachis insignis	190
Family uncertain	191
Eryops megacephalus	191
—— africanus	193
Rhytidosteus capensis	194
Incertæ sedis	195
Pholidogaster pisciformis	195
Suborder *MICROSAURIA*	196
Family UROCORDYLIDÆ	196
Urocordylus wandesfordi	196
—— scalaris	197
Ceraterpetum galvani	197
—— crassum	198
Family uncertain	199
Lepterpetum, sp.	199
—— sp.	199
Family LIMNERPETIDÆ	199
Limnerpetum elegans	200
—— laticeps	200
—— obtusatum	200
Family HYLONOMIDÆ	201
Hylonomus longicostatus	201
—— lyelli	223
—— wymani	223
—— latidens	223
Seeleya pusilla	203
Smilerpetum aciedentatum	224
Hylerpetum dawsoni	225
—— longidentatum	225

Family HYLONOMIDÆ (*continued*).

Page

Ricnodon trachylepis 203
Orthopleurosaurus microscopicus . . . 203

Family MICROBRACHIDÆ 204
Microbrachis pelicani 204
—— mollis 205

Suborder *AISTOPODA* 205

Family DOLICHOSOMATIDÆ 205
Dolichosoma longissimum 205
—— (?) angustatum 206
Ophiderpetum brownriggi 206
—— granulosum 207
—— pectinatum 207
—— vicinum 207
—— corvini 208
—— ziegleriaum 208

Suborder *BRANCHIOSAURIA* 208

Family APATEONIDÆ 208
Melanerpetum pulcherrimum 209

Family PROTRITONIDÆ 210
Protriton petrolei 210
—— salamandroides 211

Family uncertain 212
Sparodus validus 212
—— crassidens 213
Dawsonia multidens 213

Subordinal Position uncertain 214
Anthracerpetum crassosteum 214
Lepidosaurus duffi 214
Fritschia curtidentata 225
Genus *non det.* 214

Page

ICHNITES 215

Chirosaurus barthi 215

—— stortonensis 216

Genus *non det.* (*Chirotherium minus*) . . 217

Genus *non det.* 217

Chelichnus duncani 218

Genus *non det.* (*Chirotherium reiteri*) . . 218

Saurichnites rittlerianus 219

—— perlatus 219

Macropterna divaricans 219

Brontozoum giganteum 220

—— sillimani 220

—— validum 220

Anisopus gracilis 221

Anomœpus intermedius 221

—— curvatus 221

—— gracillimus 221

Grallator cuneatus 222

Tridentipes elegantior 223

SUPPLEMENT.

Class **REPTILIA** 226

Order ORNITHOSAURIA 226

Family RHAMPHORHYNCHIDÆ 226

Scaphognathus purdoni 226

Rhamphorhynchus gemmingi 226

—— jessoni 226

Order CROCODILIA 227

Family CROCODILIDÆ 227

Generically undetermined specimens . . 227

Crocodilus spenceri 228

Tomistoma (?) macrorhynchus 228

Garialis (?) dixoni 228

Page

Family GONIOPHOLIDIDÆ 229

Goniopholis crassidens 229
—— minor 229

Family TELEOSAURIDÆ 230

Geosaurus maximus 230
Suchodus durobrivensis 231
Metriorhynchus superciliosum 232
Crocodilæmus jourdani 234
Steneosaurus brevidens 234

Order PARASUCHIA 235

Order DINOSAURIA 236

Suborder *SAUROPODA* 236

Family CETIOSAURIDÆ 236

Cardiodon rugulosus 236
Morosaurus brevis 237
Pleurocœlus valdensis 237

Family ATLANTOSAURIDÆ 239

Pelorosaurus conybeari 240
—— humerocristatus 241
—— manseli 241
—— leedsi 242
Bothriospondylus suffosus 242
Hoplosaurus armatus 243

Suborder *THEROPODA* 243

Family CŒLURIDÆ 243

Calamospondylus foxi 243

Family MEGALOSAURIDÆ 244

Megalosaurus dunkeri 244
—— oweni 245
Genus *non det.* 245

Page

Family ANCHISAURIDÆ 246

Thecodontosaurus platyodon 246
Massospondylus carinatus 248
—— (?) rawesi 249

Family uncertain 250

Arctosaurus osborni 251

Suborder *ORNITHOPODA* 251

Family STEGOSAURIDÆ 251

Stegosaurus armatus 252
—— hastiger 252
Euscelesaurus browni 252
Orinosaurus capensis 253

Family SCELIDOSAURIDÆ 254

Syngonosaurus macrocercus 254
Genus *non det.* 255

Family IGUANODONTIDÆ 256

Cryptodraco eumerus 256
Camptosaurus valdensis 257
—— prestwichi 258
—— leedsi 258
Iguanodon dawsoni 259
—— fittoni 260
—— hollingtoniensis 262
—— bernissartensis 265
—— mantelli 266

Order ICHTHYOPTERYGIA 266

Family ICHTHYOSAURIDÆ 266

Ophthalmosaurus icenicus 267
—— pleydelli 268
Ichthyosaurus thyreospondylus 270

Family ICHTHYOSAURIDÆ (*continued*).

	Page
Ichthyosaurus communis	270
—— intermedius	271
Temnodontosaurus platyodon	271
—— trigonodon	271
Order SAUROPTERYGIA	271
Family PLESIOSAURIDÆ	271
Pliosaurus brachydirus	271
—— ferox	272
Peloneustes evansi	273
Cimoliosaurus valdensis	274
—— portlandicus	274
—— richardsoni	276
Plesiosaurus dolichodirus	277

LIST OF WOODCUTS.

		Page
Fig. 1.	Cyamodus laticeps. *Cranium*	8
2.	Dicynodont. *Pectoral girdle*	16
3.	——. *Pelvic girdle*	17
4.	Dicynodon, sp. *Cranium*	18
5.	——. *Occiput*	28
6.	Ptychosiagum declive. *Occiput*	34
7.	—— murrayi. *Humerus*	38
8.	—— orientale. *Occiput*	41
9.	—— ——. *Pectoral girdle*	43
10.	—— ——. *Pelvic girdle*	44
11.	—— ——. *Vertebra*	44
12.	Cistecephalus microrhinus. *Cranium*	45
13.	Dicynodont. *Humerus*	53
14.	Galesaurus planiceps. *Skull*	69
15.	(?) Cynosuchus suppostus. *Mandible and innominate*	72
16.	Ælurosaurus felinus. *Skull*	76
17.	(?) Tapinocephalus atherstonei. *Pelvis*	82
18.	—— ——. *Vertebra*	84
19.	(?) Deuterosaurus biarmicus. *Tooth*	100
20.	Empedias molaris. *Tooth*	101
21.	—— ——. *Cranium and mandible*	102
22.	—— ——. *Cranium*	103
23.	Stereorhachis dominans. *Humerus*	105
24.	Naosaurus claviger. *Vertebra*	106
25.	Embolophorus dolloverianus. *Vertebra and intercentrum*	109
26.	(?) Propappus omocratus. *Innominate*	120
27.	(?) Bufo melanostictus. *Humerus*	128
28.	Cryptobranchus scheuchzeri. *Skeleton*	138
29.	Protriton salamandroides. *Skull*	140
30.	*Rhachitomous vertebra*	141

		Page
Fig. 31.	Mastodonsaurus giganteus. *Cranium*	143
32.	—— ——. *Cranium*	143
33.	—— indicus. *Thoracic plate*	146
34.	Capitosaurus robustus. *Cranium*	150
35.	Metoposaurus diagnosticus. *Cranium*	152
36.	Genus *non det.* *Jaw*	153
37.	Loxomma allmani. *Cranium*	163
38.	—— ——. *Tooth*	164
39.	Nyrania trachystoma. *Cranium*	167
40.	—— ——. *Cranium*	168
41.	Bothriceps huxleyi. *Cranium*	173
42.	Cricotus heteroclitus. *Scutes and cranium*	176
43.	Archegosaurus decheni. *Cranium*	177
44.	Actinodon latirostris. *Thoracic buckler*	183
45.	—— ——. *Cranium*	184
46.	Trimerorhachis insignis. *Vertebræ, &c.*	190
47.	Eryops megacephalus. *Vertebræ*	192
48.	Seeleya pusilla. *Skeleton*	202
49.	Melanerpetum pusillum. *Cranium*	209
50.	Protriton petrolei. *Skeleton*	211
51.	Pleurocœlus valdensis. *Tooth*	238
52.	—— nanus. *Vertebra*	238
53.	(?) Pelorosaurus conybeari. *Tooth*	240
54.	Calamospondylus foxi. *Vertebra*	244
55.	Genus *non det.* *Axis vertebra*	246
56.	Massospondylus hislopi. *Tooth*	248
57.	——(?) rawesi. *Tooth*	248
58.	—— carinatus. *Vertebra*	249
59.	Arctosaurus osborni. *Vertebra*	250
60.	Syngonosaurus macrocercus. *Vertebra*	255
61.	Camptosaurus leedsi. *Femur*	258
62.	Ophthalmosaurus icenicus. *Pectoral limb*	267
63.	—— pleydelli. *Humerus*	268
64.	—— ——. *Vertebra*	269
65.	Ichthyosaurus intermedius. *Pectoral limb*	270
66.	Cimoliosaurus richardsoni. *Pectoral limb*	276

ABBREVIATIONS OF SERIALS QUOTED IN THIS VOLUME

AND NOT MENTIONED IN THE LISTS GIVEN IN PARTS I., II., & III.

[Where not otherwise stated, the works are in 8vo.]

Ann. Hofmus. Wien.—Annalen des k.-k. naturhistorischen Hofmuseums. *Vienna.*

Bericht. Naturf. Aertzte.—Ämtliche Bericht deutscher Naturforscher und Aertzte.

Froriep's Neue Notizen.—Neue Notizen aus dem Gebiete der Natur- und Heilkunde. 4to. *Erfurt* and *Weimar.*

Jahrb. nat. Ver. Magdeburg.—Jahrbuch des naturwissenschaftlichen Vereins zu Magdeburg. *Magdeburg.*

Jahresber. nat. Ver. Halle.—Jahresbericht naturwissenschaftlichen Verein für Sachsen und Thüringen. *Halle.*

Jahresber. schles. Ges.—Jahresbericht der schlesischen Gesellschaft fur vaterländischen Cultur. 4to. *Breslau.*

Mem. Amer. Acad.—Memoirs of the American Academy of Arts and Sciences. 4to. *Boston.*

Mém. Soc. Hist. Nat. d'Autun.—Mémoires de la Société d'Histoire Naturelle d'Autun. *Autun.*

Mém. Soc. Neuchâtel.—Mémoires de la Société des Sciences Naturelles de Neuchâtel. *Neuchâtel.*

Nat. Hist. Rev.—The Natural History Review. *London.*

Nat. Hist. Trans. Northumb. and Durham.—Natural History Transactions of Northumberland, Durham, and Newcastle-on-Tyne. *London* and *Newcastle.*

Proc. Liverpool Geol. Soc.—Abstracts of the Proceedings of the Liverpool Geological Society. *Liverpool.*

Rep. Amer. Assoc.—Proceedings of the American Association for the Advancement of Science. *Various towns.*

Trans. Roy. Irish Acad.—Transactions of the Royal Irish Academy. 4to. *Dublin.*

Verh. nat. Ver. preuss. Rheinl.—Verhandlungen des naturhistorischen Vereins der preussischen Rheinlande und Westphalen. *Bonn.*

CATALOGUE

OF

FOSSIL REPTILIA

AND

AMPHIBIA.

Class **REPTILIA.**

Ordinal position uncertain.

Group PLACODONTIA.

This group is at present only known by the skull and teeth, and its ordinal position cannot be certainly determined without the evidence of the vertebræ or the bones of the pectoral or pelvic girdles or limbs. When the reptilian nature of *Placodus* was proved, it was referred by Owen [1] to the Sauropterygia. Subsequently, however, the same writer [2] appeared inclined to regard *Placodus* as allied to the Anomodontia; and Seeley [3] ranks the Placodontia as a suborder of Anomodontia. It may be observed that all the described vertebræ and bones of the pectoral and pelvic girdles from the Muschelkalk, except those of Dinosauria, appear to be of a Sauropterygian type.

[1] See 'Palæontology,' 2nd ed. p. 236 (1861).
[2] Quart. Journ. Geol. Soc. vol. xxxv. p. 563 (1879).
[3] Phil. Trans. for 1889, p. 292.

Skull with a deep temporal arch, apparently consisting of a coalesced upper and lower arcade[1]; and the palate completely closed by the union of the pterygoids with one another and with the basisphenoid[2]. No bony floor to the nasal passage, and the posterior nares consequently opening by slits on the surface of the palate. Dentition mainly in the form of pavement-like teeth, occurring both in the alveolar margins of the jaws and also on the surface of the palate; teeth apparently inserted in very shallow sockets. The skull has a parieto-squamosal (post-temporal) and a postorbital bar.

Family PLACODONTIDÆ.

The whole of the known forms may be included in this family. Skull more or less broad and depressed, with double nares, which may be approximated to the orbits. Upper teeth arranged in an outer or maxillary series of small ones, and an inner or palatine series of large ones; mandibular teeth in a single row. Anterior teeth, in some cases, of a prehensile type.

Genus **PLACODUS**, Agassiz[3].

Cranium longer than broad, with a distinct premaxillary rostrum carrying three chisel-like teeth on either side; premaxillæ separate; three palatine teeth of polygonal contour; and four or five maxillary teeth. Mandible with produced rostrum containing two chisel-like teeth on either side of the extremity.

Placodus gigas, Agassiz[4].

Syn. *Placodus pachygnathus*, Owen[5].
Placodus bathygnathus, Owen[6].
? *Placodus hypsiceps*, Meyer[7].

The type species. Of large size, with a vaulted cranium, in which there are four comparatively large maxillary teeth on either side.

[1] Shown in the imperfect skull figured in the 'Palæontographica,' vol. xi. pl. xxvii.

[2] This is a character common to the Chelonia and to the typical *Nothosauridæ* among the Sauropterygia.

[3] Recherches sur les Poissons Fossiles, vol. ii. pt. i. p. 15 (1833).

[4] *Loc. cit.*

[5] Phil. Trans. for 1858, p. 179 (1859).

[6] *Ibid.* p, 182.

[7] 'Palæontographica,' vol. xi. pt. iv. p. 199 (1863).

One of the undermentioned specimens suggests that *P. hypsiceps* is founded on a young cranium of this species. It will also be shown below that the mandibular rami described as *P. pachygnathus* and *P. bathygnathus* present no characters by which they can be distinguished from this form.

Hab. Europe (Germany).

41139. Cast of the imperfect cranium, showing the occipital region and the greater portion of the palate. The original is the type, and was obtained from the Muschelkalk of Baireuth, Bavaria. It is figured by Agassiz in his 'Recherches sur les Poissons Fossiles,' vol. ii. pl. lxx. fig. 14; and also by Meyer in the 'Palæontographica,' vol. xi. pl. xxv. fig. 1. *Purchased*, 1868.

41096. The crushed cranium; from the Muschelkalk of Baireuth.
(*Fig.*) Noticed by Meyer in the 'Neues Jahrb.' 1868, pp. 48–52; and described, and the occipital region figured, by Seeley in the 'Phil. Trans.' for 1889, p. 281, pl. xxiv. figs. 5, 6. This specimen closely accords with the more imperfect one figured in the 'Palæontographica,' vol. xi. pl. xxviii. With the exception of the second of the right maxillary series, all the teeth of the palate are preserved, but three of the premaxillary teeth are wanting; there is a minute palatal tooth in each maxilla. Seeley considers that this specimen indicates the absence of a basioccipital condyle; but it seems doubtful whether this is not due to its imperfection, since such a condyle is shown in the 'Palæontographica,' vol. xi. pl. xxvii. fig. 3. When entire this skull evidently had the vaulted contour characteristic of *P. hypsiceps*. *Purchased*, 1868.

33071. Fragment of the left side of the posterior half of a palate probably referable to a small individual of this species; from Baireuth. Some of the teeth have been reset, and the second maxillary one appears to be placed at right angles to its true position. *Purchased*, 1858.

35868. Part of the left side of the hinder portion of the palate of an immature individual; from Baireuth. This specimen shows the three palatine teeth in use, beneath the hindmost of which is seen the germ of a much larger replacing permanent tooth in its alveolus. The germ-tooth agrees in size with the hinder palatine teeth of the type specimen, but the teeth in use agree precisely with the palatine teeth of the type specimen of *P. hypsiceps* figured by Meyer in the 'Palæontographica,' vol. xi. pl. xxiv., and suggest that the latter merely belongs to a young individual of *P. gigas*. This is confirmed by the distinctness of the sutures in the latter specimen. *Purchased*, 1860.

R. 1642. Part of a right maxilla probably belonging to an immature (*Fig.*) individual of this species; from Baireuth. Figured by Owen in the 'Phil. Trans.' for 1858, pl. x. figs. 2–5. The last three maxillary teeth are preserved, beneath the first of which is seen the germ of a replacing tooth.
Purchased.

32781. Portion of a maxilla, showing sections of two teeth, with their replacing germs; from Baireuth. *Purchased*, 1854.

19677. The posterior portion of the left ramus of the mandible, in (*Fig.*) a flattened and imperfect condition; from Baireuth. The type of *P. bathygnathus*; figured by Owen in the 'Phil. Trans.' for 1858, pl. xi. figs. 1–3. The two hinder molariform teeth are in use, while beneath the anterior one there is the crown of a successional tooth in its alveolus. As remarked by Meyer in the 'Palæontographica,' vol. xi. p. 208, this specimen presents no characters by which it can be specifically distinguished from the mandible figured in plate xxxii. figs. 1, 2 of that volume, although the last tooth is somewhat smaller. The absence of a ledge on the outer side is due partly to a chipping away of this portion of the jaw and partly to its compression.
Purchased, 1845.

33070. The imperfect posterior portion of a similar left mandibular ramus, containing the last two teeth; from Baireuth. Here the external ledge is well-preserved.
Purchased, 1858.

R. 1641. An imperfect mandible of similar type; from Baireuth. (*Fig.*) Figured by Owen in the 'Phil. Trans.' for 1858, as the type of *P. pachygnathus*. On the left side the first and third molariform teeth, and on the right the first and second are shown. The crown of the second tooth is more nearly square than in the preceding specimens, and thereby resembles the corresponding tooth on the right side of the mandible figured by Meyer. The last tooth precisely resembles that of No. 19677. *Purchased.*

32781 a. Fragment of the left ramus of a rather larger mandible; from Baireuth. The three molariform teeth are preserved, the first being in its alveolus, and the third imperfect. *Purchased*, 1854.

Specimens from the Muschelkalk of Baireuth of which a considerable number probably belong to this species.

1323. A large palatine tooth, apparently the last of the left side, in matrix. *Purchased. About* 1836.

32781 c. A large imperfect palatine or mandibular tooth, in matrix. *Purchased*, 1854.

1332. A palatine or mandibular tooth of irregular contour, in matrix. *Purchased. About* 1836.

35680. A palatine or mandibular tooth, apparently the last of the left side, in matrix. *Purchased*, 1859.

32781 d. A palatine or mandibular tooth, in matrix. *Purchased*, 1854.

32781 e. A palatine or mandibular tooth, in matrix. This specimen may belong to *P. andriani*. *Purchased*, 1854.

32781 f. The crowns of a number of teeth belonging either to the premaxillæ or to the symphysis of the mandible. *Purchased*, 1854.

48204. The crowns of a number of anterior teeth; from Luneville. One of these specimens is of very large size, while others are interesting, as showing the summits more or less abraded by wear. *Purchased*, 1877.

Placodus andriani, Münster[1].

Syn. *Placodus bombidens*, Owen[2].

Smaller than the type species, with the palatine teeth relatively shorter antero-posteriorly and the parietal foramen very large.

The undermentioned mandibular ramus described as *P. bombidens* agrees in size with the cranium of the present species, and exhibits no characters by which it can be distinguished therefrom.

Hab. Europe (Germany).

[1] In Agassiz's 'Recherches sur les Poissons Fossiles,' vol. ii. pt. ii. p. 219 (1844).

[2] Phil. Trans. for 1858, p. 179 (1859).

48204 a. A second and third palatine tooth, agreeing in size with those of the skull figured by Meyer in the 'Palæontographica,' vol. xi. pl. xxx., and therefore probably belonging to this species; from the Muschelkalk of Baireuth, Bavaria. *Purchased*, 1877.

32781 b. A first palatine tooth, agreeing in relative size with the preceding; from Baireuth. *Purchased*, 1854.

R. 1643. (*Fig.*) The imperfect left ramus of the mandible, wanting the symphysial portion; from Baireuth. Figured by Owen in the 'Phil. Trans.' for 1858, pl. ix. figs. 3–6, as the type of *P. bombidens*. The three lateral teeth are shown, the first having a replacing tooth below it. This specimen agrees precisely in relative size with the cranium of *P. andriani*, and the size of the replacing tooth shows that it belonged to an adult individual. *Purchased*.

Placodus quinimolaris, Braun[1].

Of the approximate size of *P. gigas*, but with five maxillary teeth.

Hab. Europe (Germany).

Unrepresented in the Museum.

Placodus impressus, Agassiz[2].

Founded upon teeth from the Bunter (Lower Trias) of Zweibrücken, Bavaria.

Hab. Europe (Germany).

Unrepresented in the Museum.

Specifically undetermined Specimens from the Muschelkalk of Baireuth, some of which may be referable to the next genus.

28451. A palatine or mandibular tooth, in matrix. *Purchased*, 1853.

1330. A palatine or mandibular tooth, in matrix. *Purchased. About* 1836.

32781 g. Two palatine or mandibular teeth. *Purchased*, 1854.

32781 h. Two maxillary teeth, in matrix. *Purchased*, 1854.

1328–1329. Five small maxillary teeth. *Purchased. About* 1836.

1335. The crowns of three imperfect anterior teeth. *Purchased. About* 1836.

[1] Jahreshber. Lands- u. Gewerbschule z. Bayreuth for 1862–63, *teste* Meyer, See also 'Palæontographica,' vol. xi. p. 203.

[2] Recherches sur les Poissons Fossiles, vol. ii. pt. i. p. 15 (1833).

Genus **CYAMODUS**, Meyer[1].

Cranium (fig. 1) not longer than broad, with the premaxillæ fused together and not forming a distinct rostrum; typically two rounded premaxillary teeth on either side; two or three palatine teeth, of which the crowns are rounded, and the last is much the largest; two or three maxillary teeth. Mandible with a comparatively short triangular symphysis, which was not improbably edentulous.

In the skull No. R. 1644 (p. 9) it is probable that the prominence seen on the occiput is the stapes; while the groove in the zygomatic arch probably indicates the dual origin of this arcade.

Cyamodus rostratus (Münster[2]).

Syn. *Placodus rostratus*, Munster[3].

The type species. Of comparatively small size, with three palatine and two maxillary teeth; second palatine tooth very small.

Hab. Europe (Germany).

32781. The imperfect crown of a worn palatine tooth, which is apparently the third of either this or the next species; from the Muschelkalk of Baireuth, Bavaria. So far as can be seen this specimen resembles the third palatine tooth of the type cranium figured by Meyer in the 'Palæontographica,' vol. xi. pl. xxiii. *Purchased*, 1854.

Cyamodus muensteri (Agassiz[4]).

Syn. *Placodus muensteri*, Agassiz[5].

Distinguished from the type species by the presence of three maxillary teeth, which are placed relatively farther back, and by the proportionately larger size of the second palatine tooth.

Hab. Europe (Germany).

Unrepresented in the Museum, unless the specimen entered under the preceding specific heading belongs to it.

---

[1] 'Palæontographica,' vol. xi. pt. iv. p. 179 (1863).

[2] Beiträge zur Petrefactenkunde, vol. i. p. 119 (1839).—*Placodus.*

[3] *Loc. cit.*

[4] Recherches sur les Poissons Fossiles, vol. ii. pt. ii. p. 220 (1844).—*Placodus.*

[5] *Loc. cit.*

Cyamodus laticeps (Owen [1]).

Syn. *Placodus laticeps,* Owen [2].

Of large size; two palatine and three maxillary teeth, the last palatine tooth being very large; premaxillary teeth placed externally to the line of the maxillary teeth; posterior nares small.

Hab. Europe (Germany).

Fig. 1.

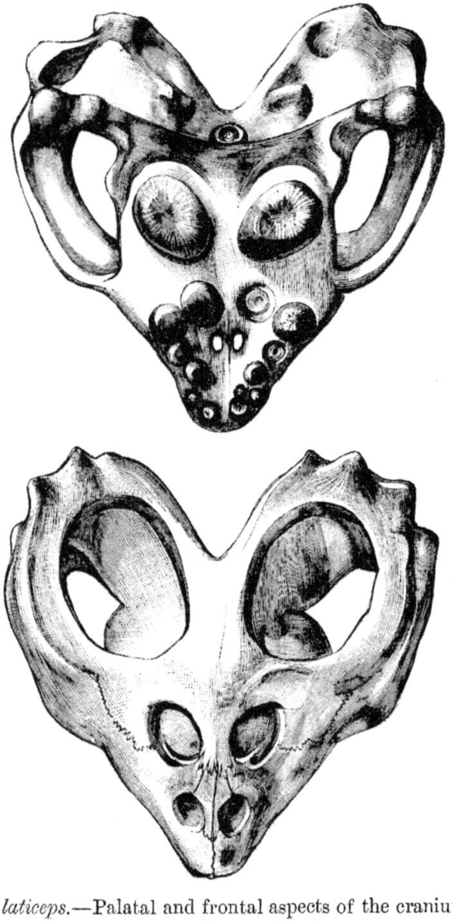

Cyamodus laticeps.—Palatal and frontal aspects of the cranium; from the Muschelkalk of Baireuth. $\frac{1}{3}$.

[1] Phil. Trans. for 1858, p. 169 (1859).—*Placodus.* [2] *Loc. cit.*

R. 1644. The imperfect cranium; from the Muschelkalk of Baireuth, Bavaria (fig. 1). The type specimen; figured by Owen in the 'Phil. Trans.' for 1858, pl. ix. figs. 1, 2, and pl. x. fig. 1. Since the original figures were drawn the matrix has been chiselled away, in order to exhibit the passage connecting the anterior and posterior nares.
(*Fig.*)

Purchased, 1857.

R. 1644 a. A restored plaster-model of the preceding specimen.
Made in the Museum, 1888.

35711. The imperfect mandible, with the oral surface buried in matrix; from Baireuth. This specimen, which shows successional teeth with the plane of their crowns placed nearly vertically in the ramus, agrees so well in relative size and contour with the cranium that there can be no hesitation in referring it to the same species. The symphysis is triangular, with a length of 0,070 (2·73 inches) and a width of 0,095 (3·73 inches); it may have had no anterior teeth. *Purchased*, 1860.

Cyamodus tarnowitzensis, Gürich [1].

Distinguished from the preceding species by the one premaxillary tooth being placed in the line of the three maxillary teeth, with which it forms a continuous series; and by the larger size of the posterior nares.

Hab. Europe (Siberia).

Unrepresented in the Museum.

Pleurodus bicolor, Gürich [2].

This name has been applied to teeth from the Muschelkalk of Silesia, which are considered to indicate a member of this group; the generic name is, however, preoccupied (*vide suprà*, Pt. I. p. 53) by *Pleurodon*.

---

[1] Zeitschr. deutsch. geol. Ges. vol. xxxvi. p. 136 (1884).

[2] Jahresber. Schles. Ges. for 1884, p. 220 (1885).

THEROMOROUS BRANCH.

The forms included in this branch or alliance are arranged in a single order, under which heading the characteristic features are noticed. Baur[1] regards the Dicynodontia (Anomodontia) and Theriodontia (Pelycosauria) as entitled to rank as distinct orders of a subclass Theromora (Theromorpha[2]). The leading features of the members of this branch are their resemblances on the one hand to the Labyrinthodont Amphibians and on the other to the Monotreme Mammals.

Order ANOMODONTIA.

Body lacertiform, the limbs adapted for walking, the tail comparatively short, and the head frequently also short. Skull with a fixed quadrate, which articulates with the pterygoid, either one or two temporal arcades, and large nasals; in the typical forms the pterygoids meeting in front of the basisphenoid, which they also join, but diverging anteriorly (fig. 4); palatines generally small, and placed internally to the pterygoids. When the temporal arcade consists of only a single chain of bones, it is, at least usually, a squamoso-maxillary one. Dentition thecodont, but the teeth frequently anchylosed to the jaws. Vertebræ amphicœlous and often notochordal; dorsals with longer or shorter transverse processes; anterior ribs usually double-headed, with the tubercle articulating with the transverse process of the arch, and the capitulum generally attached to a facet on the anterior border of the centrum. Pectoral girdle usually with a distinct precoracoid, suturally united with the coracoid and the acromial process of the scapula, and entering into the glenoid cavity. Clavicles, a T-shaped interclavicle, and apparently a sternum, present. Humerus (fig. 7) with distinct distal condyles, an entepicondylar (ulnar) foramen, and a more or less expanded delto-pectoral (radial) crest. Pelvis (fig. 3) with the component bones frequently anchylosed into an innominate, the ischium and pubis united by a long suture, in which there may be a small obturator foramen; body of ilium more or less in advance of acetabulum. Tarsus (when known) with one centrale, and the

[1] Journ. Morphol. vol. i. p. 102 (1887).

[2] This term was withdrawn by its founder Prof. Cope, on account of having been previously applied to the Amphibia Ecaudata.

phalangeals of the manus and pes 2, 3, 3, 3, 3 in number; and the structure of the foot approximating to a Mammalian type.

In the Dicynodontia and the typical Theriodontia the skull has both parieto-squamosal (post-temporal) and postorbital bars; but the former unites with the supraoccipital so as to leave no post-temporal fossa (fig. 6), and all the bones of the occiput tend to anchylose together into a large plate (fig. 5); this arrangement producing a remarkably Mammalian type of occiput[1]. Secondary posterior nares may be formed in the above-mentioned suborders by the development of palatal plates to floor the nasal passage.

The precoracoid appears to correspond with the bone in the pectoral girdle of Monotremes usually termed the epicoracoid. The acromial process of the scapula is strictly comparable with that of Monotremes, although in the latter it articulates only with the clavicle. The coracoid and precoracoid are relatively small in comparison with the scapula.

As a rule, abdominal ribs appear to have been absent[2], while there seem to be no traces of a dermal armour in most forms. Sclerotic plates may be developed in the orbit. In the cervical region it appears that, as a general rule, the ribs articulate to one process on the centrum, and to another on the arch, as in Crocodiles. In most of the dorsal vertebræ the capitular articulation usually forms a distinct facet on the anterior border of the centrum[3], but in some American forms there is no distinct facet, and occasionally all the ribs seem to have had single heads; in *Embolophorus* the capitular articulation is placed on the intercentrum, from which it appears to have been transferred to the anterior border of the centrum in other forms by the diminished size and final loss of the intercentrum.

[1] It may be observed that in the occiput of the Dicynodonts (figs. 5, 6) Owen considered that the exoccipital met above the foramen magnum, as in Crocodiles, and that the bone marked *ip* in fig. 6 represented the supraoccipital. Huxley ('Quart. Journ. Geol. Soc' vol. xv. pp. 652–653) pointed out that the supraoccipital is represented in the upper part of the compound occipital plate, forming the upper border of the foramen. The three bones marked *ip* and *pa* in fig. 6 were regarded by Huxley as the parietals; but the distinctness of the median element was indicated by Seeley in the 'Phil. Trans.' for 1889, pp. 230–235, where it is identified with the interparietal. The same writer would regard the bones here termed parietals as consisting of three elements, the part marked *pa* in fig. 6 being regarded as probably an epiotic. The correctness of Huxley's interpretation of the supraoccipital is shown by *Cistecephalus* (fig 12), where the supraoccipital, exoccipitals, and basioccipital remain separate.

[2] Present in *Theropleura*, Cope.

[3] Cope, Proc. Amer. Assoc. vol. xxxiii. p. 475 (1885) remarks on the approximation to a Mammalian type presented by this mode of articulation.

A few words are advisable to justify the sense in which the term Anomodontia is employed. In his 'Palæontology'[1] Owen made the order Anomodontia to include the "families" of the Dicynodontia, Cryptodontia (*Udenodon*), and Cynodontia (*Galesaurus*, &c.), although the definition of the order as thus constituted was incorrect. Subsequently[2] the name Anomodontia was restricted to the Bidentalia (Dicynodontia), Cryptodontia, and Endothiodontia; the Cynodontia being raised to the rank of an order under the name of Theriodontia. Although, on account of the definition, there may be some doubt whether the inclusion of the Cynodontia in the Anomodontia was not due to an error, yet the fact that such a classification was published, together with the subsequent use of the latter term in this sense by later English writers, seems to justify its retention.

It may be mentioned in respect of the African Anomodonts, that the term Karoo system is taken to embrace the four divisions known as the Stormberg, Beaufort, Koonap, and Ecca beds, of which the latter are the lowest[3]. It is probable that the majority of the undermentioned African specimens are from the Beaufort beds, but when there is no decisive evidence on this point they are entered simply as the Karoo system.

Suborder *PROCOLOPHONIA*.

The one known genus of this group appears to present an approximation in several points to the Rhynchocephalia, in which order it was at one time placed by Seeley[4], although subsequently transferred to the Anomodontia[5]. The skull is characterized by the approximation of the quadrate to the postorbital bar, so as to leave no distinct temporal arcade, and by the roofing over of the temporal fossa. The pterygoids extend forwards to join the vomer, and thus widely separate the palatines; and the dentition is fully developed. Vertebræ notochordal. Interclavicle T-shaped, like that of the Rhynchocephalia.

[1] Second edition, pp. 255–270 (1861).
[2] Cat. Foss. Rept. S. Africa (1873).
[3] See Quart. Journ. Geol. Soc. vol. xxiii. pp. 142–144. In vol. xliv. p. 240, of the same serial, A. H. Green proposes to restrict the term Karoo to the beds underlying the Stormberg, thus making it approximately equivalent to the Beaufort beds of the earlier scheme.
[4] Quart. Journ. Geol. Soc. vol. xxxiv. p. 807 (1878).
[5] Proc. Roy. Soc. vol. xliv. p. 383 (1888).

Seeley (Phil. Trans. for 1889, p. 272, pl. ix. figs. 7–9) regards the large bone identified by Owen (Cat. Foss. Rept. S. Africa, pl. xx fig. 6) with the squamosal (27 in figure) as the quadratojugal. The extension of this bone behind the quadrate, which Seeley remarks is so peculiar in a quadratojugal, is characteristic of a squamosal (as in Chelonia), and it seems more probable that the bone in question represents a squamosal or supratemporal rather than a quadratojugal.

Family PROCOLOPHONIDÆ.

Dentition of a carnivorous type, but not differentiated into incisors, tusks, and cheek-teeth; marginal teeth completely anchylosed to the bone, and teeth borne upon the vomer and pterygoids.

Genus **PROCOLOPHON**, Owen [1].

The type and only described genus. Skull short, triangular, and somewhat depressed, with a straight posterior border; its general contour approximating somewhat to that of *Sphenodon*, but with a much shorter postorbital portion. Premaxillæ projecting in advance of mandibular symphysis, which is but loosely united.

Procolophon trigoniceps, Owen [2].

Syn. *Procolophon griersoni*, Seeley [3].

The type species. Of the approximate size of *Sphenodon punctatus*; skull somewhat convex, with a comparatively narrow interorbital bar, and a broad lateral plate of bone in the quadratic region.

Hab. South Africa.

R. 1726. The imperfect skull; from the Karoo system of Tafelberg.
(*Fig.*) The type specimen; figured by Owen in his 'Catalogue of the Fossil Reptilia of S. Africa,' pl. xx. figs. 4–7. From the imperfection of this specimen it was considered that the genus had a single narial aperture; the notch marked as the temporal fossa in the figure is really the notch at the side of the occiput. The skull figured by

[1] Cat. Foss. Rept. S. Africa, p. 25 (1876).
[2] *Loc. cit.*
[3] Quart. Journ. Geol. Soc. vol. xxxiv. p. 797 (1878).

Seeley in the 'Quart. Journ. Geol. Soc.' vol. xxxiv., pl. xxxii. figs. 1–3, as the type of *P. griersoni* does not appear to be specifically distinct from the present specimen; the parietal foramen being, if anything, larger than the latter. The narrowness of the bone figured as the quadrate is apparently due to imperfection.

Presented by W. G. Atherstone, Esq., M.D.

R. 794. The imperfect skull, much concealed by matrix; from Tafelberg. This specimen appears to be identical in size and contour with the preceding. *Purchased*, 1886.

R. 518. A smaller and less imperfect skull; from Tafelberg. This specimen cannot apparently be specifically distinguished from the type; neither is it apparent how the skull figured by Seeley in the 'Quart. Journ. Geol. Soc.' vol. xxxiv. pl. xxxii. figs. 7–8, as the type of *P. cuneiceps*, differs from this specimen.

By exchange with the Blomfontein Museum, 1884.

R. 795. An imperfect skull, with traces of other bones, much concealed by matrix; from Tafelberg. This specimen seems to agree exactly with the preceding. *Purchased*, 1886.

Procolophon minor, Owen[1].

Founded upon a skull which, if adult, indicates a smaller species than the preceding. This skull appears to be somewhat more depressed than that of *P. trigoniceps*, with a relatively wider interorbital bar, and without a lateral expansion of bone in the quadratic region.

Hab. South Africa.

R. 1727. (*Fig.*) The somewhat imperfect skull; from the Karoo system of the Tafelberg. The type; figured by Owen in his 'Catalogue of the Fossil Reptilia of S. Africa,' pl. xx. figs. 8–12. Since the specimen was figured part of the left side of the occiput has been cleared from matrix.

Presented by W. G. Atherstone, Esq., M.D., 1875.

[1] Cat. Foss. Rept. S. Africa, p. 26 (1876).

Procolophon laticeps, Seeley [1].

Of larger size than the type species, the skull being wider and more depressed, with a broader interorbital bar. It seems highly probable that this form may be the adult of the preceding, in which case the present name should be retained, since that of *P. minor* would be inappropriate.

Hab. South Africa.

49427. The imperfect skull; from the Karoo system of Tafelberg. Agrees closely with the type skull figured by Seeley in the 'Quart. Journ. Geol. Soc.' vol. xxxiv. pl. xxxii. figs. 4–6.
Presented by John Dunn, Esq.

SPECIFICALLY UNDETERMINED SPECIMENS.

R. 514. Fragment of rock showing the outer aspect of a right mandibular ramus; from the Karoo system of Vogelstruisfontein, Bethulie district.
Presented by Herr H. S. Viljoen, 1884.

R. 796. Fragment of sandstone showing the external aspect of part of a right mandibular ramus; from the Cape Colony.
Purchased, 1886

Suborder *DICYNODONTIA.*

Skull with a single squamoso-maxillary temporal arcade and secondary posterior nares; premaxillæ united to form a single beak-like bone; mandibular symphysis deep, laterally compressed, and anchylosed: a lateral vacuity in the mandibular ramus. Anterior branches of the pterygoids separated by the palatines and vomer. Never more than one pair of teeth in alveolar borders of the upper jaws, and none in those of the mandible. Vertebræ fully ossified and without intercentra; four or five in sacrum. Humerus much expanded at the extremities, with a narrow bar over the entepicondylar foramen, of which the distal aperture opens high up on the palmar aspect (fig. 13); no supinator flange on preaxial border of distal extremity. Pelvis (fig. 3) with an obturator foramen; the ilium expanded in an antero-posterior direction, and the expanded plate lying nearly parallel to the sacrum.

[1] Quart. Journ. Geol. Soc. vol. xxxiv. p. 801 (1878).

Family DICYNODONTIDÆ.

Intra-alveolar surfaces of palate and mandible without teeth; alveolar margins of jaws trenchant; nares lateral; temporal arcade very slender; posterior nares more backward than in the *Endothiodontidæ*, and the palatines less incurved.

The prefrontals are very small, and the lachrymal has its duct situated within the cavity of the orbit. The quadrate has no antero-posterior perforation.

Fig. 2.

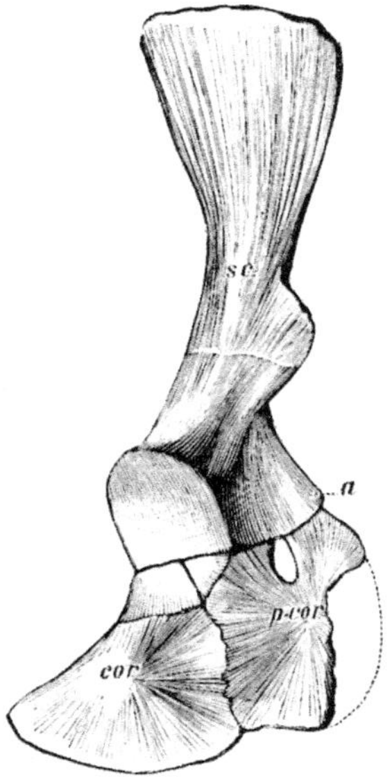

Dicynodont (?*Ptychosiagum*, sp.).—Dorsal aspect of the cartilage bones of the right side of the pectoral girdle; from the Karoo system of the Cape. $\frac{1}{2}$. *sc*, scapula; *a*, acromial process of ditto; *cor*, coracoid; *p.cor*, precoracoid.

Genus **DICYNODON**, Owen [1].

The type genus. Skull vaulted, with a pair of tusk-like maxillary teeth, growing from persistent pulps; a more or less narrow parietal bar; orbits of variable direction; and the palate carrying one ridge on the vomer, and a pair of ridges on the premaxillæ. General contour of cranium of a normal type, the parieto-frontal plane being continued by a more or less marked curve into that of the naso-premaxillary region; supratemporal fossæ more or less elongated; preorbital region comparatively short, with the nares near the

[1] Quart. Journ. Geol. Soc. vol. i. p. 318 (1845); and Trans. Geol. Soc. ser. 2, vol. vii. p. 59 (1845).

muzzle, and the facial portion of the moderately long premaxillæ devoid of lateral ridges; supraoccipital with a deep bar above foramen magnum; inferior aspect of mandibular symphysis rounded. Vertebral centra (when known) with deep terminal cups. Scapula (when known) with the acromial process separated by a notch from the glenoidal portion, and a long emargination on the preaxial border, which is but slightly reflected towards the dorsal aspect (see p. 21).

Fig. 3.

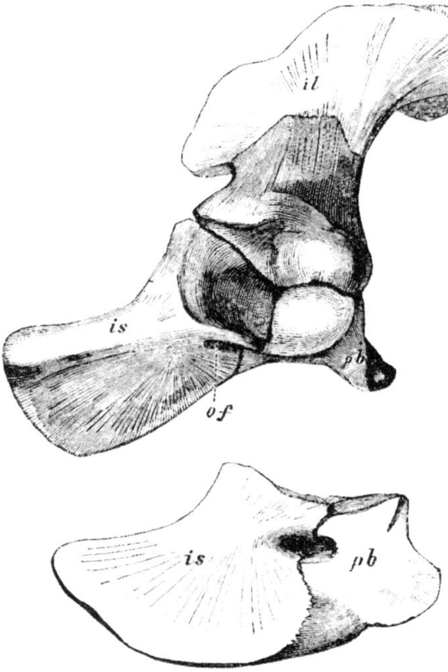

Dicynodont.—Dorsal aspect of the right side of the pelvic girdle; from the Karoo system of the Cape. ½. *il*, ilium; *is*, ischium; *pb*, pubis; *of*, obturator foramen. In the upper figure the pubis and ischium are represented in their natural position, while in the lower one their whole contour is shown.

A diagrammatic restoration of the palatal aspect of the Dicynodont skull is given in fig. 4[1].

It is probable that at least the majority of the species attained very large dimensions.

[1] In the restoration given by Seeley in the 'Phil. Trans.' for 1889, p. 246, fig. 3, the median processes of the pterygoids are represented as uniting and extending forwards to join the vomer.

Dicynodon lacerticeps, Owen[1].

The type species. Typically of comparatively small size. Cranium with produced muzzle and nearly straight fronto-parietal profile; orbits apparently directed partly frontally and partly laterally, and a comparatively narrow interorbital bar; parietal bar of moderate length and comparatively narrow; temporal fossæ of medium width and length; frontal aspect of bones of temporal arcade comparatively narrow; occiput nearly vertical; and root of tusk inclined forwards.

Hab. South Africa.

Fig. 4.

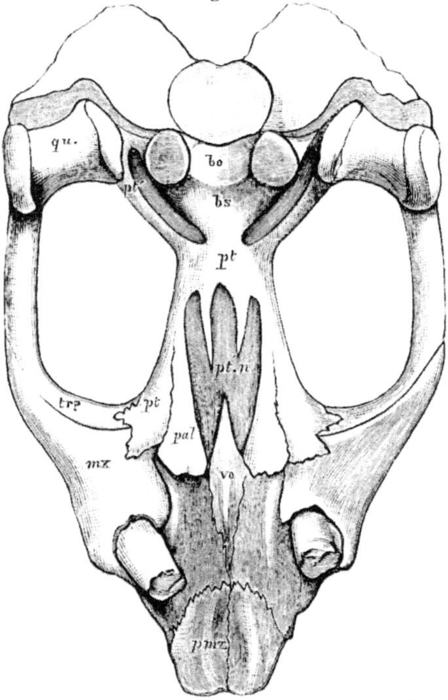

Dicynodon, sp.—Palatal aspect of the cranium; from the Karoo system of the Cape. ½. *pmx*, premaxilla; *mx*, maxilla; *vo*, vomer; *pal*, palatine; *pt*, pterygoid; *bs*, basisphenoid; *bo*, basioccipital; *qu*, quadrate; *tr*?, transverse bone (?); *pt.n*, posterior nares. The anterior portion is drawn from No. R. 860 (p. 27), and the posterior region from Owen's figure of *D. leoniceps*.

36233. The imperfect skull; from the Beaufort beds of the Karoo (*Fig.*) system of the Winterberg Peak, north of Fort Beaufort[2]. The type of the genus and species; figured by Owen in

[1] Trans. Geol. Soc. ser. 2, vol. vii. p. 62 (1845).
[2] See the maps in 'Trans. Geol. Soc.' ser. 2, vol. vii. p. 55, and pl. ii.

the 'Trans. Geol. Soc.' ser. 2, vol. vii. pls. iii., iv., and also in his 'Catalogue of the Fossil Reptilia of South Africa,' pl. xxiii. The tusks are considerably damaged, and the borders of the orbits are broken, so that the orbital contour and the width of the interorbital bar cannot be accurately determined. The suture introduced into the occiput of the first figure as dividing the exoccipitals from the supra-occipital is really a fracture across the former.

Presented by A. G. Bain, Esq., 1853.

R. 859. The imperfect skull of a rather larger individual; from the Karoo system of the Gouph district near Beaufort West[1], south of the Nieuwveldt range. Of the cranium only the preorbital portion remains, but the mandible is nearly entire. The base of the tusk has the same forward inclination as in the type.

Purchased from T. Bain, Esq., 1880.

Dicynodon leoniceps, Owen[2].

Syn. (?) *Dicynodon recurvidens*, Owen[3].

Of very large dimensions. Apparently allied to *D. lacerticeps*, but the orbits directed mainly laterally, and not distinctly triangular in contour; interorbital bar very wide; the parietal bar long and narrow; and the temporal fossæ also elongated and narrow. The occiput seems to be inclined forwards.

If the young skull on which *D. recurvidens* was founded really belong to the present form there will be no question but that the latter cannot be the adult of *D. lacerticeps*.

Hab. South Africa.

47047. The imperfect cranium of an adult; from the Karoo system (*Fig.*) of the Gat river, on the southern flank of the Sneewberg range in the Graaf-Reinet district. The type; figured by Owen in his 'Catalogue of the Fossil Reptilia of S. Africa,' pls. xxiv., xxvi. The temporal arcades have been somewhat crushed inwardly, but the left orbit is nearly entire. The tusk approximates in its inclination to that of the type, but is perhaps somewhat less oblique. The larger and narrower parietal bar and temporal fossæ are well shown. The articular surface of the quadrate and the posterior portion of the palate are well displayed. The bones marked maxillæ (20) in Owen's figure (pl. xxvi.)

[1] Beaufort West is on the flank of the Nieuwveldt range, and must not be confounded with Fort Beaufort, which lies considerably to the eastward.

[2] Cat. Foss. Rept. S. Africa, p. 32 (1876).

[3] *Ibid.* p. 46.

are really the anterior portions of the pterygoids, the apparent division between the bars and the body of the pterygoids being due to fracture. The quadrate shows the absence of an antero-posterior perforation.

Presented by W. G. Atherstone, Esq., M.D., 1875.

47098. (*Fig.*) A small crushed and somewhat imperfect skull, which may apparently belong to a young individual of this species; from the Beaufort beds near Fort Beaufort. The type of *D. recurvidens*; figured by Owen in his 'Catalogue,' pl. lxix. figs. 3, 4. In its wide interorbital bar, long and narrow temporal fossæ, and elongated parietal bar this specimen agrees with *D. leoniceps* and differs from *D. lacerticeps*, and if referable to the former indicates its specific distinctness from the latter. The position of the tusk may be accounted for by its having been thrust backwards out of its socket.

Presented by A. G. Bain, Esq., 1853.

Dicynodon feliceps, Owen[1].

In the young skull the root of the tusk more nearly vertical than in the type of *D. lacerticeps*, and the muzzle shorter. In the older skulls probably referable to this species the orbits have a distinctly triangular contour.

Hab. South Africa.

47052. (*Fig.*) The imperfect skull of a young individual; from the Beaufort beds of the Karoo system near Fort Beaufort. The type specimen; figured by Owen in his 'Catalogue of the Fossil Reptilia of S. Africa,' pl. xliii. The external surface of the bone is lost. *Presented by A. G. Bain, Esq.*, 1853.

R. 857. A somewhat smaller imperfect skull; from the Karoo system of the Gouph district, near Beaufort West, south of the Nieuwveldt range. The contour is precisely the same as in the preceding specimen.

Purchased from T. Bain, Esq., 1880.

R. 858. The imperfect skull of a still smaller individual; from the Gouph district. *Purchased from T. Bain, Esq.*, 1880.

47099. A small imperfect skull probably referable to this form; from near Fort Beaufort. The tusk is not visible.

Presented by A. G. Bain, Esq., 1853.

47079. The anterior portion of a cranium not improbably referable to this species; from the Graaf-Reinet district, north-west

[1] Cat. Foss. Rept. S. Africa, p. 45 (1876).

of Fort Beaufort. This specimen indicates a somewhat larger individual than the type.

Presented by W. G. Atherstone, Esq., M.D., 1875.

47103. An imperfect cranium probably belonging to a nearly adult individual of this species; from the Cape Colony. This specimen, which comprises the middle region, apparently accords well in contour with the type. The orbit is distinctly triangular. The narrowness of the parietal bar at once distinguishes this specimen from *D. leoniceps*.

Presented by A. G. Bain, Esq., 1853.

47056. The anterior extremity of a larger cranium, specifically (*Fig.*) identical with the preceding specimen; from the Gat river, Graaf-Reinet district. Noticed by Owen in his 'Catalogue of the Fossil Reptilia of S. Africa,' p. 47 (No. 80), as *D. leoniceps*; and the occipital region figured by Seeley in the 'Phil. Trans.' for 1889, pl. x. fig. 3. In profile this specimen differs from the type of that species by the regularly triangular form of the orbit, the shorter and more deflected muzzle, and the more nearly vertical direction of the tusks; in all of which respects it accords with the type of the present species. There is no reasonable doubt that it really belongs to the latter, and it is certainly specifically distinct from the type of *D. leoniceps*.

Presented by W. G. Atherstone, Esq., M.D., 1875.

47056*. A mass of rock showing several bones belonging to the (*Fig.*) same individual as the preceding specimen. Portions of dorsal vertebræ, and the ventral aspect of the right scapula and a fragment of the precoracoid are shown. The scapula and precoracoid are figured by Owen in his 'Catalogue,' pl. lxx. fig. 1. In the scapula the acromial process (*e* of Owen's figure) is separated by a distinct notch from the glenoidal region; and the preaxial border of the bone has a long emargination, above which there is only a slight torsion towards the dorsal aspect.

Presented by W. G. Atherstone, Esq., M.D., 1875.

Dicynodon pardiceps, Owen[1].

Syn. (?) *Dicynodon dubius*, Owen[2].

Cranium with a comparatively short and abruptly deflected muzzle, concave fronto-parietal profile, and the orbits triangular and

[1] Cat Foss. Rept. S. Africa, p. 42 (1876).
[2] *Ibid.* p. 46.

directed somewhat frontally; parietal bar long and narrow, with large foramen; temporal fossæ of moderate width, and the bone of the temporal arcade comparatively narrow; occiput nearly vertical; tusks directed mainly downwards. On the palate the premaxillary ridges are inclined downwards and forwards.

The young skull which was made the type of the provisional species *D. dubius* agrees so well in characters with the type of the present species that it is probably not separable.

Dicynodon simocephalus, Weithofer [1], is founded on an imperfect cranium having a marked resemblance to the type of *D. pardiceps*, but with a more inclined tusk.

Hab. South Africa.

47045. (*Fig.*) The nearly entire cranium; from the Beaufort beds of the Karoo system near Fort Beaufort. The type; figured by Owen in his 'Catalogue of the Fossil Reptilia of S. Africa,' pls. xxxviii., xxxix., and also by Seeley in the 'Phil. Trans.' for 1889, p.. 244, fig. 2. The bones marked maxillæ (20) in Owen's figure are the anterior prolongations of the pterygoids, with the palatines on their inner sides. On both sides the suture between the premaxilla and maxilla is distinctly shown; but the sulcus on the left side of the premaxilla, which in Seeley's figure is regarded as a suture forming the anterior border of an 'infranasal' bone, if anything definite, is probably the impression formed by the junction of two horny shields.

Presented by A. G. Bain, Esq., 1875.

47048. The imperfect and crushed muzzle of a larger cranium probably referable to this species; from the Karoo system of the Graaf-Reinet district, on the southern flank of the Sneewberg range. Noticed by Owen on p. 35 (No. 57) of his 'Catalogue,' as *D. leoniceps*. The contour of the muzzle appears to be exactly the same as in the type of the present species, but the alveolar margins of the jaws are broken away. The section of the tusk is slightly elliptical. *Presented by A. G. Bain, Esq.*, 1853.

47054. (*Fig.*) A young skull probably referable to this species; from the Graaf-Reinet district. The type of *D. dubius*, figured by Owen, *op. cit.* pl. lxix. figs. 1, 2. A fracture and dislocation have disturbed the relative positions of the occipital and parietal planes. The tusk, which has only just pierced the jaw, has an elliptical section. The general contour

[1] Ann. Hofmus. Wien, vol. iii. p. 1 (1888).

of this specimen resembles that of the type, although the parietal bar is relatively wider. The parietal foramen is very large, and although proportionately much smaller in the type is still comparatively large.

Presented by A. G. Bain, Esq., 1853.

** Dicynodon rectidens, Owen[1].

A doubtful species, apparently only separable from the preceding by the cylindrical section of the tusk.

Hab. South Africa.

47057. (*Fig.*) The imperfect skull; from the Beaufort beds of the Karoo system near Fort Beaufort. The type specimen; figured by Owen in his 'Catalogue of the Fossil Reptilia of S. Africa,' pl. xl. The whole of the upper surface of the cranium is wanting. If placed in the same position as the type skull of *D. pardiceps* (which is not the case in the figures), the contour of the two appears to be identical.

Presented by A. G. Bain, Esq., 1853.

47058. (*Fig.*) Mass of rock containing five imperfect dorsal vertebræ apparently associated with the preceding specimen. Figured by Owen in his 'Catalogue,' pl. xxxv. fig. 3. Two of the vertebral centra have been longitudinally bisected. *Presented by A. G. Bain, Esq.*, 1853.

47058 a. Mass of rock containing some imperfect cervical (?) vertebræ and some flat bones, probably belonging to the same individual as the preceding specimen.

Presented by A. G. Bain, Esq., 1853.

** Dicynodon curvatus, Owen[2].

Founded upon a very imperfect cranium which appears to approximate in contour to that of *D. pardiceps*.

Hab. South Africa.

R. 1655. Cast of the imperfect cranium. The original, which is the type, and was formerly in the collection of the late Prof. John Morris, was obtained from the Beaufort beds of the Karoo system at Cradock[3], on the upper part of the Great Fish river, north-west of Fort Beaufort.

Made in the Museum.

[1] Cat. Foss. Rept. S. Africa, p. 44 (1876).
[2] *Ibid.* p. 44 (1876).
[3] See Trans. Geol. Soc. ser. 2, vol. vii. p. 55.

Dicynodon testudiceps, Owen[1].

Founded upon a small and imperfect cranium characterized by the extreme shortness and abrupt deflection of the muzzle, the relatively large size of the nares, and the almost vertical direction of the ridges on the palatal aspect of the premaxillæ. The parietal region was probably of the type obtaining in the next species; but it does not appear that the latter is the adult of the present species, which is from a higher horizon.

Hab. South Africa.

47051. (*Fig.*) The anterior two-thirds of the cranium in an imperfect condition; from the Stormberg[2] beds of the Karoo system on the Modder tributary of the Orange river[3]. The type; described and figured by Owen in the 'Trans. Geol. Soc.' ser. 2, vol. vii. p. 71, pl. v., and also in his 'Catalogue of the Fossil Reptilia of S. Africa,' p. 46, pl. xliv. figs. 1, 2, where it is said to be from the Tarka (Tacka) prolongation of the Winterberg. In the former figure the palatines (21) are rightly determined, but in the latter they are not marked, and the anterior prolongations of the pterygoids are lettered as the maxillæ (20). A transverse section has been made through the right tusk.

Presented by A. G. Bain, Esq., 1853.

Dicynodon tigriceps, Owen[4].

Syn. *Dicynodon baini*, Owen[5].

Cranium with comparatively short and deflected muzzle, large triangular and lateral orbits, short and wide parietal bar, very wide temporal fossæ, upper part of occiput inclining forwards, bone of posterior portion of temporal arcade expanding into a broad horizontal plate, and the tusks directed mainly downwards.

The nasals form prominent projections above the comparatively small nares, and the postorbital bar is inclined backwards.

Hab. South Africa.

36235. (*Fig.*) The cranium of an adult individual, imperfect anteriorly; from the Beaufort[6] beds of the Karoo system of the Gonzia

[1] Trans. Geol. Soc. ser. 2, vol. vii. p 71 (1845).

[2] See Quart. Journ. Geol. Soc. vol. xxiii. p. 144.

[3] See Trans. Geol. Soc. ser. 2, vol. vii. p. 58.

[4] Trans. Geol. Soc. ser. 2, vol. vii. p. 233 (1856—read 1855).

[5] Cat. Foss. Rept. of S. Africa, p. 36 (1876). The name also occurs in Trans. Geol. Soc. ser. 2, vol. vii. p. 76 (1845), but with insufficient description.

[6] See Quart. Journ. Geol. Soc. vol. xxiii. p. 143.

river. The type; figured by Owen in the 'Trans. Geol. Soc.' ser. 2, vol. vii. pls. xxix.–xxxii., and also in his 'Catalogue of the Fossil Reptilia of S. Africa,' pl. xxx. fig. 2, pl. xxxi. fig. 1, pl. xxxii. fig. 1, and pl. xxxiii. In the view of the palate given in the former memoir the maxillæ are lettered palatines (21); while in the latter the bar comprising the anterior extremity of the pterygoids and palatine is lettered as the maxilla (20). The contour of the orbits appears to be natural, but the effect of distortion is shown by the left temporal fossa being wider than the right. *Presented by A. G. Bain, Esq.*, 1853.

36238. (*Fig.*) An imperfect adult cranium; from the Gonzia river. Figured by Owen in his 'Catalogue of the Fossil Reptilia of S. Africa,' pl. xxix., pl. xxx. fig. 1, pl. xxxi. fig. 2, and pl. xxxiii. fig. 2, as *D. baini*, of which it may be regarded as the type[1]. Although there are differences in the contour of the temporal fossæ and orbits, yet the present specimen agrees so closely in general characters with the preceding that there is every probability that the two belong to the same species, and that the differences are due to the effects of crushing and fracture. Thus it has been already mentioned that the preceding specimen is unsymmetrical; while in the present one it is evident that on the left side the prefrontal has been thrust laterally over the lachrymal region of the temporal arcade, by which the size of the orbit has been reduced, and the position of the postorbital bar altered. The slight lateral compression of the tusk does not seem to be a feature of much importance.

In Owen's 'Catalogue' the locality of this specimen is given as Fort Beaufort.

Presented by A. G. Bain, Esq., 1853.

R. 1660. (*Fig.*) The imperfect anterior portion of an adult cranium; from the Cape Colony. Figured by Seeley in the 'Phil. Trans.' for 1889, pl. xiii. This specimen is broken off immediately behind the parietal foramen, and the extremity of the muzzle is wanting. The boundaries of the various bones on the frontal aspect are well shown, the small size of the prefrontals and the position of the lachrymal

[1] There is no evidence to show that this cranium is specifically identical with the fragment of a maxilla described in the 'Trans. Geol. Soc.' *loc. cit.* under the same name.

duct being very conspicuous. The right tusk is preserved and has a cylindrical section.

Purchased from T. Bain, Esq.

** **Dicynodon parvidens,** Owen[1].

A very doubtful undescribed species based upon the following specimen.

Hab. South Africa.

47062. (*Fig.*) An imperfect small skull enveloped in matrix, which has been longitudinally and vertically bisected; from the Karoo system of Styl-Krantz, in the Sneewberg range. Figured by Owen in his 'Catalogue of the Fossil Reptilia of S. Africa,' pl. xxviii. figs. 3–4, and mentioned on p. 62 (no. 106) under the heading *Udenodon megalops*.

Presented by A. G. Bain, Esq., 1853.

Specifically Undetermined Specimens.

Some of the following Skulls are probably referable to Udenodon.

47341. (*Fig.*) An imperfect and laterally flattened skull, together with other bones, of a large *Dicynodon*, partially embedded in matrix; from the Karoo system of the Graaf-Reinet district on the southern flanks of the Sneewberg range. Figured by Owen in the 'Phil. Trans.' for 1862, pl. xxii., and in his 'Catalogue of the Fossil Reptilia of S. Africa,' pl. xxxiv. fig. 1, as *D. tigriceps*; and also by Seeley in the 'Phil. Trans.' for 1889, p. 222, fig. 1. This specimen seems to be too imperfect for specific determination. An imperfect dorsal vertebra is shown on one side of the block.

Presented by H.R.H. the Duke of Edinburgh, K.G., 1862.

36236. (*Fig.*) The crushed and imperfect symphysial extremity of the mandible of a large individual; from the Karoo system of Styl-Krantz, on the flank of the Sneewberg range. Figured by Owen in the 'Trans. Geol. Soc.' ser. 2, vol. vii. pl. xxx., as *D. tigriceps*, and noticed on p. 34 of his 'Catalogue' (as no. 64) under the same name. This specimen, which affords no evidence as to its species, shows the lateral vacuity very clearly.

Presented by A. G. Bain, Esq., 1853.

[1] Cat. Foss. Rept. S. Africa, pl. xxviii. (1876).

R. 1651. The imperfect mandible of a still larger individual; from the Karoo system of the Cape Colony. The extremity of the symphysis, the hinder portion of the left ramus, and the whole of the alveolar borders are wanting. The right articular bone is well preserved, and shows that the undetermined bone from the Gondwanas of India, figured by the writer in the 'Palæontologia Indica' (Mem. Geol. Surv. Ind.), ser. 4, vol. i. pt. 3, pl. ii. figs. 12, 13, is also an articular. The shortness and depth of the symphysis suggests reference to *D. tigriceps*.

Presented by W. G. Atherstone, Esq., M.D., 1875.

R. 860. (*Fig.*) The imperfect anterior extremity of the cranium of a comparatively small individual; from the Karoo system of the Gouph district, near Beaufort West, on the southern flank of the Nieuwveldt range. The palatal aspect is figured, with a restoration of the posterior half, in fig. 4 (p. 18). This specimen is important as being the only one clearly showing the small palatines lying on the inner sides of the anterior bars of the pterygoids; the articulation of the latter with the maxillæ is also exhibited, while portions of the maxillo-premaxillary suture are likewise shown.

Purchased from T. Bain, Esq., 1880.

R. 1671. (*Fig.*) A water-worn mass containing an imperfect skull; from the Cape Colony. This specimen, which has been bisected in a longitudinal and vertical plane, is figured by Seeley in the 'Phil. Trans.' for 1889, pl. ix. fig. 1. The cavity of the right tusk is exposed, and the contour of the mandibular symphysis is well preserved. *No history.*

36239. The lower portion of the occiput of an individual of the size of *D. tigriceps*, and not improbably referable to that species; from near Fort Beaufort. The condyle and the portion below is shown.

Presented by A. G. Bain, Esq., 1853.

R. 1021. (*Fig.*) The occiput of an individual of the size of the type of *Dicynodon lacerticeps*; from the Rhenosterberg. Figured by Seeley in the 'Phil. Trans.' for 1889, pl. x. figs. 1, 2; and also in woodcut fig. 5 of this volume. A trace of the suture between the basi- and exoccipitals can be detected on the external surface, but no other is visible.

Presented by Sir George Grey, K.C.B., 1858.

R. 1654. A rather smaller occiput, in a broken condition; from the Rhenosterberg.

Presented by Sir George Grey, K.C.B., 1858.

R. 861. The anterior portion of the mandible of a comparatively small individual; from the Karoo system of the Gouph district, near Beaufort West.

Purchased from T. Bain, Esq., 1880.

Fig. 5.

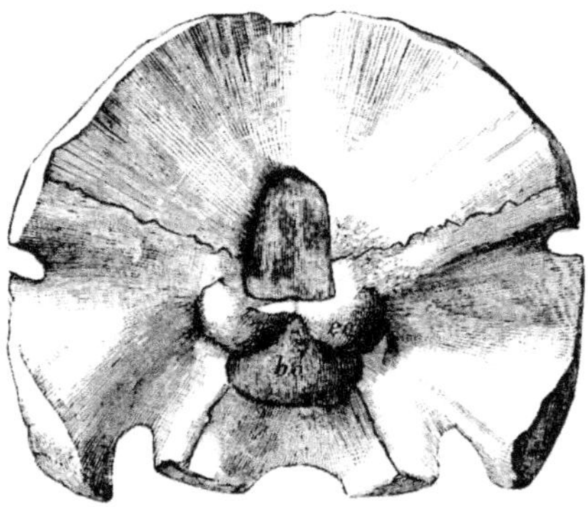

Dicynodon, sp.—Occipital plate of cranium; from the Karoo system of the Cape. ⅔. *bo*, basioccipital; *eo*, exoccipital.

47053. The anterior portion of a rather smaller mandible; from the Karoo system of the Winterberg range. Noticed by Owen on page 31 (no. 55) of his 'Catalogue of the Fossil Reptilia of S. Africa,' and provisionally referred to *D. lacerticeps*. *Presented by A. G. Bain, Esq.*, 1853.

R. 862. The anterior extremity of a mandible; from the Gouph district. *Purchased from T. Bain, Esq.*, 1880.

Genus **UDENODON**, Owen [1] (*ex* Bain).

Both jaws (at least in the adult) edentulous; maxillæ with an external caniniform ridge; palate with a vomerine ridge; general contour of skull as in *Dicynodon*.

It has been suggested that *Udenodon* may be founded upon female skulls of *Dicynodon*, but, as observed by Owen [2], there does not appear to be sufficient resemblance between the skulls of species

[1] Quart. Journ. Geol. Soc. vol. xvi. p. 54 (1860).—*Oudenodon*.

[2] Cat. Foss. Rept. S. Africa, p. 60.

of *Dicynodon* and *Udenodon* to justify this view. If the specimens described as *Platypodosaurus* prove to belong to *Udenodon*, its right to generic distinctness will be certain.

The specimens from the Upper Permian of Russia, described and figured by Trautschold in the 'Mém. Soc. Imp. Nat. Moscou,' vol. xv. pt. i. p. 35, pl. viii., as *Udenodon rugosus*, seem too imperfect to admit of definite determination.

Udenodon baini, Owen [1].

Syn. *Udenodon brevirostris*, Owen [2].
(?) *Udenodon raniceps*, Owen [3].

The type species. Of large size. Cranium with a short muzzle, in which the caniniform ridge is nearly vertical; orbits small and directed frontally; parietal and interorbital bars comparatively short and wide in the young; in the adult the parietal bar forming a sharp crest. Plane of occiput sloping backwards.

There appears every reason for regarding *U. brevirostris* as the adult of this form.

Hab. South Africa.

36232. (*Fig.*) The somewhat imperfect cranium of a young individual; from the Beaufort beds of the Karoo system, near Fort Beaufort. The type specimen; figured by Owen in the 'Quart. Journ. Geol. Soc.' vol. xvi. pl. i. fig. 1 (with a restoration of the mandible), and also in the 'Catalogue of the Fossil Reptilia of S. Africa,' pl. lvi. fig. 2, and pl. lx. The occiput and hinder portion of the palate are concealed by matrix; and the everted inner borders of the orbits are partly broken. *Presented by A. G. Bain, Esq.*, 1853.

R. 1649. (*Fig.*) The imperfect cranium of an adult individual; from the Karoo system of the Graaf-Reinet district, on the southern flank of the Sneewberg range. Figured by Owen in his 'Catalogue of the Fossil Reptilia of S. Africa,' pl. lviii., as *U. brevirostris*, of which it is the type. So far as can be seen, and allowing for difference of age, this specimen agrees in all essential characters with the type, the production of a sagittal crest being evidently due merely to age. The orbits appear to be relatively wider, but their borders have been extensively broken away. The canini-

[1] Quart. Journ. Geol Soc vol. xvi. p. 55 (1860).
[2] Cat. Foss Rept. S. Africa, p. 57 (1876).
[3] *Ibid.* p. 61.

form ridge of the maxilla has approximately the same subvertical position as in the type.

Presented by A. G. Bain, Esq., 1853.

R. 1649 a. (*Fig.*) The imperfect mandible; from Graaf-Reinet. Apparently associated with the preceding specimen; figured by Owen, *op. cit.* pl. lix., as *U. brevirostris.* The symphysis is well preserved, and exhibits the curved extremity of the beak. *Presented by A. G. Bain, Esq.*, 1853.

R. 1650. Two fragments of rock containing an imperfect skull which may belong to a young individual of this species; from the Beaufort beds of East London, at the mouth of the Buffalo river. The type of *U. raniceps*; noticed on p. 61 of Owen's 'Catalogue of the Fossil Reptilia of S. Africa.' The greater part of the cranium is restored in putty. The occipital condyle is formed almost entirely by the exoccipitals.

? *Presented by the Director of the Museum at Harrisberg.*

Udenodon strigiceps, Owen[1].

Syn. *Dicynodon* (?) *strigiceps*, Owen[2].

The muzzle still shorter and wider than in the type species, with its alveolar portion incurved; internarial and interorbital bars very broad; orbits extending immediately over the nares, with strongly everted anterior and lateral borders.—Imperfectly known; type specimen small.

Hab. South Africa.

47060. (*Fig.*) The anterior extremity of the cranium; from the Beaufort beds of the Karoo system in the Tarka branch of the Winterberg range[3]. The type specimen; figured by Owen in the 'Trans. Geol. Soc.' ser. 2, vol. vii. pl. vi. figs. 2, 3 (as *Dicynodon*), and in the 'Catalogue of Fossil Reptilia of S. Africa,' pl. xliv. fig. 2.

Presented by A. G. Bain, Esq., 1853.

Udenodon megalops, Owen[4].

Muzzle of cranium apparently more produced than in the type species, and the orbits relatively much larger; caniform ridge of maxilla nearly vertical; and the occipital plane also approaching

[1] Trans. Geol. Soc. ser. 2, vol. vii. p. 75 (1845).—*Dicynodon.*
[2] *Loc. cit.* [3] *Ibid.* p. 57.
[4] Cat. Foss. Rept. S. Africa, p. 62 (1876).

the vertical. The known specimens are of comparatively small size.

Hab. South Africa.

47061. (*Fig.*) The imperfect cranium; from the Beaufort beds of the Karoo system at Styl-Krantz, Sneewberg range. The type specimen; figured by Owen in his 'Catalogue of the Fossil Reptilia of S. Africa,' pl. lxiii. figs. 4, 5. The greater part of the palate is well preserved, but the aperture of the posterior nares is obscure. The bone marked palatine (20) in Owen's figure also includes the anterior extension of the pterygoid. *Presented by A. G. Bain, Esq.*, 1853.

R. 1646. The imperfect and laterally crushed cranium of a rather larger individual; from the Karoo system of the Cape. The nearly vertical plane of the occiput and the large size of the orbits are well shown.

Presented by A. G. Bain, Esq., 1853.

Udenodon prognathus, Owen[1].

Syn. *Udenodon magnus*, Owen[2].

The largest species. Cranium with the muzzle somewhat produced, and the caniniform ridge of the maxilla inclining obliquely forwards; orbits of moderate size, and to a great extent lateral; interorbital bar long and wide; parietal bar elongated; and the plane of the occiput nearly vertical.

There appear to be no characters by which the large cranium on which *U. magnus* was founded can be specifically distinguished from the type of this species; the length of the former cranium is 0,482 (19 inches).

Hab. South Africa.

47059. (*Fig.*) The imperfect skull of a young individual; from the Beaufort beds of the Karoo system at Mildenhalls, near Fort Beaufort. The type: described by Owen in the 'Quart. Journ. Geol. Soc.' vol. xvi. p. 56, and figured in his 'Catalogue of the Fossil Reptilia of S. Africa,' pl. lxi. The mandible is nearly entire; and the cranium is fairly well preserved, but has lost the occipito-squamosal region and the postorbital bar. The horizon of this specimen is stated[3] to be the "Fort Beaufort Grit," at the base of the Beaufort beds. *Presented by A. G. Bain, Esq.*, 1853.

[1] Quart. Journ. Geol. Soc. vol. xvi. p. 55 (1860).
[2] Cat. Foss. Rept. S. Africa, p. 56 (1876).
[3] Quart. Journ. Geol. Soc. vol. xxiii. p. 143.

R. 1720. The frontal region of a somewhat larger cranium apparently referable to this species; from the Karoo system of the Cape Colony. This specimen exhibits very clearly the marked eversion of the anterior border of the orbit and the prominent tuberosity above the nares, which form such striking features in the type skull.

Presented by A. G. Bain, Esq., 1853.

36252. (*Fig.*) The imperfect cranium of an adult individual; from the Beaufort beds of the Karoo system on the Brak river, near Fort Beaufort. The type of *U. magnus*; figured by Owen in his 'Catalogue,' pls. liv., lv., lvi. This specimen agrees so exactly in contour with the preceding that there is no reasonable doubt as to their specific identity. The parietal bar in the present specimen is indeed constricted into a sagittal crest; but this appears to be due solely to age.

Presented by A. G. Bain, Esq., 1853.

36237. (*Fig.*) The symphysial extremity of the mandible, associated with the preceding. Figured by Owen, *op. cit.* pl. lvii. The extremity of the beak has been cut and polished to show the absence of teeth.

Presented by A. G. Bain, Esq., 1853.

Udenodon greyi, Owen [1].

Known by a small and probably immature skull. Distinguished from the young of *U. prognathus* by the narrower parietal and interorbital bars.

Hab. South Africa.

36231. (*Fig.*) The imperfect and laterally crushed skull, showing on the left side several of the hyoid bones; from the Beaufort beds of the Karoo system of the Rhenosterberg branch of the Sneewberg range [2]. The type; figured by Owen in the 'Quart. Journ. Geol. Soc.' vol. xvi. pl. iii. fig. 5, and also in his 'Catalogue of the Fossil Reptilia of S. Africa,' pl. lxii., and pl. lxv. fig. 8.

Presented by Sir George Grey, K.C.B., 1858.

Udenodon (?), sp.

The undermentioned specimen, which is provisionally referred to this genus, is characterized by the extreme width of the parietal and interorbital bars.

Hab. South Africa.

[1] Quart. Journ. Geol. Soc. vol. xvi. p. 56 (1860). [2] *Ibid.* vol. xv. p. 94.

47098. The imperfect posterior portion of a small cranium; from the Karoo system of the Cape Colony. The specimen is broken off near the anterior border of the orbits.

Presented by A. G. Bain, Esq., 1853.

Genus **PTYCHOSIAGUM**, Lydekker[1].

Syn. *Ptychognathus*, Owen[2].

Cranium, so to speak, bent upon itself, with more or less strongly marked angles at the junction of the occipital with the parieto-frontal plane, and of the latter with the naso-premaxillary plane. Supratemporal fossæ wider than long; preorbital region much elongated, and the facial portion of the long premaxillæ with strongly marked lateral ridges, bordering a flat anterior surface: nares far behind muzzle; nasals short; orbits lateral; supraoccipital with only a very narrow bar over foramen magnum, above which is a notch for the reception of a descending process of the interparietal (fig. 6); inferior aspect of mandibular symphysis flattened, with lateral angulation.

The posterior extremities of the tusks are situated immediately below the nares; and the tusks extend downwards and forwards parallel with the naso-premaxillary plane, not leaving the sockets until they have passed beyond the level of the posterior end of the mandibular symphysis.

In some cases at least (*P. orientale*) the centra of the dorsal vertebræ long and much less deeply cupped than in *Dicynodon*. Scapula (when known) with the acromial process separated only by a groove on the ventral aspect from the glenoidal region, and a short notch-like emargination on the preaxial border, which above this notch is twisted towards the dorsal aspect.

None of the species attained dimensions at all approaching those reached by the larger species of *Dicynodon* and *Udenodon*.

Specimens of the pectoral girdle probably belonging to this genus are noticed on pages 52 and 53.

[1] In Nicholson and Lydekker's 'Manual of Palæontology,' 3rd ed. vol. ii. p. 1063 (1889).

[2] Quart Journ. Geol Soc. vol. xvi. p 49 (1860).—Preoccupied by Stimpson, Proc. Ac. Nat. Sci. Philad. 1858, p. 104, for a Crustacean genus.

Ptychosiagum declive (Owen[1]).

Syn. *Dicynodon (Ptychognathus) declivis*, Owen[2].
Ptychognathus declivis, Owen[3].

The type species. Cranium with the occipital and fronto-parietal planes forming an acute angle with one another, and the fronto-parietal and premaxillary planes a very obtuse one; fronto-parietal plane not convex antero-posteriorly; premaxillæ much elongated. Parietal ridges of interparietal separated by a considerable interval, which is deeply channelled; frontals carrying a pair of tubercles; maxillæ laterally emarginate for a long distance in front of orbits; occiput flattened above foramen magnum; muzzle of moderate width.

Hab. South Africa.

Fig. 6.

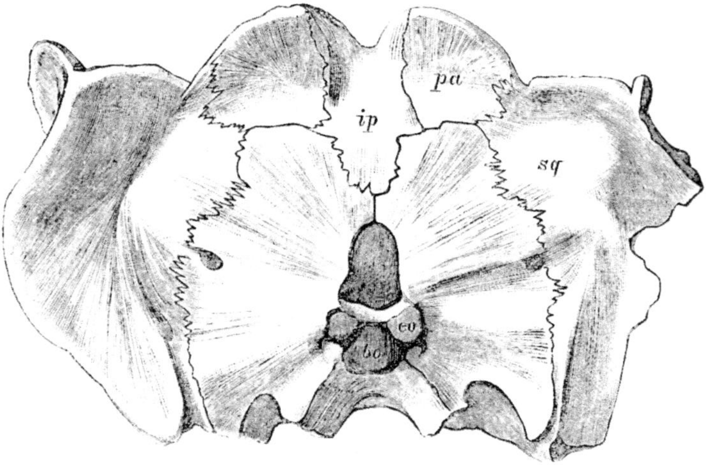

Ptychosiagum declive.—Occipital aspect of the cranium; from the Karoo system of the Cape. ½. *pa*, parietal; *ip*, interparietal; *sq*, squamosal; *bo*, basioccipital; *co*, exoccipital.

36221. The imperfect cranium; from the Beaufort beds of the (*Fig.*) Karoo system of the Rhenosterberg branch of the Sneewberg range[4]. The type; figured by Owen in the 'Quart. Journ. Geol. Soc.' vol. xvi. pl. i. figs. 3–5, and also in his

[1] Quart. Journ. Geol. Soc. vol. xvi. p. 49 (1860).—*Dicynodon (Ptychognathus)*. [2] *Loc. cit.*

[3] Cat. Foss. Rept. S. Africa, p. 48 (1876).

[4] See Quart. Journ. Geol. Soc. vol. xv. p. 194.

'Catalogue of the Fossil Reptilia of S. Africa,' pls. xlv., xlvi. fig. 1. The orbits have been somewhat flattened by crushing. The occipital region of this specimen is shown in woodcut, fig. 6.

Presented by Sir George Grey, K.C.B., 1858.

Ptychosiagum latirostris (Owen[1]).

Syn. *Ptychognathus latirostris*, Owen[2].
(?) *Ptychognathus alfredi*, Owen[3].
(?) *Ptychognathus depressus*, Owen[4].

Allied to the type species, but the parietal ridges of the interparietal separated by a wider interval, which is not distinctly channelled; frontals apparently without distinct tubercles: maxillæ laterally emarginate for a short distance in front of the orbits; muzzle very wide.

Hab. South Africa.

36222. (*Fig.*) The imperfect skull; from the Beaufort beds of the Karoo system of the Rhenosterberg branch of the Sneewberg range. The type; described by Owen in the 'Quart. Journ. Geol. Soc.' vol. xvi. p. 51, and figured by the same writer in the 'Phil. Trans.' for 1862, pl. xxi. fig. 1, and also in his 'Catalogue of the Fossil Reptilia of S. Africa,' pl. xlvi. fig. 2, pl. xlvii., and pl. xlviii. fig. 2. The shape of this specimen has been but little altered by crushing. The flattened inferior surface of the mandibular symphysis characteristic of the genus is well shown.

Presented by Sir George Grey, K.C.B., 1858.

47064. (*Fig.*) A flattened skull, apparently referable to this species; from the Karoo system of Styl-Krantz[5], on the flanks of the Sneewberg range. The type of *Ptychognathus depressus*; figured by Owen in his 'Catalogue,' pl. li. This specimen has been flattened from above downwards, its contour being precisely what might be expected if the preceding specimen had been subjected to the same process.

Presented by A. G. Bain, Esq., 1853.

[1] Quart. Journ. Geol. Soc. vol. xvi. p. 51 (1860).—*Ptychognathus.*

[2] *Loc. cit.*

[3] Phil. Trans. for 1862, p. 456.

[4] Cat. Foss. Rept. S. Africa, p. 53 (1876).—Preoccupied by Fischer, see p. 37.

[5] Some of the higher beds at Styl-Krantz may belong to the Stormberg group, see Quart. Journ. Geol. Soc. vol. xxiii. p. 143.

47342. The imperfect skull of a younger individual, apparently (*Fig.*) referable to this species; from the Beaufort beds of the Rhenosterberg. The type of *P. alfredi*; figured by Owen in the 'Phil. Trans.' for 1862, pls. xix., xx., and also in his 'Catalogue,' pl. 50. Allowing for the difference of age and its somewhat crushed condition, there appear to be no characters by which the present specimen can be specifically distinguished from the type. The absence of the frontal tubercles found in the type of *P. declive* is well shown.

Presented by H.R.H. the Duke of Edinburgh, K.G., 1862.

Ptychosiagum microtrema (Seeley[1]).

Syn. *Dicynodon microtrema*, Seeley[2].
Dicynodon (*Tropidostoma*) *dunni*, Seeley[3].

Imperfectly known. Apparently allied to the type species, but the parietal bar longer; the parietal ridges of the interparietal either closely approximated or united, and the occiput deeply depressed above the foramen magnum. The angle formed by the planes of the occiput and parieto-frontal region is more acute than in the type.

This species appears to have attained a larger size than either of the preceding.

Hab. South Africa.

R. 868. The imperfect occipital region of the cranium; from the (*Fig.*) Karoo system of the Gouph district, near Beaufort West, on the southern flank of the Nieuwveldt range. The type; figured by Seeley in the 'Phil. Trans.' for 1889, pl. xi. figs. 1, 2. This specimen agrees approximately in size with the type of *P. declive*, from which it is at once distinguished by the longer parietal bar and the extremely narrow channel between its two lateral elements. In *P. declive* the length of that bar is 0,036 (1·38 inches), whereas in the present specimen the corresponding length to the line of fracture is 0,060 (2·35 inches). Both quadrates are wanting.

Purchased from T. Bain, Esq., 1880.

R. 1662. The nearly entire occiput of a larger individual; from the Gouph district. The parietal ridges of the inter-

[1] Phil. Trans. for 1889, p. 228.—*Dicynodon.*
[2] *Loc. cit.*
[3] *Ibid.* p. 232.

parietal have come into contact. The deep depression in the supraoccipital above the foramen magnum is well shown; and the left quadrate is in position.

Purchased from T. Bain, Esq., 1880.

R. 866. (*Fig.*) The crushed occipital portion of a nearly similar cranium, with the anterior cervical vertebræ; from the Gouph district. Figured by Seeley, *op. cit.* pl. xii., as *Dicynodon (Tropidostoma) dunni*, of which it is the type. The two sides of the occiput have been crushed in towards the middle line. The right quadrate is entire. The cervical ribs articulate by one head to a process on the arch and by another on the centrum of the vertebræ. This specimen has received a lateral crush, so that the squamosals are approximated to one another and deflected backwards; the root of the temporal arcade is likewise thrust towards the parietal bar. Making allowance for this difference, the specimen apparently presents no characters by which it can be specifically distinguished from the preceding.

Purchased from T. Bain, Esq.

Ptychosiagum murrayi (Huxley[1]).

Syn. *Dicynodon murrayi*, Huxley[2].
Ptychognathus verticalis, Owen[3].
Ptychognathus boöpis, Owen[4].
Dicynodon copei, Seeley[5].

Cranium with the occipital and fronto-parietal planes forming a largely obtuse angle with one another, and the fronto-parietal and premaxillary planes a slightly obtuse one; fronto-parietal region convex antero-posteriorly, and thus tending to form a gradual passage from the occipital to the naso-premaxillary plane. Attains dimensions equal to the preceding species.

The differences between the specimens to which the above names have been applied appear to be solely due to age or to distortion by pressure and crushing. If such points of difference be regarded as specific, nearly every specimen will form the type of a species.

It is not improbable that *Ptychognathus depressus*, Fischer[6], is based upon a much flattened skull of this species.

[1] Quart. Journ. Geol. Soc. vol. xv. pp. 555 and 649 (1859).—*Dicynodon.*
[2] *Loc. cit.* [3] *Ibid* vol. xvi. p. 54 (1860).
[4] Cat. Foss. Rept. S. Africa, p. 50 (1876).
[5] Phil. Trans. for 1889, p. 241.
[6] Nouv. Arch. du Muséum, sér. 1, vol. vi. p. 179, pl. x. (1870).

One of the specimens referred to this species shows sclerotic plates in the orbit. The humerus [1], represented in woodcut, fig. 7, was found with the type remains, and has been referred by Huxley (Quart. Journ. Geol. Soc. vol. xv. pl. xxiii. fig. 3) to this species; if rightly determined, it indicates a small individual.

Hab. South Africa.

Fig. 7.

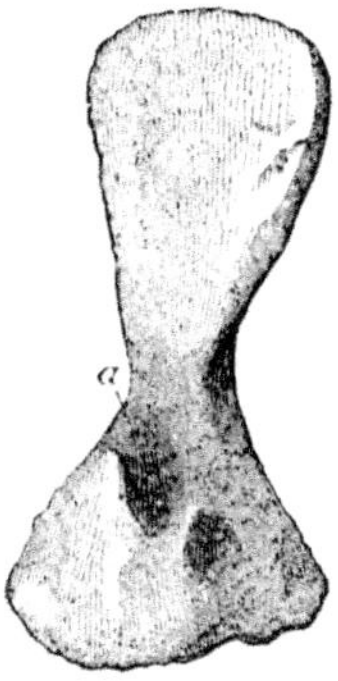

? *Ptychosiagum murrayi.*—Palmar aspect of the imperfect left humerus; from the Karoo system of Colesberg. ½ nat. size. *a*, entepicondylar foramen, with the bridge broken away. (From the 'Palæontologia Indica.')

R. 1291. The imperfect skull of an adult individual; from the Beaufort beds [2] of the Karoo system near Colesberg, to the north of the Sneewberg range. The type specimen; figured by Huxley in the 'Quart. Journ. Geol. Soc.' vol. xv. pl. xxiii. figs. 1, 2. The muzzle is wanting. In the figure of the occipital region it appears that the line separating the interparietal (parietal of Huxley) from the supraoccipitals is a fracture, and that in reality the interparietal gave off a process descending into the supraoccipital, as in other examples.

Presented by Prof. T. H. Huxley, 1888.

R. 1292. (*Fig.*) An imperfect skull, which has been cut into transverse segments; from near Colesberg. Described, and the sections figured by Huxley, *op. cit.* p. 654, pl. xxii. figs. 3–6. *Presented by Prof. T. H. Huxley*, 1888.

[1] Now in the Indian Museum, Calcutta.

[2] See Quart. Journ. Geol. Soc. vol. xxiii. p. 143.

R. 1294. The imperfect middle region of an immature cranium, showing impressions of the sclerotic plates in the orbit; from near Colesberg. Noticed by Huxley in the 'Quart. Journ. Geol. Soc.' vol. xv. p. 657. There appear to have been four or five bones in the sclerotic ring; and there is an azygous "Wormian bone" in front of the parietal foramen, as in No. 36253.

Presented by Prof. T. H. Huxley, 1888.

33224. (*Fig.*) The imperfect and somewhat distorted cranium of a half-grown individual; from the Beaufort beds of the Karoo system of the Rhenosterberg branch of the Sneewberg range. The type of *P. verticalis*; figured by Owen in the 'Quart. Journ. Geol. Soc.' vol. xvi. pl. i. fig. 2, and also in his 'Catalogue of the Fossil Reptilia of S. Africa,' pl. xlix. figs. 1, 2. This specimen agrees very closely in contour with the type, but the convexity of the frontal region is less marked, which may be due partly to crushing and partly to immaturity.

Presented by Sir George Grey, K.C.B., 1858.

36253. (*Fig.*) The imperfect skull of a slightly larger individual, attached to a mass of matrix containing vertebræ and other bones; from the Rhenosterberg. The type of *P. boopis*; figured by Owen in his 'Catalogue,' pl. xlviii. fig. 1, and pl. xlix. fig. 3. The orbit is not perceptibly larger than in the preceding specimen; the alleged larger size of the nares and their approximation to the orbit is a feature of the right side of the specimen due to imperfection, the left nares being precisely the same as in the preceding specimen. The contour of the frontal region is similar to that of the type. This specimen is important, as showing the composition of the temporal arcade; the squamosal (27 of Owen's figure) is seen overlying the upper part of the quadrate (=tympanic, 28, of Owen), and is itself connected both with the postfrontal and the maxilla, without any sign of a distinct jugal. Since the quadrate has no connecting chain with the maxilla, the one temporal arcade may be termed squamoso-maxillary, and thus corresponds with the Mammalian zygomatic arcade[1]. In

[1] This interpretation of the relations of the bone of the temporal arcade agrees with that given by Seeley in the 'Phil. Trans.' for 1889, p. 244, fig. 2.

advance of the parietal foramen there is a large azygous "Wormian bone," separating the posterior portion of the frontals; this feature is noticed by Huxley in the 'Palæontologia Indica' (Mem. Geol. Surv. Ind.), ser. iv. vol. i. pt. i. p. 11, where this specimen (then labelled *Ptychognathus declivis*) is referred to the present species.
Presented by Sir George Grey, K.C.B., 1858.

47073. The very imperfect and distorted skull of a slightly larger individual; from the Cape Colony.
Presented by A. G. Bain, Esq., 1853.

47074. (*Fig.*) An imperfect and distorted skull, agreeing closely in size with the preceding; from the Cape Colony. Figured by Seeley in the 'Phil. Trans.' for 1889, pl. xiv., as *Dicynodon copei*, of which it is the type. There are no characters by which this specimen can be specifically distinguished from the preceding. *Presented by A. G. Bain, Esq.*, 1853.

36223. An imperfect skull of the same size as the preceding specimen; from the Rhenosterberg. The frontal and left lateral aspects are fairly well shown.
Presented by Sir George Grey, K.C.B., 1858.

Specifically Undetermined Specimens from the Karoo system of the Cape Colony.

47075. A very imperfect skull; from the Cape Colony. The occipital region and the greater portion of the mandible are preserved. *Presented by A. G. Bain, Esq.*, 1853.

R. 1652. The imperfect occipital region of the cranium; from the Cape Colony. The quadrates are entire. *No history.*

R. 1652 a. The anterior portion of a mandible; from the Cape Colony. The flattening of the inferior surface of the symphysis characteristic of this genus is well shown; and part of the lateral vacuity is also exhibited. *No history.*

36284. The symphysial portion of a mandible; from the Rhenosterberg branch of the Sneewberg range.
Presented by Sir George Grey, K.C.B., 1858.

Ptychosiagum orientale (Huxley[1]).

Syn. *Dicynodon orientalis*, Huxley[2].
Ptychognathus orientalis, Lydekker[3].

Imperfectly known; of the approximate size of *P. declive*. This species was founded upon fragments of the skull and teeth, and imperfect bones of the limbs. The undermentioned specimen of the occipital region, which agrees in relative size with the larger limb-bones, and therefore probably belongs to the same species, affords the grounds of the generic reference.

The bones of the pectoral and pelvic girdles (figs. 9, 10) were at first incorrectly determined, and are redescribed by the writer in a paper published in the 'Rec. Geol. Surv. Ind.' vol. xxiii. pt. i. (1890). The difference in the size of the bones of the limb-girdles and limbs may be merely sexual or individual.

Hab. India.

Fig. 8.

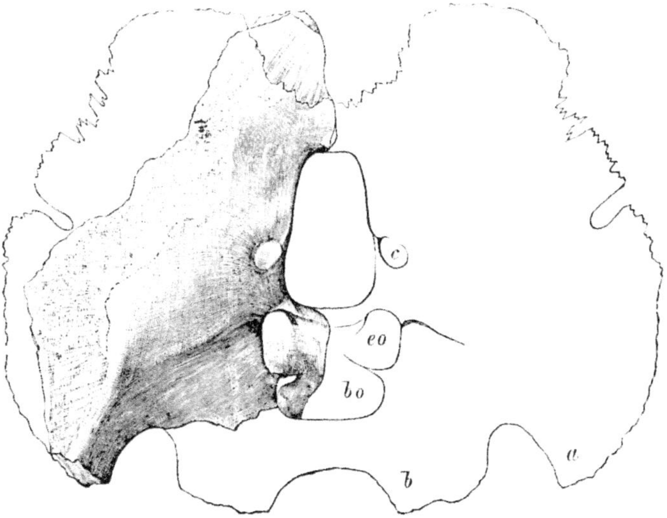

Ptychosiagum orientale.—The occiput, restored; from the Panchet group of Bengal. $\frac{1}{1}$. *bo*, basioccipital; *eo*, exoccipital; *c*, tubercle at junction of exoccipital with supraoccipital; *a*, *b*, descending processes; the imperfect bone above the bar over the foramen magnum is the interparietal. (From the 'Rec. Geol. Surv. Ind.')

---

[1] Palæontologia Indica (Mem. Geol. Surv. Ind.), ser. 4, vol. i. pt. i p. 8 (1865).—*Dicynodon.* [2] *Loc cit.*
[3] Rec. Geol. Surv. Ind. vol xx. p 68 (1887).

R. 1026. Cast of the imperfect left side of the occipital region. The original was obtained from the Panchet stage of the Lower Gondwana system at Panchet, near Raniganj, Bengal; and is preserved in the Indian Museum, Calcutta. It is figured by the writer in the 'Palæontologia Indica,' ser. 4, vol. i. pt. iii. pl. i. fig. 2. In the figure the upper part of the bone should have been inclined towards the middle line, thus narrowing the upper part of the foramen magnum, and the notch for the descending plate of the interparietal; an amended restoration is given in fig. 8. In the parietal region of a cranium figured by Huxley, *op. cit.* pt. i. pl. i. fig. 1, there is an azygous bone in advance of the parietal foramen, as in *P. murrayi*.

Made in the Museum, 1889.

R. 1026 a. Cast of the imperfect glenoidal extremity of the right scapula. The original was obtained from Panchet, and is preserved in the Indian Museum. It is figured by the writer in the 'Palæontologia Indica,' ser. 4, vol. i. pt. iii. pl. i. fig. 5. This specimen agrees in size with the scapula of the African pectoral girdle represented in fig. 2, and if, as is probable, that specimen belongs to *P. declive*, the present one would agree in relative size with the occiput above mentioned. *Made in the Museum*, 1889.

R. 1026 b. Cast of the imperfect right scapula. The original was obtained from Panchet, and is preserved in the Indian Museum. It is figured by Huxley, *op. cit.* pt. i. pl. v. fig. 5, as a Saurian scapula, and referred by the present writer, *op. cit.* pt. iii. p. 8, to the Dicynodonts. This specimen, which is much smaller than the preceding, is also figured by the writer in the 'Proc. Zool. Soc.' 1889, pl. lv. fig. 2, and in the 'Rec. Geol. Surv. Ind.' vol. xxiii. pt. i. (from which fig. 9 is reproduced), in conjunction with the next specimen. *Made in the Museum*, 1889.

R. 1026 c. Cast of the right coracoid. The original was obtained with the preceding, and is in the Indian Museum. It is figured by the writer in the 'Pal. Ind.' *op. cit.* pl. ii. fig. 8, from the ventral aspect, and provisionally referred to the Dinosaurian genus *Epicampodon* (*Ancistrodon*). It is refigured in the 'Rec. Geol. Surv. Ind.' vol. xxiii. pt. i., in conjunction with the scapula and a restoration of the pro-coracoid, the woodcut being reproduced in fig. 9. This specimen precisely resembles the larger coracoid of the pectoral girdle represented in fig. 2, and clearly shows the articular surface for the precoracoid.

Made in the Museum, 1889.

Fig. 9.

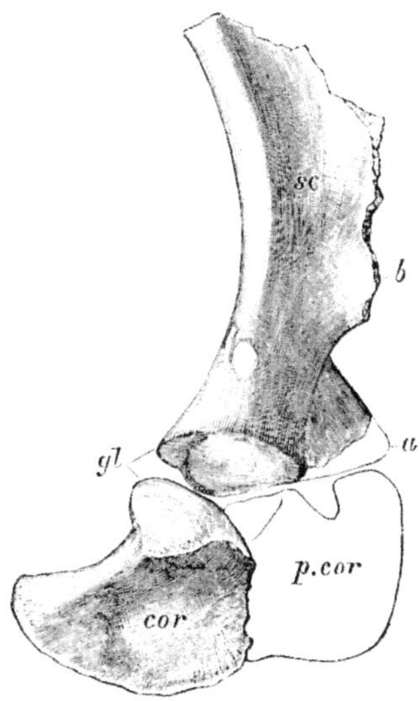

Ptychosiagum orientale.—The imperfect right side of the pectoral girdle; from the Panchet group of Bengal. $\frac{1}{1}$. *sc*, scapula; *a*, acromial process of do.; *b*, supra-acromial process of do.; *cor*, coracoid; *p.cor*, precoracoid; *gl*, glenoid cavity. (From the 'Rec. Geol. Surv. Ind.')

R. 1026 d. Cast of the right humerus, wanting the bridge over the entepicondylar foramen. The original was obtained with the preceding, and is figured by the writer in the 'Palæontologia Indica,' *op. cit.* pt. iii. pl. i. figs. 3, 6, 11.

Made in the Museum, 1889.

R. 1026 e. Three imperfect dorsal vertebræ; from the Panchet group. These specimens agree with those figured by Huxley in the memoir cited, one of those figures being reproduced in figure 11. The nearly flat terminal faces of the centra are well shown.

Presented by the Director of the Geological Survey of India, 1888.

Fig. 10.

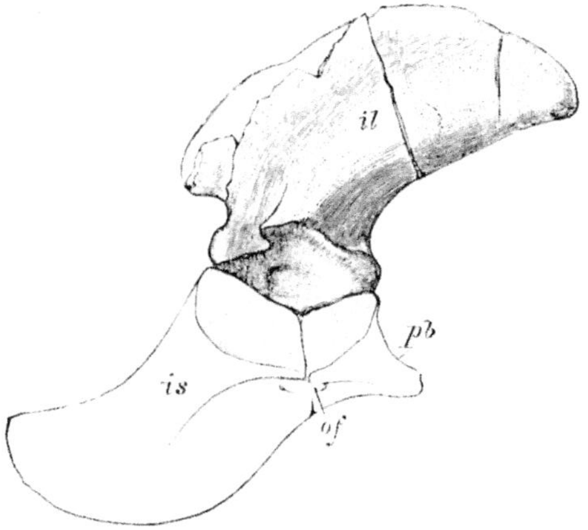

Ptychosiagum orientale.—The imperfect right ilium, with a restoration of the pubis and ischium; from the Panchet group of Bengal. ½. *il*, ilium; *pb*, pubis; *is*, ischium; *of*, obturator foramen. (From the 'Rec. Geol. Surv. Ind.')

R. 1026 f. The anterior portion of the centrum and right rib of a sacral vertebra; from Panchet. Resembles the specimen figured by Huxley, *op. cit.* pt. i. pl. ii. fig. 7.

Presented by the Director of the Geological Survey of India, 1888.

Fig. 11.

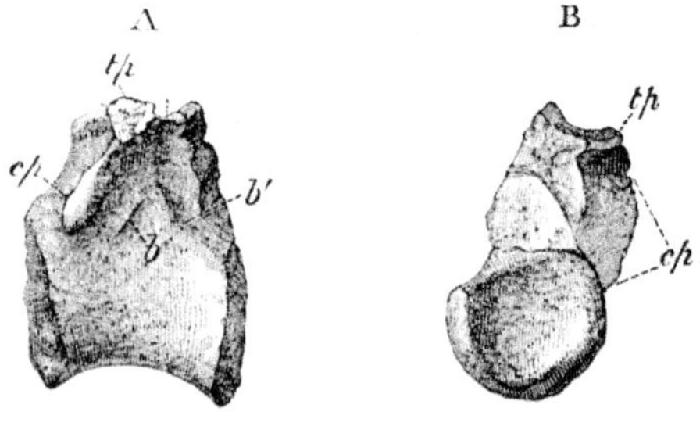

Ptychosiagum orientale.—An imperfect dorsal vertebra; from the Panchet group of Bengal. ⅟₁. A, left lateral; B, posterior aspect; *tp*, transverse process; *cp*, capitular facet; *b*, *b'*, ridges. (From the 'Palæontologia Indica.')

Genus **CISTECEPHALUS,** Owen[1].

Typically the skull much depressed, and in some instances, at least, with a pair of tusk-like maxillary teeth; a very wide parietal bar; and the orbits directed to a great extent frontally. Nares subterminal, and considerably in advance of orbits; præmaxillæ very short; nasals longer.

The width of the parietal appears always to exceed that of the interorbital bar. All the species appear to have been of comparatively small size. In the type species the contour of the skull recalls that of the Labyrinthodont *Brachyops*.

Fig. 12.

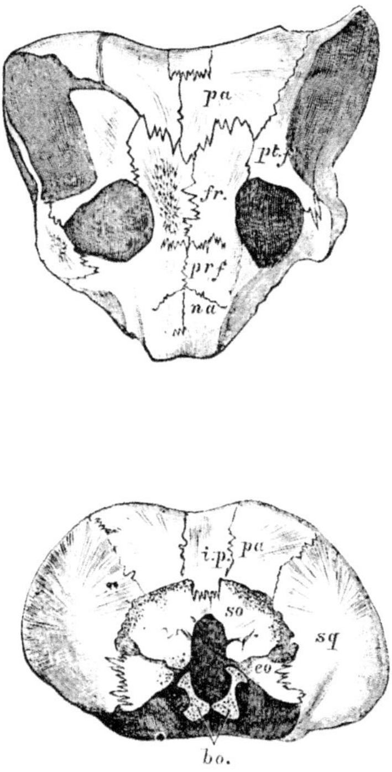

Cistecephalus microrhinus.—Frontal and occipital aspects of the imperfect cranium; from the Karoo system of the Cape. $\frac{2}{3}$. *pa*, parietal; *ip*, interparietal; *pt f*, postfrontal; *fr*, frontal; *prf*, prefrontal; *na*, nasal; *sq*, squamosal; *so*, supraoccipital; *co*, exoccipital; *bo*, section of basioccipital.

[1] Cat. Foss. Rept. S. Africa, p. 63 (1876).—*Kistecephalus.*

Cistecephalus microrhinus, Owen[1].

Syn. *Cistecephalus chelydroides*, Owen[2].

The type species. Skull (fig. 12) very broad and flat, of triangular contour, and with very wide parietal and interorbital bars. The presence of tusks is shown in one of the undermentioned specimens.

Hab. South Africa.

47066. (*Fig.*) The imperfect cranium, partly embedded in matrix; from the Karoo system of Styl-Krantz, on the flanks of the Sneewberg range. The type specimen; figured by Owen in his 'Catalogue of the Fossil Reptilia of S. Africa,' pl. lxiv. figs. 4–7. *Presented by A. G. Bain, Esq.*, 1853.

47071. (*Fig.*) The imperfect skull, with portions of the vertebral column, pectoral girdle and limbs, partly embedded in matrix; from Styl-Krantz. Part of the occipital region of the cranium, the two scapulæ, the right side of the (?) clavicle, the proximal half of the right humerus, and a section of two cervical vertebræ are figured by Owen in his 'Catalogue,' pl. lxix. figs. 7–10; and the frontal and occipital aspects of the cranium are shown in fig. 12. This specimen is noticed by Seeley in the 'Phil. Trans.' for 1888, pp. 489 (*errore* as No. 17071) and 496, where the characters of the scapula and humerus are discussed. The bones marked *eo* in the figure, which impinge on the borders of the foramen magnum, appear to be certainly the exoccipitals, while the one marked *so* cannot be anything but the supraoccipital. This small size of the exoccipitals is a feature often found in Mammals. The bone regarded by Owen as the interclavicle or episternum (*h* of figure) is referred by Seeley, *op. cit.* p. 491, note, to the clavicle.

Presented by W. G. Atherstone, Esq., M.D., 1875.

47068. (*Fig.*) The imperfect cranium of an individual of rather larger size than the two preceding specimens; from Styl-Krantz. The type of *C. chelydroides*; figured by Owen in his 'Catalogue,' pl. lxiv. figs. 8, 9. The alleged difference in the proportionate size of the orbits and temporal fossæ as compared with the preceding specimens is certainly not a specific one. *Presented by A. G. Bain, Esq.*, 1853.

[1] Cat. Foss. Rept. S. Africa, p. 63 (1876). [2] *Ibid.* p. 64.

R. 1689. The imperfect anterior portion of a still larger cranium; from Styl-Krantz. This specimen is important as showing sections of a pair of tusk-like maxillary teeth on the palatal aspect. It may belong to the next form.

Presented by A. G. Bain, Esq., 1853.

47082. The imperfect cranium; from the Karoo system of the Graaf-Reinet district, on the southern flank of the Sneewberg range. The bones forming the upper borders of the supratemporal fossæ are wanting.

Presented by R. N. Rubidge, Esq., M.D.

Cistecephalus leptorhinus, Owen [1].

In the type of this species the muzzle appears to be more deflected than in *C. microrhinus*, in consequence of which the large nares are placed considerably below the level of the orbits. The type skull is larger than that of *C. microrhinus*.

Hab. South Africa.

47067. (*Fig.*) The imperfect cranium; from the Karoo system of Styl-Krantz, in the Sneewberg range. The type; figured by Owen in his 'Catalogue of the Fossil Reptilia of S. Africa,' pl. lxiv. figs. 1–3. This specimen shows a large azygous bone in advance of the parietal foramen comparable to the one noticed in *Ptychosiagum* (p. 39).

Presented by A. G. Bain, Esq., 1853.

Cistecephalus planiceps, Owen [2].

Syn. *Cistecephalus bathygnathus*, Owen [3].

Smaller than the type species, with a relatively longer cranium, in which the temporal bar is proportionately longer and narrower.

Hab. South Africa.

47070. (*Fig.*) The imperfect cranium, with the palatal surface concealed by matrix; from the Karoo system of Styl-Krantz, on the flanks of the Sneewberg range. The type; figured by Owen in his 'Catalogue of the Fossil Reptilia of S. Africa,' pl. lxiv. figs. 10–13. *Presented by A. G. Bain, Esq.*, 1853.

[1] Cat. Foss. Rept. S. Africa, p. 64 (1876).

[2] *Loc. cit.*

[3] *Ibid.* p. 65.

47069. The imperfect skull in a laterally crushed condition; from
(*Fig.*) Styl-Krantz. The type of *C. bathygnathus*; figured by Owen in his 'Catalogue,' pl. lxv. fig. 7. The points in which this specimen differs from the preceding are solely due to the effects of pressure and crushing.

Presented by A. G. Bain, Esq., 1853.

Cistecephalus (?) arctatus, Owen[1].

Skull still longer and narrower than in the preceding species, with a more elongated and narrower parietal bar, which is, however, wider than the interorbital bar.

This species seems to connect the more typical representatives of the genus with *Dicynodon* and *Udenodon*, the skull having a marked resemblance to that of the young skull described as *D. recurvidens* (*suprà*, p. 20). In the latter the parietal bar is, however, narrower than the interorbital, the orbits are more lateral, and the proper contour of the skull was probably more vaulted. No traces of teeth are shown in the undermentioned specimens.

Hab. South Africa.

R. 1690. A small and slightly imperfect skull, with the palatal
(*Fig.*) surface largely concealed by matrix; from the Karoo system of the Cape Colony. The type; figured by Owen in his 'Catalogue of the Fossil Reptilia of S. Africa,' pl. lxv. figs. 2–6. The extremity of the muzzle is wanting; projections as if for tusks are seen on the maxillæ, but no traces of tusks are visible; the anterior portion of the mandible is exposed.

Presented by the Director of the Museum at Albany.

47083. A similar, but more imperfect and somewhat crushed skull, partly embedded in matrix; from the Karoo system of the Graaf-Reinet district, on the southern flank of the Sneewberg range.

Presented by W. G. Atherstone, Esq., M.D., 1875.

47088. Slab of rock showing the outlines of the parietal aspect of
(*Fig.*) a larger skull referred by its describer to this species; from Styl-Krantz, on the flank of the Sneewberg range. Figured by Owen in his 'Catalogue,' pl. lxv. fig. 1.

Presented by W. G. Atherstone, Esq., M.D., 1875.

[1] Cat. Foss. Rept. S. Africa, p. 65 (1876).

Specimens of Uncertain Generic Position.

a. (*Ptychognathus* [?] *pusillus*, Owen[1].)

R. 1691. (*Fig.*) Slab showing the dorsal aspect of six lumbar vertebræ and of the sacrum and ilia of a small Dicynodont; from the Karoo system of the Cape Colony. Described by Owen in his 'Catalogue of the Fossil Reptilia of S. Africa,' pp. 54, 55, without name, and figured in pl. liii. fig. 1 of the same, under the name of *Ptychognathus* (?) *pusillus*, of which it must be regarded as the type.

Presented by the Directors of the Museum at Cape Town.

b. (*Cirognathus cordylus*, Seeley[2].)

The undermentioned specimen probably belongs to a young *Dicynodon*, the alleged difference in the characters of the humerus from other Dicynodonts being due to a comparison of its preaxial aspect with the palmar aspect of the other forms and there being no evidence to support the alleged reduction in the number of the phalangeals.

49413. (*Fig.*) Slab of rock showing the imperfect bones, or their impressions, of the skull and the anterior region of the postcephalic skeleton; from the Karoo system of Klipfontein, Fraserburg, near the northern frontier of the Cape Colony. Described and figured by Seeley in the 'Phil. Trans.' for 1888, pp. 487–501, pl. lxxv., a restoration of the skeleton being given in pl. lxxvi. This specimen is the type of *Cirognathus*. The fragment of the scapula is described as closely resembling the corresponding bone of *Cistecephalus macrorhinus*, No. 47071 (p. 46); but the humerus, which is mainly indicated by the impression, is stated (*l. c.* p. 496) to be distinguished from the corresponding bone of the latter specimen "in a much less development of the radial crest, which is more proximal in position, and continuous with the articular head of the bone." In the humerus of *Cistecephalus* it is, however, the palmar or ventral aspect which is exposed (as in fig. 13 A, p. 53), whereas in the humerus of the present specimen the impression of the preaxial aspect is preserved. The contour of

[1] Cat. Foss. Rept. S. Africa, pl. liii. (1876).

[2] Phil. Trans. for 1888, p. 487.—*Keirognathus.*

this impression corresponds exactly with fig. 13 B; and the difference between figs. A and B is precisely the same as that obtaining between the humerus of the present specimen and that of *Cistecephalus*, which is solely due to the different aspects from which the bones are viewed. There do not appear to be any adequate grounds for the restoration of the hinder part of the skeleton. The contour of the lateral aspect of the skull closely resembles that of a young *Dicynodon*, *e. g.* the skull No. 47098 (*suprà*, p. 20) described as *D. recurvidens*. In regard to the alleged reduction of the number of the phalangeals, which is described as being 2, 2, 3, 2, 2, in the manus; it appears that the right manus shows the prepollex and the first, second, and third digits in connection with the radius, and what is probably the fifth digit attached to the ulna, the fourth being missing. The prepollex is suggested by Seeley to be part of the 1st digit. The 1st digit (2nd of Seeley) shows two phalangeals; the 2nd digit (3rd of Seeley) has three phalangeals; the 3rd (4th of Seeley) has two phalangeals remaining, the terminal one being apparently missing; while in the 5th digit there are clearly three digits, of which the basal one was regarded in the description as a metacarpal.

Purchased from T. Bain, Esq., 1878.

c. (*Eurycarpus oweni*, Seeley[1].)

46075. (*Fig.*) Slab of rock showing the impression of part of the vertebral column and portions of the left limb-bones of a comparatively small form; from the Karoo system of the Sneewberg range. The type of *Eurycarpus*. Parts of the scapula and humerus and the whole of the remaining portion of the pectoral limb are shown; while in the pelvic limb only the distal extremity of the femur is indicated. The palmar aspect of the manus is shown; this together with the radius and ulna being figured by Owen in his 'Catalogue,' pl. lii. fig. 3. The specimen is noticed by Seeley in the 'Phil. Trans.' for 1888, p. 500, and figured in the volume for 1889, pl. xviii. It is there regarded as generically distinct from *Dicynodon*, since the humerus is stated to be of different contour and shorter than the radius, and it is suggested that it may

[1] Phil. Trans. for 1888, p. 500.

belong to a Pariasaurian. The imperfect state of the mpression of the humerus renders it, however, impossible to draw any safe conclusions on this point, and the serial position of the specimen must for the present remain undecided. *Presented by Sir H. Barkly, K.C.B.*, 1874.

46075 a. A plaster cast taken from the preceding specimen to show the bones in relief. The vertebral column and pectoral limb are figured by Owen, *l. c.* figs. 1, 2; the manus being represented from the palmar aspect. The prepollex is omitted in the figure. *Made in the Museum.*

d. *The undermentioned specimen is probably referable to the present family.*

47095. Fragment of rock showing the impression of the frontal surface of the temporal arcade, and a natural cast of the inferior aspect of the anterior and middle regions of the brain; from the Karoo system of Styl-Krantz, in the Sneewberg range. The cast of the brain shows a marked general resemblance to that of a Permian Anomodont from the United States, described and figured by Cope in the 'Proc. Amer. Phil. Soc.' vol. xxiii. pp. 234–238 (1886), and plate. Thus there is the same remarkable narrowness in the region of the olfactory lobes and hemispheres, and the prominences on the lateral borders of the under surface of the mid-brain are precisely similar to those shown in fig. 2 of Cope's plate, which are considered to be the casts of vacuities in the cranial walls. In the present specimen the olfactory lobes appear to be more distinctly defined from the hemispheres than in the American specimen. *Presented by A. G. Bain, Esq.*, 1853.

47095 a. A plaster cast taken from the preceding specimen. This cast shows the hinder part of the temporal arcades in relief, and the concavity of the lower part of the brain-case. *Made in the Museum*, 1889.

e. *Of the following specimens of the pectoral and pelvic girdles and limb-bones, it is probable that while the majority of the larger ones are referable to* Dicynodon, *several of the smaller ones may belong to* Ptychosiagum.

47104. Mass of rock showing the symphysis of the mandible, the glenoidal region of the right pectoral girdle, and the right radius of a very large Dicynodont; from the Karoo system of Tepid-spring, near Fort Beaufort. The pectoral girdle includes the adjacent portions of the scapula, coracoid, and precoracoid, and is precisely similar in structure to the undermentioned small specimen No. 36287, thus proving the association of that type of pectoral girdle with undoubted Dicynodont remains, and also that the precoracoid and coracoid persisted as separate bones till maturity. *Presented by A. G. Bain, Esq.*, 1853.

36287. (*Fig.*) The imperfect right half of the pectoral girdle of a comparatively small form, probably referable to a species of *Ptychosiagum*; from the Karoo system of the Rhenosterberg branch of the Sneewberg range. Of the scapula only the glenoidal extremity remains; the precoracoid has lost its antero-internal angle, but the coracoid is nearly entire. Figured by Owen in his 'Catalogue,' pl. lxix. figs. 5, 6, and also by Seeley in the 'Phil. Trans.' for 1888, p. 492, fig. 1. It is also represented in woodcut fig. 2 (p. 16) of the present volume, with a restoration of the scapula. From the close resemblance of the scapula to that of *Ptychosiagum orientale*, it is probable that this specimen is referable to that genus, and it would agree approximately in relative size with the skull of *P. declive*[1]. In the restoration of the precoracoid given by Seeley there is no justification for the obliteration of the notch at the junction of the precoracoid with the acromion, which seems to be a natural one. Owen identified the process marked *a* in woodcut fig. 2 with the acromion, correlating it with the similar process in the scapula of *Dicynodon* No. 47056* (p. 21); but Seeley regarded the reflected projection of the preaxial border (*b* of fig. 9) as the acromion. That the former identification is the true one is shown by No. 47056*, where the process corresponding to *b* is situated near the distal extremity of the bone.
Presented by Sir George Grey, K.C.B., 1858.

36272 x. A split nodule of rock showing the imperfect right half of a similar pectoral girdle; from the Rhenosterberg. Both sides of the glenoidal portion of the scapula are shown,

[1] From the proportions obtaining in the specimen described as *Cirognathus* (p. 49), it was suggested by Seeley (*op. cit.* p. 491) that this pectoral girdle indicated a skull of some 12 inches in length, the larger relative size of the skull in young animals being overlooked.

but of the coracoid and precoracoid only the inner aspect is visible. This specimen shows that the notch at the junction of the anterior border of the precoracoid with the scapula is a natural one.

Presented by Sir George Grey, K.C.B., 1858.

36272 y. Fragment of rock showing the inner surface of an imperfect right precoracoid closely resembling that of the preceding specimen; from the Rhenosterberg.

Presented by Sir George Grey, K.C.B., 1858.

36272 z. Mass of rock showing the imperfect left scapula and a portion of the precoracoid, together with imperfect vertebræ and other bones, of a rather larger individual; from the Rhenosterberg.

Presented by Sir George Grey, K.C.B., 1858.

Fig. 13.

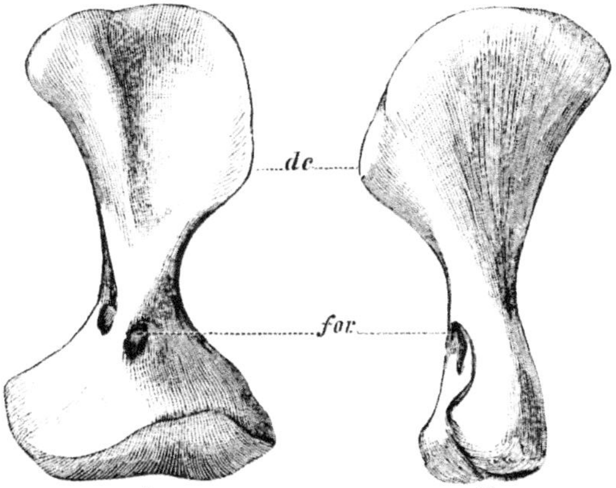

Dicynodont.—Palmar (A) and preaxial (B) aspects of the left humerus; from the Karoo system of the Rhenosterberg. ½. *dc*, deltopectoral (radial) crest; *for*, entepicondylar foramen.

36259. (*Fig.*) Mass of rock, showing some imperfect dorsal vertebræ and ribs, and a series of associated bones of the right side of the appendicular skeleton of a large form: from the Karoo system of the Gonzia river, Kaffraria. Figured by Owen in his 'Catalogue,' pl. xxxv., as *Dicynodon tigriceps*. The vertebræ have been longitudinally bisected. The bones of the appendicular skeleton, which are more or less imperfect, comprise the scapula, humerus, radius, and the proximal half of the ulna.

Presented by A. G. Bain, Esq., 1853.

47049. The proximal half of the left humerus of a large individual,
(*Fig.*) in a somewhat imperfect condition; from the Karoo system of Kaffraria. Figured by Owen in his 'Catalogue,' pl. xxvii., and referred to *Dicynodon leoniceps*. The specimen is broken off across the entepicondylar foramen.
Presented by A. G. Bain, Esq., 1853.

47108. The entire right humerus of an individual of the same approximate size as the preceding specimen; from the Karoo system of the Cape Colony.
Presented by A. G. Bain, Esq., 1853.

50120. The distal extremity of a rather smaller left humerus; from the Karoo system of Catelomds, Claremont.
Presented by E. J. Dunn, Esq., 1879.

47101. The imperfect distal extremity of a nearly similar left humerus; from the Cape Colony.
Presented by A. G. Bain, Esq., 1853.

47046. A smaller imperfect left humerus, with the delto-pectoral
(*Fig.*) crest bent over towards the ulnar side; from the Karoo system near Fort Beaufort. Figured by Owen in his 'Catalogue,' pls. xli., xlii., and referred to *Dicynodon pardiceps*. *Presented by A. G. Bain, Esq.*, 1853.

49378. The imperfect distal extremity of a right humerus, agreeing approximately in size with the preceding; from the Cape Colony. *Purchased from T. Bain, Esq.*, 1878.

47072. A considerably smaller right humerus, wanting the proximal extremity; from the Cape Colony. Noticed on p. 53 of Owen's 'Catalogue' as No. 88.
Presented by A. G. Bain, Esq., 1853.

36299. The proximal half of a comparatively small left humerus;
(*Fig.*) from the Karoo system of the Rhenosterberg. Figured by Owen in his 'Catalogue,' pl. xix. figs. 3–6, in conjunction with No. 36289, and referred to *Galesaurus*. There is no reason for regarding this specimen as not referable to a Dicynodont, and it is certainly far too large for *Galesaurus*. It closely resembles the smaller humerus of *Ptychosiagum orientale*, figured by the writer in the 'Palæontologia Indica,' ser. 4, vol. i. pt. iii. pl. i., and is figured with a restoration of the lower extremity in fig. 13.
Presented by Sir George Grey, K.C.B., 1858.

R. 1700. An imperfect right humerus, agreeing in size with the preceding, but belonging to a distinct form; from the Cape Colony. The distal extremity is wanting, and there is a very sudden expansion below the foramen on the ulnar border. *No history.*

36301. The water-worn proximal half of a smaller left humerus; from the Rhenosterberg.
Presented by Sir George Grey, K.C.B., 1858.

36292. The distal half of a right humerus agreeing in relative size with No. 36299; from the Rhenosterberg.
Presented by Sir George Grey, K.C.B., 1858.

36293. A nearly similar specimen; from the Rhenosterberg.
Presented by Sir George Grey, K.C.B., 1858.

36289. A nearly similar specimen of the right side; from the
(*Fig.*) Rhenosterberg. Figured by Owen in his 'Catalogue,' pl. xix. figs. 3–6, in conjunction with No. 36299, and referred to *Galesaurus*.
Presented by Sir George Grey, K.C.B., 1858.

36288. The imperfect distal portion of a slightly smaller right humerus; from the Rhenosterberg.
Presented by Sir George Grey, K.C.B., 1858.

R. 469. The distal portion of a left humerus, agreeing approximately in size with the preceding; from the Cape Colony.
Presented by Sir R. Owen, K.C.B., 1884.

49389. A radius of a large form; from the Karoo system of the
(*Fig.*) Gouph district, Beaufort West. Figured by Seeley in the 'Phil. Trans.' for 1889, pl. xxiii., without generic determination, as an ulna. *Purchased from T. Bain, Esq.*, 1878.

36249. A rather smaller imperfect radius of the same side; from the Karoo system of the Brak river, near Fort Beaufort. Closely resembles the radius in No. 36259 (*suprà*, p. 53).
Presented by A. G. Bain, Esq., 1853.

50124. A considerably smaller and somewhat imperfect radius of the opposite side; from the Karoo system of Catelomds, Claremont. *Presented by E. J. Dunn, Esq.*, 1879

47078. Fragment of rock showing most of the bones of a manus or pes of a comparatively small form; from the Karoo system of the Graaf-Reinet district. Noticed by Owen in his 'Catalogue,' p. 54, no. 91, where it is suggested that it may be the manus. The terminal phalangeals are imperfect, but the proximal ones are of the massive type characteristic of the African and Indian Dicynodonts.
Presented by R. N. Rubidge, Esq., M.D.

R. 515. Natural sandstone cast of an imperfect footprint of a larger form; from the Karoo system of Rouxville, Orange Free State. *By exchange with the Blomfontein Museum*, 1884.

36243. The right ilium of a large form; from the Karoo system
(*Fig.*) near "Tola's Kraal" [1], Kaffraria. Figured by Owen in the 'Trans. Geol. Soc.' ser. 2, vol. vii. pl. xxiv. figs. 2, 3 (the former figure in conjunction with No. 36244, *infrà*), and provisionally regarded as a scapula of *Dicynodon*. Described by the same writer in his 'Catalogue,' p. 70, no. 122, as the left ilium of *Pariasaurus*. Mentioned by Seeley in the 'Phil. Trans.' for 1888, p. 94, where it is shown to be a right ilium probably referable to *Dicynodon*; a reversed figure is given on p. 107 of the same memoir, where it is described as the ilium of an Anomodont. The contour of this specimen is almost identical with that of the much smaller ilium represented in woodcut, fig. 3 (p. 17). *Presented by A. G. Bain, Esq.*, 1853.

36245. The right ilium of a rather larger and specifically distinct individual; from the road to Block Drift, Kaffraria. The supra-acetabular portion is more elevated, and thereby approximates to the ilia described as *Platypodosaurus* (p. 63). *Presented by A. G. Bain, Esq.*, 1853.

R. 1699. The right ilium and sacral ribs of a comparatively small
(*Fig.*) form; from the Karoo system of the Cape Colony. Figured in woodcut 3 in conjunction with an ischium and pubis. This bone precisely resembles the ilium of the opposite side figured by Huxley in the 'Palæontologia Indica' (Mem. Geol. Surv. Ind.), ser. 4, vol. i. pt. i. pl. v. fig. 1, and regarded as rather a scapula or coracoid, but referred to its true position by the writer in the 'Rec. Geol. Surv. Ind.' vol. xxiii. pt. i. (see fig. 10). *No history.*

[1] Incorrectly "Sola's" in Owen's 'Catalogue.'

36265. Mass of rock exhibiting the inner aspect of a nearly similar right ilium and other bones; from the Rhenosterberg.
Presented by Sir George Grey, K.C.B., 1858.

36267. An imperfect left ilium of similar type, partially concealed by matrix; from the Rhenosterberg.
Presented by Sir George Grey, K.C.B., 1858.

36263. Mass of matrix exhibiting an imperfect ilium, ischio-pubis, femur, tibia, several trunk-vertebræ, and other bones, of an individual of the same approximate size as the preceding; from the Rhenosterberg.
Presented by Sir George Grey, K.C.B., 1858.

R. 1698. (*Fig.*) The right innominate of a comparatively small form, wanting the distal portion of the ilium; from the Cape Colony. This well-preserved specimen is represented in fig. 3 (p. 17), the missing portion of the ilium being restored from No. R. 1699. The suture between the ischium and pubis is still visible. Three imperfect vertebræ are attached by matrix to the ventral surface of the specimen. *No history.*

36285. (*Fig.*) The somewhat flattened and slightly imperfect right ischium of a comparatively small form; from the Rhenosterberg. This bone closely resembles the ischium of No. R. 1698, but the acetabular portion is altered by distortion. The channel leading to the obturator foramen is preserved, and the thickening below the foramen is also shown. Figured by Seeley in the 'Phil. Trans.' for 1889, pl. xv. fig. 5, as a coracoid.
Presented by Sir George Grey, K.C.B., 1858.

36286. (*Fig.*) A flattened left ischium, together with some centra of dorsal vertebræ, apparently associated with the preceding specimen; from the Rhenosterberg. Figured by Seeley, *op. cit.* pl. xv. figs. 6, 7, the ischium being described as a coracoid. The foramen and inferior thickening are shown. From the length of the vertebral centra it is possible that this specimen belongs to *Ptychosiagum.*
Presented by Sir George Grey, K.C.B., 1858.

36244. (*Fig.*) The imperfect innominate of a large form; from the Karoo system, on the road to Block Drift, Kaffraria. These specimens comprise a fragment of the acetabular articulation of the ilium, and the greater portion of the ischium and pubis of either side. The right ischio-pubis is figured by Owen in the 'Trans. Geol. Soc.' ser. 2, vol. vii. pl. xxxiv. fig. 2, in connection with the ilium, No. 36243 (p. 53), where it is provisionally regarded as a coracoid. The specimens are noticed by Owen in his 'Catalogue,' p. 35, no. 58, where they are referred to the pectoral girdle of *Dicynodon leoniceps*, their true position being indicated by Seeley in the 'Phil. Trans.' for 1888, p. 94. The line of suture between the ischium and pubis is still visible, and the general contour of these bones is very similar to that obtaining in the specimens described as *Platypodosaurus*.

Presented by A. G. Bain, Esq., 1853.

47877. The imperfect right pubis of an individual agreeing closely in size with the preceding; from the Cape Colony. This specimen is fractured through the obturator foramen on a line which probably indicates the original sutural union between the pubis and ischium.

Presented by A. G. Bain, Esq., 1853.

47050. The right femur of a very large individual; from the Karoo system of the Great Karroo, south of Beaufort West. Noticed by Owen in his 'Catalogue,' p. 36, no. 61, and referred to *Dicynodon leoniceps*. The proximal extremity is crushed and imperfect, but when entire was probably very similar to that of the femur of *Platypodosaurus*.

Presented by A. G. Bain, Esq., 1853.

R. 1701. The flattened and somewhat imperfect right femur of a considerably smaller form; from the Cape Colony. This specimen indicates a rather larger individual than the type of *Platypodosaurus*, and when entire would appear to have had the same general contour.

Presented by A. G. Bain, Esq., 1853.

50127. The imperfect proximal extremity of the left femur of an individual of slightly smaller size than that of the type of *Platypodosaurus*; from Catelomds, Claremont.

Presented by E. J. Dunn, Esq., 1879.

36296. A small left femur, wanting the distal extremity; from the Rhenosterberg.

Presented by Sir George Grey, K.C.B., 1858.

36290. The proximal extremity of a tibia of a comparatively small form; from the Rhenosterberg.

Presented by Sir George Grey, K.C.B., 1858.

36297. The proximal half of a still smaller right tibia; from the Rhenosterberg.

Presented by Sir George Grey, K.C.B., 1858.

The following bones are not determined.

36272. A somewhat imperfect bone; from the Karoo system of the
(*Fig.*) Rhenosterberg. Figured by Seeley in the 'Phil. Trans.' for 1889, pl. xv. fig. 4, as a scapula.

Presented by Sir George Grey, K.C.B., 1858.

36302. The expanded extremity of a somewhat larger bone of the same general type as the preceding; from the Rhenosterberg. Closely resembles a smaller specimen figured by the writer in the 'Palæontologia Indica,' ser. 4, vol. i. pt. iii. pl. iii. fig. 10, and suggested to be a pubis, which it certainly is not. *Presented by Sir George Grey, K.C.B.*, 1858.

Of the following vertebræ it is probable that, while a large proportion are referable to Dicynodonts, a moiety may belong to Theriodonts.

R. 1704. Three imperfect anterior dorsal or late cervical vertebræ of a medium-sized individual, cemented together by matrix; from the Karoo system of the Cape Colony. The costal articulations are situated as in the cervicals of Crocodilia.

No history.

R. 1704 a. Five smaller anterior dorsal or late cervical vertebræ, with some of the ribs attached, in matrix; from the Cape Colony. *No history.*

R. 1702. Fragment of rock showing the right lateral aspect of five
(*Fig.*) small dorsal vertebræ, together with several imperfec ribs; from the Rhenosterberg. Figured by Seeley in the 'Phil. Trans.' for 1889, pl. xvi. fig. 1. The transverse processes are well preserved, and there appears to be a rib-facet on the anterior border of the fourth centrum.

Presented by Sir George Grey, K.C.B., 1858.

R. 1703. Fragment of rock showing a lateral aspect of three imperfect dorsal vertebræ, probably belonging to the same individual as the preceding; from the Rhenosterberg.
Presented by Sir George Grey, K.C.B., 1858.

36304. Four imperfect dorsal vertebræ in matrix, not improbably associated with the preceding.
Presented by Sir George Grey, K.C.B., 1858.

R. 1704 b. Four larger imperfect dorsal vertebræ, cemented together by matrix; from the Cape Colony. *No history.*

R. 1704 c. Portions of three dorsal vertebræ cemented together by matrix; from the Cape Colony. These specimens agree approximately in size with the preceding. *No history.*

R. 725 x. Four somewhat larger dorsal vertebræ, in matrix; from the Karoo system of Natal.
Presented by Prof. T. Rupert Jones, 1886.

R. 725 y. The imperfect centra of two associated dorsal vertebræ of nearly the same size as the preceding; from Natal. The extreme depth of the terminal cups is well shown.
Presented by Prof. T. Rupert Jones, 1886.

R. 1704 d. A nearly similar dorsal centrum; from the Cape Colony.
No history.

36240. The imperfect anterior dorsal vertebra of a larger form;
(*Fig.*) from Kaffraria. Figured by Owen in the 'Trans. Geol. Soc.' ser. 2, vol. vii. pl. xxxviii. figs. 1–3, and also in his 'Catalogue,' pl. liii. figs. 2, 3, in the latter instance being referred to *Dicynodon pardiceps*. The costal facet is seen on the centrum, and the deep cupping of the terminal faces is conspicuous. *Presented by A. G. Bain, Esq.*, 1853.

50126. An imperfect dorsal centrum of similar type; from Catelomds, Claremont. *Presented by E. J. Dunn, Esq.*, 1879.

R. 1704 e. An imperfect dorsal centrum of the same general type and size; from the Cape Colony. *No history.*

36242. Portions of the first two sacral vertebræ of a large form;
(*Fig.*) from the Brak river, near Fort Beaufort. Figured by Owen in the 'Trans. Geol. Soc.' ser. 2, vol. vii. pl. xxxiii figs. 4, 5, and also in his 'Catalogue,' pl. xxxvii. figs. 2, 3

(as *Dicynodon tigriceps*). The presence of a distinct rib to each vertebra is clearly shown, whereby the specimen differs from the entire sacrum figured in the latter plate under the same name (see pp. 117, 118).

Presented by A. G. Bain, Esq., 1853.

36241. Two later sacral vertebræ, apparently belonging to the same (*Fig.*) individual as the preceding: from the same locality. Figured by Owen in the 'Trans. Geol. Soc.' ser. 2, vol. vii. pl. xxxiii. figs. 6, 7, and noticed in his 'Catalogue,' p. 42, no. 69, as *Dicynodon tigriceps*. The expansion of the ribs for attachment to the ilium is well shown, and the specimen agrees in all essential characters with the sacra described as *Platypodosaurus* (p. 63).

Presented by A. G. Bain, Esq., 1853.

47105. Portion of the imperfect sacral region of a large form; from the Karoo system of the Cape Colony.

Presented by A. G. Bain, Esq., 1853.

R. 725 z. The centrum of a caudal vertebra of a comparatively small form; from Natal. Probably associated with No. R. 725 *y*.

Presented by Prof. T. Rupert Jones, 1886.

The following vertebræ have more compressed centra than the preceding, and some of them may be referable to the Theriodont Galesauridæ.

R. 869. Two imperfect dorsal vertebræ of medium size; from the Cape Colony. These specimens were found with skulls of *Ptychosiagum*, and not improbably belong to that genus. The centra are long and much compressed, with moderately deep terminal cups. On the left side a portion of a rib remains, with its head resting on the capitular articular surface of the centrum.

Purchased from T. Bain, Esq., 1880.

R. 1704 f. An imperfect lumbar (?) vertebra, approximating in size to the preceding; from the Cape Colony. The centrum is comparatively short. *No history.*

R. 1704 g. Four conjoint imperfect dorsal vertebræ of a smaller form; from the Cape Colony. The centra approximate in contour to No. R. 869. *No history.*

R. 1704 h. Two small and imperfect trunk vertebræ; from the Cape Colony. The centra are much compressed. *No history.*

R. 1704 i. Three imperfect and somewhat larger vertebral centra; from the Cape Colony. These specimens are of a broad and uncompressed type. *No history.*

Family uncertain.

**PLATYPODOSAURUS, Owen[1].

Founded upon portions of the axial and appendicular skeleton, which do not include the skull, but which appear to be decidedly Dicynodont. The chief distinctive feature of these specimens is the great development of the inferior portion of the delto-pectoral crest of the humerus, which descends lower than in *Dicynodon* and forms a distinct projection; and also the presence of a smaller process on the opposite or postaxial border. Scapula approximating to that of *Ptychosiagum*.

From their general Dicynodont character it would appear probable that these specimens are referable either to *Udenodon* or to *Endothiodon*.

**Platypodosaurus robustus, Owen[2].

The type and only described species; of large size.

Hab. South Africa.

50121. (*Fig.*) Part of the vertebral column, the right humerus, and the imperfect pelvis; from the Karoo system[3] of Catelomds, Claremont. These, with the undermentioned associated specimens, are the types. Three of the vertebræ and the humerus are figured by Owen in the 'Quart. Journ. Geol. Soc.' vol. xxxvi. pl. xvi. figs. 1–3, 7, the figure of the humerus being reversed. Three of the vertebræ have been longitudinally bisected, and show the great depth of the terminal cups. In the figure of the humerus there does not appear to be any justification for the prominent angle at the upper extremity of the delto-pectoral crest. When entire the ilium would appear to be of the general type of that of *Dicynodon*. *Presented by E. J. Dunn, Esq.*, 1879.

[1] Quart. Journ. Geol. Soc. vol. xxxvi. p. 414 (1880). [2] *Loc. cit.*

[3] There is no evidence to show whether these specimens are from the Beaufort or Stormberg beds.

50122. (*Fig.*) The nearly entire right scapula, with the terminal phalangeals of four digits of the manus attached to the ventral surface; associated with the preceding specimens. Figured by Owen, *op. cit.* pl. xvii. figs. 1, 2. In the description the process in the figure marked *e* is correlated with the one so marked (*a* of fig. 2) in Owen's figures of Nos. 47056* (p. 21) and No. 36287 (p. 52), whereas this really corresponds with the process above *a* in fig. 2 or that marked *b* in fig. 9, and there termed the supra-acromial process of the preaxial border. That the identification of the lower process with the acromion of Monotremes is probably correct has been already mentioned under the head of No. 47056*, the question being alluded to by the writer in the 'Proc. Zool. Soc.' 1889, p. 575.

Presented by E. J. Dunn, Esq., 1879.

50123. (*Fig.*) The proximal portion of the right femur; associated with the preceding specimens. Figured by Owen, *op. cit.* pl. xvii. figs. 6, 7. *Presented by E. J. Dunn, Esq.*, 1879.

50124. An imperfect bone, which may be a portion of a tibia, associated with the preceding specimens.

Presented by E. J. Dunn, Esq., 1879.

50125. (*Fig.*) An azygous plate-like bone, which is probably the presternum; associated with the preceding. Figured by Owen, *op. cit.* pl. xvi. fig. 5.

Presented by E. J. Dunn, Esq., 1878.

50126. An imperfect vertebra; associated with the preceding.

Presented by E. J. Dunn, Esq., 1879.

47090. (*Fig.*) The sacrum and imperfect pelvis and ten caudal vertebræ; from the Beaufort beds of the Karoo system, near Fort Beaufort. Noticed by Owen in his 'Catalogue of the Fossil Reptilia of S. Africa,' p. 73, no. 126, and the caudal vertebræ figured on p. 74, as an undetermined Dicynodont; described and the pelvis and sacrum figured by the same writer in the 'Quart. Journ. Geol. Soc.' vol. xxxvii. p. 266, pl. x., where it is referred to the present form. The caudal vertebræ are figured by Seeley in the 'Phil. Trans.' for1889, pl. xvii.; a reversed view of the right innominate being given by the same writer in the 'Phil. Trans.' for 1888, p. 107. The contour of the ilium appears to have been very similar to that obtaining in *Dicynodon*; there is no justification for the restoration given in the figure in the 'Phil. Trans.,' and it appears that what is there regarded as the upper extremity of the ilium is really matrix. The ischio-pubis closely resembles the corresponding element, which is probably referable to *Dicynodon* (No. 36244, p. 57): but the obturator foramen is less elliptical. The resemblance to the type pelvis renders it almost certain that the present specimen belongs to the same form. *Presented by A. G. Bain, Esq.*, 1853.

Family ENDOTHIODONTIDÆ.

Oral surface of palate and mandible carrying one or more longitudinal rows of cylindrical columnar teeth, those of the upper jaw being apparently borne on the maxilla. Nares terminal. Posterior nares in advance of pterygoids, and palatines deeply incurved.

Until the postcephalic skeleton is known, the serial position of this family cannot be regarded as absolutely fixed The approximation in the characters of the skull, and more especially the mandible, to that of the Dicynodontidæ, supports Owen's view that the Endothiodontidæ should be included in the same suborder; and if this be correct, it is quite probable that the specimens described as *Platypodosaurus* may prove to belong to *Endothiodon*.

Genus **ENDOTHIODON,** Owen [1].

Syn. *Theriognathus*, Owen [2].

The type and only described genus. Skull somewhat depressed, with a flattened frontal region and large nasals, separated by a deep groove from the maxillæ, and overhanging the nares; orbits small and lateral; alveolar border of maxilla and premaxilla with a caniniform swelling; interorbital region very wide; occiput unknown.

Endothiodon has been compared to *Placodus*, *Sphenodon*, and *Hyperodapedon*; but it differs fundamentally in that there is a bony floor beneath the narial passage; while, if the interpretation given below be correct, the palatal teeth are supported upon the maxilla instead of upon the palatine and vomer.

[1] Cat Foss. Rept. S. Africa, p. 66 (1876).

[2] *Ibid.* p. 62.

Endothiodon bathystoma, Owen[1].

The type species; of large dimensions. Three longitudinal rows of palatal and mandibular teeth in the adult.

Hab. South Africa.

R. 1646. (*Fig.*) The imperfect skull, wanting the whole of the occipital region; from the Beaufort beds of the Karoo system of the Sneewberg range[2]. The type; figured by Owen in his 'Catalogue of the Fossil Reptilia of S. Africa,' pls. lxvi., lxvii., and noticed by the same writer in the 'Quart. Journ. Geol. Soc.' vol. xxxv. p. 557. The cranium has been laterally crushed, and the oral surfaces have been cut and polished in order to exhibit the dentition: and the right ramus of the mandible has also been vertically and transversely cut. *Presented by the Directors of the Museum at Albany, Cape Colony.*

49415. (*Fig.*) The imperfect anterior portion of the mandible; from the Beaufort beds in the Gouph district near Beaufort West, on the southern flank of the Nieuwveldt range. Figured by Owen in the 'Quart. Journ. Geol. Soc.' vol. xxxv. pl. xxvi. fig. 1. The upper surface of the right ramus has been cut and polished, in order to exhibit the dentition, which is the portion figured by Owen. *Purchased from T. Bain, Esq.*, 1878.

Endothiodon microps (Owen[3]).

Syn. *Theriognathus microps*, Owen[4].
Endothiodon uniseries, Owen[5].

Smaller than the preceding, with very small orbits and but one row of palatal teeth; mandibular symphysis probably less deep. It is assumed that the undermentioned specimens indicate adult individuals.

Hab. South Africa.

47065. (*Fig.*) The imperfect skull; from the Beaufort beds of the Karoo system at Styl-Krantz, on the flanks of the Sneewberg

[1] Cat. Foss. Rept. S. Africa, p. 66 (1876).

[2] This is the locality given in Owen's 'Catalogue,' but in the passage cited from the 'Quart. Journ. Geol. Soc.' the specimen is stated to be from the Gouph district.

[3] Cat. Foss. Rept. S. Africa, p. 62 (1876).—*Theriognathus.*

[4] *Loc. cit.*

[5] Quart. Journ. Geol. Soc. vol. xxxv. p. 557 (1879).

range. The type of *Theriognathus*; figured by Owen in his 'Catalogue of the Fossils of S. Africa,' pl. lxiii. figs. 1–3. Nearly the whole of the bone has disappeared from the facial region of the cranium, and the mandible has also largely exfoliated. The teeth are not shown, but from a comparison with the next specimen the generic position of the specimen is quite evident. This is especially shown by the great width of the interorbital region and the natural cast of the right orbit, which evidently had a bony roof identical with that of the next specimen. *Theriognathus* was placed by its founder near *Udenodon*.

Presented by A. G. Bain, Esq., 1853.

49414. (*Fig.*) The anterior portion of the cranium; from the Beaufort beds on the flanks of the Nieuwveldt range. The type of *E. uniseries*; figured by Owen in the 'Quart. Journ. Geol. Soc.' vol. xxxv. pl. xxvii. figs. 2–9. This specimen is broken off a short distance behind the root of the temporal arcade. The aperture of the posterior nares is well shown, and is seen to be mesially divided by the vomer, and bounded laterally by deeply incurved bones which appear to be the palatines, although they are termed pterygoids by Owen. If this view be correct, the teeth will be situated on the maxillæ. The bone on the outer side of the palatine forming the anterior boundary of the palatal aperture of the temporal seems to be the extremity of the pterygoid, and evidently corresponds to the bone similarly situated in the skull of *Dicynodon*, No. R. 860 (p. 27). *Purchased from T. Bain, Esq.*, 1878.

Suborder *THERIODONTIA.*

This suborder is taken as equivalent to the Pelycosauria of Cope [1], but cannot at present be fully defined. Some of the most generalized forms referred by Cope to the Pelycosauria [2] are here classed in the undermentioned group Pariasauria, between which and the Theriodontia there was probably a more or less complete transition.

[1] See Cope, 'Proc. Amer. Phil. Soc.' vol. xvii. pp. 529, 530 (1878); and Baur, Journ. Morphol. vol. i. p. 102 (1887). The term Pelycosauria is retained by Baur for this group.

[2] Typified by *Clepsydrops* and originally regarded as a suborder of Rhynchocephalia; see 'Proc. Amer. Phil. Soc.' vol. xvii. pp. 511 and 529 (1878).

Typically the skull with open temporal fossæ and a single temporal arch, apparently consisting of a conjoint squamoso-maxillary and quadrato-maxillary arcade, and with secondary posterior nares; premaxillæ separate; mandibular symphysis shallow and uniting by suture; no lateral vacuity in ramus. Relations of pterygoids and palatines probably as in the Dicynodontia. Dentition fully developed, and frequently of a carnivorous type.

In some cases, as in the American forms, the vertebral centra are notochordal, and intercentra may be present, to which the capitular heads of the ribs articulate in *Embolophorus*. Not more than two or three sacral vertebræ. Scapula of African forms (when known) with the acromial process confluent with the glenoidal region, and only a short notch between the acromial and supra-acromial processes. Humerus in the typical forms approximating more or less closely to the Dicynodont type, with the inferior aperture of the entepicondylar foramen opening on the palmar aspect, and with a more or less distinct supinator flange on the preaxial border, and in some cases with the distal extremity much more expanded than the proximal (fig. 23).

In some of those African forms of which the pelvis is known the ilium is somewhat intermediate in structure between that of the Dicynodontia and Pariasauria, and there is a minute obturator foramen (fig. 17), but in other cases (fig. 15) the ilium is more like that of the Pariasauria.

The more generalized forms included in this suborder have no secondary posterior nares; while in the American *Chilonyx* and the African *Gorgonops*, which are provisionally classed in this suborder, there is a bony roof to the temporal fossæ.

Family GALESAURIDÆ.

The type family. Dentition of a carnivorous type, and usually differentiated into incisives, tusks, and cheek-teeth; no teeth on palate. Skull with secondary posterior nares. Humerus (when known) elongated.

The incisive teeth are simple, those of the upper jaw always exceeding by one pair those of the mandible; and the pulp-cavities of all the teeth were closed inferiorly.

Owen's division of the members of this family into Mononarialia and Binarialia will not hold good, the distinction being due, at least in many cases, to the imperfection of the specimens.

Genus **GALESAURUS**, Owen [1].

Syn. *Nythosaurus*, Owen [2].

The type genus. Incisive teeth $\frac{4}{3}$, cheek-teeth $\frac{?}{12}$; hinder cheek-teeth with laterally compressed tricuspidate crowns; incisives, if serrated, very indistinctly so. Skull depressed, with a comparatively short, shallow, and subconical muzzle, and the temporal fossæ much larger than the orbits, which are directed frontally. Inferior surface of mandibular symphysis sloping obliquely upwards.

Galesaurus planiceps, Owen [3].

Syn. *Nythosaurus larvatus*, Owen [4].

The type and only described species. Of comparatively small size, the length of the skull being some 0,090 (3·55 inches).

Hab. South Africa.

36220. (*Fig.*) The imperfect skull; from the Beaufort beds [5] of the Karoo system of the Rhenosterberg branch of the Sneewberg range. The type; figured by Owen in the 'Quart. Journ. Geol. Soc.' vol. xvi. pl. ii., and also in his 'Catalogue of the Fossil Reptilia of S. Africa,' pl. xviii. figs. 6–11. The cranium has been flattened by a vertical crush, thus causing the plane of the occiput to be more oblique than is naturally the case. The teeth are very imperfectly preserved, but in the left side of the mandible there are four teeth in advance of the tusk, and apparently some 12 cheek-teeth behind it. The narial aperture is very imperfect. *Presented by Sir George Grey, K.C.B.*, 1858.

R. 511. (*Fig.*) The nearly entire skull; from the Karoo system of the Orange Free State. Described and figured by Owen in the 'Quart. Journ. Geol. Soc.' vol. xliii. p. 1, pl. i., and by Seeley in the 'Phil. Trans.' for 1889, pl. ix. figs. 3, 4. In Owen's figure of the frontal aspect the occiput was restored from the preceding specimen; this is, however, incorrect, the occiput of the present specimen being really in its true subvertical plane. The tooth represented in the figure as the crown of the upper tusk is really the

[1] Quart. Journ. Geol. Soc. vol. xvi. p. 58 (1860).
[2] Cat. Foss. Rept. S. Africa, p. 24 (1876).
[3] Quart. Journ. Geol. Soc. vol. xvi. p. 58 (1860).
[4] Cat. Foss. Rept. S Africa, p. 24 (1876).
[5] See Quart. Journ. Geol. Soc. vol. xxiii. p. 143.

root of the lower. The four premaxillary teeth remain, but there are four cheek-teeth in the left maxilla. In the description the narial aperture is stated to be single, but it is clearly divided in the same manner as in *Lycosaurus*. On both sides the temporal arcade shows a longitudinal division indicating that it is composed of an upper and a lower moiety. The lower arcade seems to be undoubtedly a quadrato-maxillary one (as in *Sphenodon*), while the upper one would seem to be a squamoso-maxillary one (as in the Dicynodontia, p. 39). The constituent elements of the lower arcade cannot be determined; but the relations of the upper one confirm Cope's conclusion that the postorbito-squamosal arcade of *Sphenodon* is absent in this suborder. A figure of this skull is given in the accompanying woodcut.

By exchange with the Blomfontein Museum, 1884.

Fig. 14.

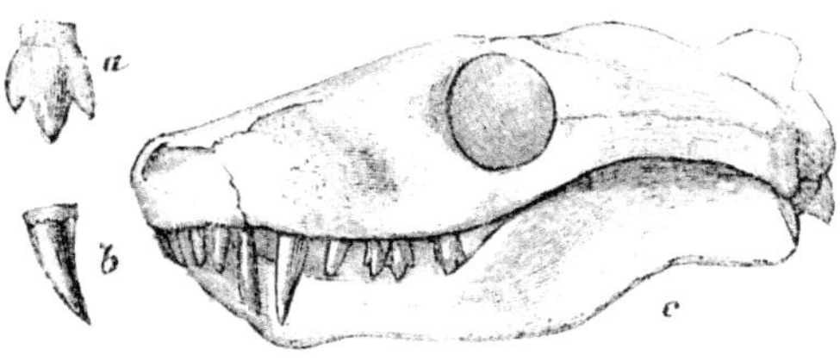

Galesaurus planiceps.—Left lateral aspect of the skull. $\frac{2}{3}$. *a*, an upper cheek-tooth, and *b* an incisive tooth. $\frac{2}{1}$. From the Karoo system of South Africa.

R. 511 a. The imperfect skull; from the Orange Free State. The contour of the facial region is well preserved, but the narial septum is broken away. Four posterior cheek-teeth remain on the left side, five cheek-teeth being shown in the opposite jaw. On the right side the lower (quadrato-maxillary) moiety of the compound temporal arcade is preserved. Neither in this nor in the preceding specimen can it be determined whether the quadrate was perforated.

By exchange with the Blomfontein Museum, 1884.

R. 845. Fragment of the middle region of the skull; from the (*Fig.*) Orange Free State. A cast is shown of the anterior extremity of the brain and of the inner aspect of the nasal region. The cusped crowns of the cheek-teeth of both the upper and lower jaws are well shown. The tusk and upper cheek-teeth are figured by Seeley in the 'Proc. Roy. Soc.' vol. xliv. p. 137, fig. 3, to show the closure of the pulp-cavity. *Presented by C. S. Orpen, Esq.*, 1886.

R. 1715. (*Fig.*) Cast of the brain-cavity and inner surfaces of an imperfect skull; from the Karoo system of Tafelberg. Figured by Owen in his 'Cat. Foss. Rept. S. Africa,' pl. xx. figs. 1, 2, and pl. xxxiv. fig. 2, as *Nythosaurus*, of which it is the type; referred to the present genus by Seeley in the 'Phil. Trans.' for 1889, p. 278. The cast of the anterior portion of the brain and of the nasal region precisely resembles that of the preceding specimen. The general contour of the upper surface of the brain approximates to that of the brain of *Diadectes*, described and figured by Cope in the 'Proc. Amer. Phil. Soc.' vol. xxiii. pp. 234–238, and plate (1886); the cast of the parietal foramen, representing the pineal gland (*o* of Owen's figure), is, however, much smaller than the latter. There is, moreover, no downward flexure immediately behind this point, this flexure not occurring till after the tubercle immediately behind *c* of the figure, which appears to represent the optic lobes, although correlated by Owen with the cerebellum. The cerebellum appears to be immediately behind this tubercle, the canals of the auditory capsule being clearly shown on the left at this point.

Presented by W. G. Atherstone, Esq., M.D., 1872.

Genus **SCALOPOSAURUS**, Owen[1].

Dentition probably very nearly the same numerically as in *Galesaurus*, but the upper teeth relatively smaller, and the cheek-teeth taller and more slender, but the contour of their crowns unknown. Skull much depressed, of a triangular form, with a long muzzle, and the supratemporal fossæ scarcely larger than the orbits, which are mainly directed frontally. Mandible very slender, with the inferior surface of the symphysis sloping obliquely upwards.

[1] Cat. Foss. Rept. S. Africa, p. 24 (1876).

Scaloposaurus constrictus, Owen[1].

The type and only described species. The smallest representative of the family, being considerably inferior in size to *Sphenodon punctatus*.

Hab. South Africa.

R. 1723. (*Fig.*) The skull, wanting the extremity of the muzzle; from the Karoo system of the Sneewberg range. The type; figured by Owen in his 'Catalogue of the Fossil Reptilia of S. Africa,' pl. xvi. figs. 10–15. The teeth are much damaged, but near the muzzle they can be distinctly seen to be free from anchylosis to the bone, by which this form is at once distinguished from the Procolophonia.

Presented by the Trustees of the Albany Museum.

Genus **CYNOSUCHUS,** Owen[2].

Known by the anterior extremity of the skull. Upper incisive teeth apparently 4 in number, and probably without serrations; upper cheek-teeth 7 or 8 in number, with short thick crowns, in which the posterior edge has a basal tubercle, but no serrations. Mandibular symphysis with a nearly vertical inferior face. Palate apparently very wide, and not suddenly contracted behind the tusk.

The absence of the internarial septum in the type is probably due to its imperfection.

Cynosuchus suppostus, Owen[3].

The type and only described species. Assuming the type specimens to be adult this species will be of the approximate size of *Ælurosaurus felinus*; upper cheek-teeth seven in number.

Hab. South Africa.

R. 1718. (*Fig.*) The anterior extremity of the skull; from the Karoo system of the Sneewberg range. The type; figured by Owen in his 'Catalogue of the Fossil Reptilia of S. Africa,' pl. xvi. figs. 1–5. The muzzle of the cranium has been crushed down upon the mandible, and has thus become abnormally widened. The presence of seven upper cheek-teeth is clearly seen on the right side. Two

[1] Cat. Foss. Rept. S. Africa, p. 24 (1876).
[2] *Ibid.* p. 21. [3] *Loc. cit.*

premaxillary teeth remain on either side, but there are indications of the presence of two other pairs of these teeth. The contour of the mandibular symphysis is well preserved; and the section of the upper tusk indicates the large size of that tooth. The disappearance of the internarial septum is doubtless due to the effects of crushing. The apparent absence of serrations in the premaxillary teeth is shown by a tooth on the right side, on which aspect the characters of the cheek-teeth are also well displayed.

Presented by the Trustees of the Albany Museum.

49404. (*Fig.*) Six fragments of rock containing portions of the skeleton of a Theriodont probably referable to a rather smaller individual of the present species; from Palimeet Fontein, Cape Colony. The specimens comprise part of the right ramus of the mandible (fig. 15); the right scapula, attached to

Fig. 15.

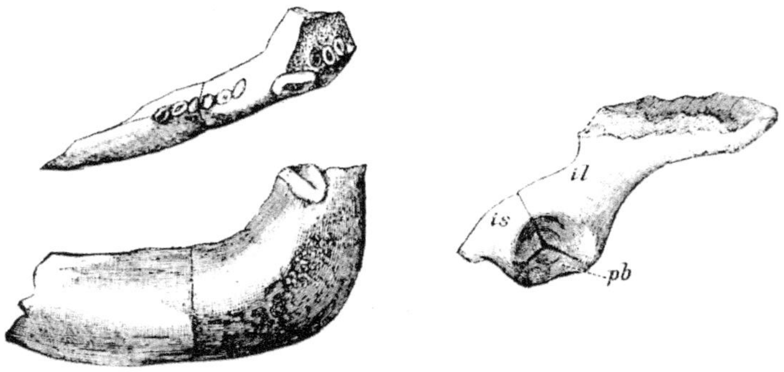

A B

? *Cynosuchus suppostus.*—The imperfect right ramus of the mandible (A), and the greater part of the right innominate (B); from the Karoo system of the Cape Colony. $\frac{1}{1}$.

the right side of six cervical vertebræ; the impression of the dorsal surface of the distal two thirds of the left humerus, and also what is apparently the head of the ulna; part of the right innominate (fig. 15) with fragments of that of the opposite side, and portions of the sacral region, and other imperfectly preserved bones. The mandibular ramus (fig. 15) closely resembles that of the type, although of somewhat smaller size; sections of three teeth in advance of the tusk are shown, and there is a space which might contain a fourth. Sections of five or six cheek-teeth are also visible. The cervical vertebræ, as shown by a cross-section, seem to have had deep terminal cups to the centra. The humerus, as shown by a wax cast taken from the mould, appears to have the same general contour as that of *Cynodraco serridens*. In the innominate (fig. 15) the ilium and the acetabular region of the ischium and pubis are preserved. The contour of the ilium appears to come nearest on the whole to that of the specimen referred to *Tapinocephalus* (fig. 17), but the preacetabular portion is relatively larger and more produced. *Purchased from T. Bain, Esq.*, 1878.

Genus **CYNOCHAMPSA,** Owen[1].

Very imperfectly known. Apparently allied to *Cynosuchus*, but the palate very narrow and suddenly constricted behind the tusks. Incisive teeth $\frac{4}{3}$ in number, but their structure unknown. Number and structure of cheek-teeth unknown.

Cynochampsa laniaria, Owen[2].

The type and only described species. Of the approximate dimensions of *Cynosuchus suppostus*, but probably with a larger and narrower skull.

Hab. South Africa.

36229. (*Fig.*) The imperfect anterior extremity of the skull; from the Beaufort beds[3] of the Karoo system of the Rhenosterberg branch of the Sneewberg range. The type; figured by Owen in the 'Quart. Journ. Geol. Soc.' vol. xvi. pl. iii. figs. 1–4, and also in his 'Catalogue of the Fossil Reptilia of S. Africa,' pl. xviii. figs. 1–5. The mandible is represented only by the superior portion of the symphysis, and exhibits the outer aspect of the incisive teeth and a section of the tusks. Portions of the four upper incisives are seen, as well as a section of the upper tusk of either side. The nasals were evidently elongated; the internarial septum is wanting.

Presented by Sir George Grey, K.C.B., 1858.

[1] Quart. Journ. Geol. Soc vol xvi. p. 61 (1860). [2] *Loc. cit.*
[3] See Quart. Journ. Geol. Soc. vol. xxiii. p. 143.

Genus **CYNODRACO,** Owen[1].

Syn. *Cynodracon,* Owen[2].

Imperfectly known. Incisive teeth $\frac{5}{7}$, with compressed and curved crowns, of which the posterior edge is strongly serrated; tusks large, and likewise serrated on the posterior border.

The apparently single narial aperture of the type specimen is probably due to its imperfection. So far as present evidence goes there appears no character by which *Ælurosaurus* (*infrà*) can be separated from this genus.

Cynodraco serridens, Owen[3].

Syn. *Cynodraco major,* Owen[4].

The type species, and one of the two largest representatives of the family. There appear to be no grounds for regarding *C. major* as specifically distinct from *C. serridens.*

Hab. South Africa.

47084. (*Fig.*) The imperfect anterior extremity of the skull, from the Karoo system near Fort Beaufort. The type; figured by Owen in his 'Catalogue of the Fossil Reptilia of S. Africa,' pl. xvii. figs. 1–4; and also, in conjunction with No. 47086, in the 'Quart. Journ. Geol. Soc.' vol. xxxii. p. 358, fig. 5. The whole series of upper incisive teeth is shown, but a great part of the outer surface of the premaxilla is wanting, so that the absence of the internarial septum may readily be accounted for.

Presented by A. G. Bain, Esq., 1853.

47086. (*Fig.*) The imperfect mandibular symphysis, showing the bases of the incisive teeth and the tusks; from near Fort Beaufort. Figured by Owen in his 'Catalogue,' pl. xvii. figs. 9, 10. *Presented by A. G. Bain, Esq.*, 1853.

47309. (*Fig.*) The imperfect mandibular symphysis, with the broken exposed portions of the upper caniniform teeth attached to it; from Mildenhalls, near Fort Beaufort. Figured by Owen in the 'Quart. Journ. Geol. Soc.' vol. xxxii. pl. xi. figs. 1–5, and also in his 'Catalogue,' pl. ii. figs. 4–6, as *Cynodraco major*, of which it is the type. There are no characters by which this specimen can be specifically

[1] Quart. Journ. Geol. Soc. vol. xxxii. p 95 (1876).
[2] Cat. Foss. Rept. S. Africa, p. 18 (1876). [3] *Loc. cit.*
[4] Quart. Journ. Geol. Soc. vol. xxxii. p. 95 (1876).

distinguished from the preceding. Sections are seen of the four incisives and also of the tusks of the mandible. The serration of the posterior edge of the upper tusks is also well shown. *Presented by A. G. Bain, Esq.*, 1853.

47065. (*Fig.*) Mass of rock, showing either portions of the bone of the inner layers of the skull or their impression; from the Karoo system of Styl-Krantz, in the Sneewberg range. Figured by Owen in his 'Catalogue,' pl. xvii. figs. 5, 7, 8. The impression of the first incisiform tooth of the left side exhibits the serrations on its posterior border very clearly, and also the greater length of this tooth as compared with the next one.

Presented by A. G. Bain, Esq., 1853.

47310. (*Fig.*) The nearly entire left humerus; from Mildenhalls. This specimen appears to have belonged to the same individual as No. 47309. It is figured by Owen of half the natural size in the 'Quart. Journ. Geol. Soc.' vol. xxxii. pl. xi. figs. 6–9, from the palmar and dorsal aspects and the two extremities; and also of the natural size from the palmar aspect in his 'Catalogue,' pl. xix. fig. 1.

Presented by A. G. Bain, Esq., 1853.

Genus **ÆLUROSAURUS,** Owen [1].

Incisive teeth $\frac{(4-5)}{(3-4)}$; cheek-teeth $\frac{5}{5}$, with compressed and backwardly curved crowns having serrated posterior edges; premaxillary teeth also serrated posteriorly. Skull vaulted, with the facial portion deep, and typically of considerable length; size of temporal fossa unknown; direction of orbits mainly lateral; postorbital bar inclined downwards and backwards. Superior surface of mandibular symphysis sloping obliquely upwards.

So far as regards the anterior dentition there appear to be no characters by which this genus can be distinguished from the preceding, so that its right to stand has yet to be proved.

Ælurosaurus felinus, Owen [2].

The type species. Considerably larger than *Galesaurus planiceps*, but entire length of skull unknown. Premaxillary teeth five in

[1] Quart. Journ. Geol. Soc. vol. xxxvii. p. 261 (1881).
[2] *Loc. cit.*

number; preorbital portion of cranium long and straight, with a long interval between the orbit and nares.

Hab. South Africa.

R. 339. (*Fig.*) The skull, wanting nearly the whole of the preorbital portion; from the Karoo system of the Gouph district, Beaufort West. The type; described and figured by Owen in the 'Quart. Journ. Geol. Soc.' vol. xxxvii. p. 261, pl. ix. The apparently single narial aperture is probably due to the imperfection of the specimen. Serration of premaxillary teeth very dense, but that of cheek-teeth invisible.

Purchased from T. Bain, Esq., 1880.

Fig. 16.

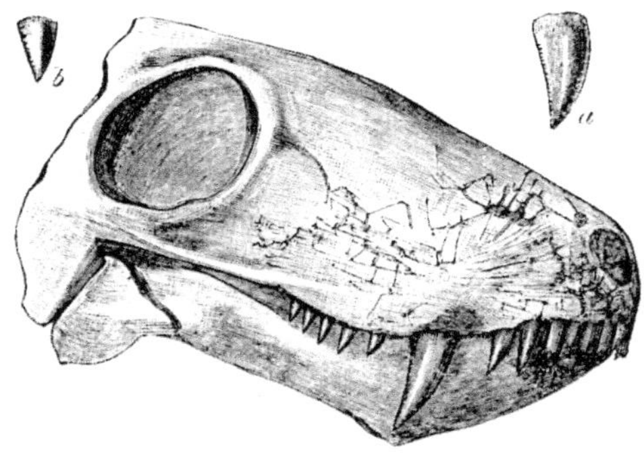

Ælurosaurus felinus.—The right lateral aspect of the imperfect cranium; from the Karoo system of the Cape Colony. $\frac{2}{3}$. An upper incisive tooth (*a*) is shown on the right, and an upper cheek-tooth (*b*) on the left side on an enlarged scale.

R. 855. The anterior portion of the cranium; from the Gouph district. The postorbital portion is wanting, and the specimen is very imperfect: but sufficient remains to show the close resemblance in contour and also in the number of the premaxillary teeth with the type specimen. The internarial septum is preserved.

Purchased from T. Bain, Esq., 1880.

Ælurosaurus, sp.

Distinguished from the type species by the shorter interval between the orbit and nares, and the somewhat more upward inclination of the inferior surface of the mandibular symphysis; apparently attains a somewhat larger size than the type species.

Hab. South Africa.

R. 855 a. The anterior extremity of the crushed skull; from the Karoo system of the Gouph district, Beaufort West. The cranium has been flattened by a vertical crush; but the short interval between the orbit and nares is well shown. There are seen to be five premaxillary teeth on either side; while in the mandible four teeth are shown in advance of the tusk. The serrations on the posterior border of both the premaxillary and cheek-teeth are clearly shown. *Purchased from T. Bain, Esq.*, 1880.

49419. The anterior extremity of a rather larger skull apparently referable to this form; from the Gouph district. The internarial septum is broken away, and the teeth are much damaged. Five premaxillary teeth can, however, be seen on either side, the summit of the fifth on the left side showing the serrated posterior border very distinctly. The upper tusk of the right side is much smaller than the remains of the corresponding tooth on the left, and would appear to be a successional tooth.

Purchased from T. Bain, Esq., 1878.

Ælurosaurus curvimola (Owen[1]).

Syn. *Lycosaurus curvimola*, Owen[2].

Considerably larger than the type species. Four upper incisive teeth; and the preorbital portion of the skull comparatively short and tapering. The small size of the upper tusk in the type specimen is not improbably indicative of a female. The resemblance of the skull of this form to that of the type of *Ælurosaurus* was pointed out by Owen in his description of the latter, and now that the division of the nares in that form is known, it seems scarcely advisable to generically separate the present form solely on account of having four in place of five premaxillary teeth.

Hab. South Africa.

47339. The imperfect skull; from the Karoo system of Kugaberg.
(*Fig.*) The type specimen; figured by Owen in his 'Catalogue of the Fossil Reptilia of S. Africa,' pl. lxviii.; a small-sized

[1] Cat. Foss. Rept S. Africa, p. 71 (1876).—*Lycosaurus*.

[2] *Loc. cit.*

reversed view of the right lateral aspect being also given in the 'Quart. Journ. Geol. Soc.' vol. xxxii. p. 359, fig. 8. The occipital region is very imperfect, and portions of the roof of the brain-case are wanting, so as to expose a cast of parts of the brain. The mandible is tightly closed against the palate, so as to conceal the lower teeth. The general contour of the cranium and mandible accords closely with that of the type species, the inferior aspect of the mandibular symphysis having precisely the same oblique upward inclination. The postorbital bar also has the same backward inclination as in the latter. The serration of the posterior borders of the cheek-teeth is well shown; but owing to this border of the premaxillary teeth being concealed in matrix the serrations are invisible.

Presented by A. G. Bain, Esq., 1853.

Genus **LYCOSAURUS**, Owen[1].

Incisive teeth apparently $\frac{4}{3}$ in number, and the cheek-teeth 5 or 6 in upper jaw; structure of teeth not shown; contour of skull very imperfectly shown in the type species, but the postorbital bar inclined downwards and forwards; if the form mentioned below as *Hyorhynchus* belongs to this genus, the orbits will be directed laterally. Inferior surface of mandibular symphysis nearly vertical, and its lateral surface with a distinct hollow for the protection of the upper tusk.

Lycosaurus pardalis, Owen[2].

Syn. *Lycosaurus tigrinus*, Owen[3].

The type species. Attains dimensions considerably exceeding those of *Ælurosaurus curvimola*.

Hab. South Africa.

R. 1717. (*Fig.*) The very imperfect and laterally flattened skull; from the Karoo system of the Cape Colony. The type specimen; figured by Owen in his 'Catalogue of the Fossil Reptilia of S. Africa,' pl. xiv., and small-sized figures given in the 'Quart. Journ. Geol. Soc.' vol. xxxii. p. 358, figs. 6, 7.

[1] Cat. Foss. Rept. S. Africa, p. 15 (1876).
[2] *Loc. cit.* [3] *Ibid.* p. 17.

The crushed condition of the specimen renders the vertical position of the inferior jaw of the mandibular symphysis not very clearly shown in the figures. In his 'Catalogue,' Owen gives the number of premaxillary teeth as three, but four are represented in the figure in the 'Quart. Journ. Geol. Soc.' There is no evidence as to the number of cheek-teeth.

Presented by the Trustees of the Cape Town Museum, 1877.

R. 1717 a. Fragment of rock split from the right side of the preced-
(*Fig.*) ing specimen. Figured by Owen in his 'Catalogue,' pl. xiv. fig. 5. Portions of the upper tusk and four cheek-teeth are shown.

Presented by the Trustees of the Cape Town Museum, 1877.

R. 1719. The anterior extremity of the skull, in an imperfect and
(*Fig.*) somewhat crushed condition; from the Karoo system of Mildenhalls, near Fort Beaufort. The type specimen of *L. tigrinus*; figured by Owen in his 'Catalogue,' pl. xv. There are no characters by which this specimen can be specifically distinguished from the preceding. The upper tusk is well shown on the right side, on which side the split crowns of five cheek-teeth are also shown. In Owen's figure a cheek-tooth is introduced in advance of these five, but this is merely a portion of the matrix; there may have been smaller cheek-teeth behind the fifth. The upright ventral surface of the mandibular symphysis is well exhibited. So far as can be determined, the premaxillary teeth do not seem to be serrated.

Presented by A. G. Bain, Esq., 1853.

49396. The anterior extremity of the imperfect skull of a larger individual apparently referable to this species; from the Karoo system of Beaufort West. The lateral aspect of the right ramus of the mandible and part of the upper jaw has been cut and polished to show a section of the tusks and anterior cheek-teeth. There must have been at least five of the latter in the upper jaw.

Purchased from T. Bain, Esq., 1878.

Genus **HYORHYNCHUS, Seeley[1].

Hyorhynchus platyceps, Seeley[2].

Known by the undermentioned specimen, which indicates an individual of the approximate size of the type of *Lycosaurus pardalis*, and does not show any characters by which it can be generically separated from *Lycosaurus*.

Hab. South Africa.

R. 872. (*Fig.*) The imperfect cranium; from the Karoo system of the Gouph district, Beaufort West. Figured by Seeley in the 'Phil. Trans.' for 1889, pl. xv. figs. 1, 3. The occipital region and muzzle are wanting, but the remaining portion retains its original contour. The direction of the post-orbital bar serves to distinguish the specimen from *Ælurosaurus*. Compared with the flattened type skull of *Lycosaurus pardalis* there appears to be, so far as can be seen, a marked resemblance between the two. Thus in *L. pardalis* the forward direction of the postorbital bar is clearly visible on the left side, and the orbits appear to have been lateral. The contour of the anterior border of the left temporal fossa is also seen to have been similar in the two specimens; and the width of the interorbital bar seems to have been likewise the same.

Purchased from T. Bain, Esq., 1880.

Genus **TIGRISUCHUS**, Owen[3].

Known only by the imperfect anterior extremity of the cranium. Perhaps allied to *Cynodraco*, but with only three upper incisive teeth.

Tigrisuchus simus, Owen[4].

The type and only described species. Fully equal in size to *Cynodraco serridens*, but with relatively larger incisive teeth.

Hab. South Africa.

R. 1721. (*Fig.*) The anterior extremity of the cranium, in a damaged condition; from the Karoo system of the Sneewberg range. The type; figured by Owen in his 'Catalogue of the Fossil Reptilia of S. Africa,' pl. xvi. figs. 6–9. Fragments of the tusks and sections of the roots of the three pairs of

[1] Phil. Trans. for 1889, p. 242.
[2] *Loc. cit.*
[3] Cat. Foss. Rept. S. Africa, p. 17 (1876).
[4] *Loc. cit.*

incisives are shown. The latter have very small pulp-cavities, and are thereby readily distinguished from the corresponding teeth of the following family.

Presented by the Director of the Albany Museum.

GENERICALLY UNDETERMINED SPECIMENS.

R. 1722. The proximal half of the right humerus of a large form; from the Karoo system of the Cape Colony. This specimen is broken off at the entepicondylar foramen. Although of smaller size, it approximates in contour to the humerus of *Cynodraco.*

Presented by Sir George Grey, K.C.B., 1858.

36300. The distal portion of a right humerus, agreeing in size with the preceding specimen, and probably referable to the present suborder; from the Karoo system of the Rhenosterberg branch of the Sneewberg range.

Presented by Sir George Grey, K.C.B., 1858.

R. 510. Fragment of rock containing numerous imperfect bones of a small Anomodont probably referable to the present suborder; from the Karoo system of the Orange Free State. This specimen shows a portion of the dorso-lumbar and caudal regions of the vertebral column, portions of the pelvis, the femur of either side, and the left tibia.

By exchange with the Blomfontein Museum.

Family TAPINOCEPHALIDÆ.

This family is taken to include the undermentioned imperfectly known large forms. The teeth always appear to have had large pulp-cavities, but in the type genus the nature of the dentition is very imperfectly known, although there were apparently no tusk-like teeth. In the other genus, however, the dentition was of a carnivorous type and differentiated into groups. The type genus appears to have no secondary posterior nares.

Vertebræ notochordal, with flattened terminal faces to the centra and moderately tall neural spines; no intercentra in dorsal region. Humerus relatively short and wide. with the proximal extremity much expanded, and a sudden angulation at the point of origin of the delto-pectoral crest.

Genus **TAPINOCEPHALUS**, Owen[1].

Syn. (?) *Phocosaurus*, Seeley[2].

The type genus; founded upon the imperfect cranial muzzle, with which a vertebra and tibia were apparently associated. No tusks in upper jaw; muzzle of cranium forming a depressed rounded projection, above which the frontal region probably rose suddenly; apparently no secondary posterior nares. Lumbar vertebra (fig. 18)

Fig. 17.

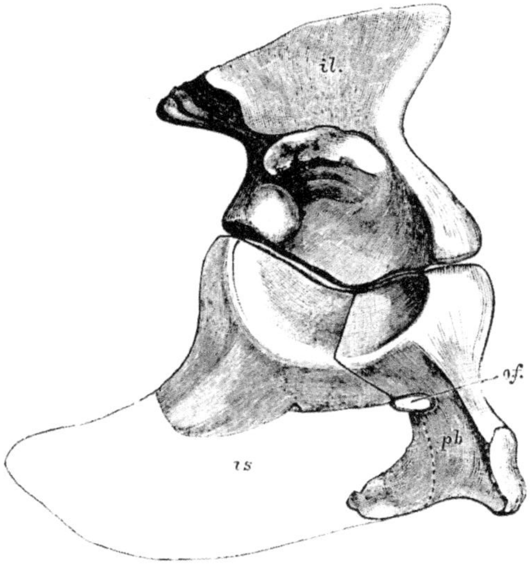

(?) *Tapinocephalus atherstonei*.—Reversed view of the left side of the pelvis, with the ischium restored; from the Karoo system of the Cape Colony. $\frac{1}{8}$. *il*, ilium; *pb*, pubis; *is*, ischium; *o.f*, obturator foramen.

with a short centrum, having angulated terminal faces, and distinct surfaces for free ribs or tranverse processes at the junction of the arch and centrum.

The undermentioned series of vertebræ, which are most probably referable to this genus, are characterized by their broad and more or less shortened centra and the absence of intercentra in the post-cervical region. The anterior dorsals have the hæmal surface of the centrum rounded, but in the lumbar and sacral region, where there are autogenous transverse processes or ribs, the hæmal and lateral

[1] Cat. Foss. Rept. S. Africa, p. 1 (1876). [2] Phil. Trans. for 1888, p. 91.

faces become flattened and form distinct angles at their junction with one another. The rib-facets on the anterior dorsals are sessile.

The series of bones of the limb-girdles and limbs, upon certain of which the genus *Phocosaurus* was based, are provisionally included under the present generic heading, since there is no evidence to show that they are generically distinct. In this series the coracoid and precoracoid remain distinct; the humerus has the head placed high up, the distal extremity moderately expanded, the radial condyle of normal size, and no ectepicondylar foramen; the pelvis (fig. 17) is characterized by the large acetabulum and the comparative lowness of the ilium, which has distinct pre- and postacetabular processes.

Tapinocephalus atherstonei, Owen[1].

Including *Phocosaurus megischion*, Seeley[2].

The type species. Of huge dimensions, perhaps somewhat exceeding those of *Titanosuchus*. It is quite probable that the under-mentioned specimens may indicate two distinct species.

Hab. South Africa.

R. 1705. (*Fig.*) The imperfect anterior extremity of the cranium; from the Karoo system near Jan Willem's Fontein, Prince Albert district. The type; figured by Owen in his 'Catalogue of the Fossil Reptilia of S. Africa,' pl. i. The broken posterior surface and part of the alveolar surface have been cut and polished. The bones forming the free surfaces of the facial aspect are in great part or entirely wanting, and no trace of the position of the nares remains. The determination of the bones exhibited in the transverse section by Owen appears to be correct. In the section of the alveolar border of the maxilla an outer series of fully developed teeth and an inner one of replacing germs are distinctly visible. The hindmost tooth, of which a section is shown, has a large pulp-cavity, like the teeth of *Titanosuchus*. In advance of this tooth there are several empty dental alveoli, but none of them appear large enough to have held a tusk answering to the lower tusk of *Titanosuchus*. This specimen apparently indicates the absence of secondary posterior nares; and it evidently

[1] Cat. Foss. Rept. S. Africa, p 1 (1876).
[2] Phil. Trans. for 1888, p. 91.

formed a flat and projecting muzzle, above which the region of the brain was probably much elevated.

Presented by W. G. Atherstone, Esq., M.D., 1872.

R. 1706. (*Fig.*) A nearly entire lumbar vertebra; from the same locality as the type, and from the nature of its mineralogical condition and peculiar colour, as well as from its somewhat rolled condition, doubtless belonging to the same individual. This specimen (woodcut, fig. 18) is noticed by Owen in his 'Catalogue,' p. 10, no. 19, as a late dorsal or lumbar of *Pariasaurus*. Facets for a free rib, or transverse process, at the junction of the arch with the centrum, show the presence of autogenous lateral elements in the hinder part of the presacral series. The centrum has a flattened hæmal surface, with the lateral surfaces also flattened and having an angulated junction with the former. The width of the centrum is 0,138 (5·5 inches).

Presented by W. G. Atherstone, Esq., M.D., 1872.

Fig. 18.

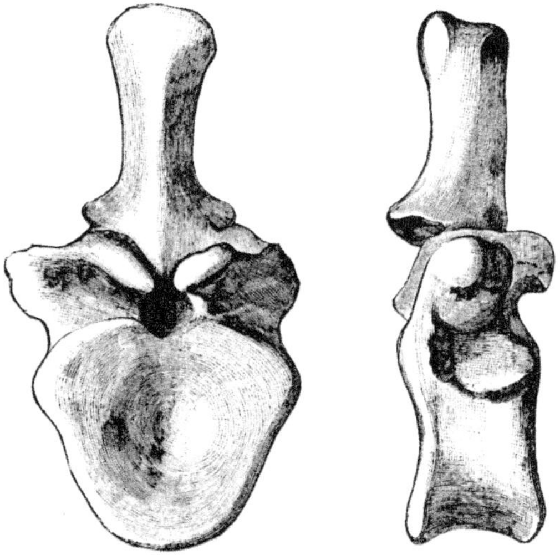

Tapinocephalus atherstonei.—Anterior and right lateral aspects of a lumbar vertebra; from the Karoo system of the Cape Colony. About $\frac{1}{5}$.

R. 1707. (*Fig.*) The right tibia; from the same locality as the type, and, for the reasons stated under the head of the preceding specimen, in all probability referable to the same individual. Figured by Seeley in the 'Phil. Trans.' for 1889, pl. xxv., without generic determination, and noticed by Owen in his 'Catalogue,' p. 13, no. 29, as *Pariasaurus*. The distal extremity is described by Seeley as indicating Mammalian affinities. The femur, No. 43525 *t* (*infrà*, p. 89), may belong to the same individual as the present specimen.

Presented by W. G. Atherstone, Esq., M.D., 1872.

R. 1706 a. An imperfect anterior dorsal vertebra; from the same locality as the type, and perhaps therefore referable to the same individual. This specimen has been longitudinally bisected in a vertical plane to show the notochordal canal. Noticed by Owen in his 'Catalogue,' p. 3, no. 4, and referred to *Tapinocephalus*. The rib-facet is partly on the arch and partly on the centrum; and the specimen closely approximates to the under-mentioned dorsal No. 43525 *d*, although of somewhat smaller dimensions. If the series to which the latter belongs is referable to the present genus, this specimen will likewise be referable to the same, although perhaps specifically distinct.

Presented by W. G. Atherstone, Esq., M.D., 1872.

The following associated series of vertebræ, of which the majority were referred by Owen to this species, appear generically inseparable, but may indicate a distinct species.

43525 a. The centrum and base of the arch of a cervical vertebra,
(*Fig.*) probably belonging to the same individual as the under-mentioned dorsals, and accordingly provisionally referred to the same form; from the Karoo system, apparently of Gats-Plaatz, Spreuw Fontein, Prince Albert district. Figured by Owen in his 'Catalogue of the Fossil Reptilia of S. Africa,' pl. x., and also in the 'Quart. Journ. Geol. Soc.' vol. xxxii. pl. v., as *Pariasaurus bombidens*. This specimen is there stated to have been obtained from Vers Fontein, but its mineralogical condition is precisely similar to that of the undermentioned dorsals. The centrum has been longitudinally bisected in a horizontal plane. Large facets for intercentra are shown.

Presented by W. G. Atherstone, Esq., M.D., 1872.

R. 854. An imperfect and somewhat crushed later cervical vertebra, agreeing in general character with the preceding; from the Karoo system of the Gouph district, near Beaufort

West. The transverse process of the left side is nearly entire, and is situated at a higher level than in the preceding specimen. The large intercentral facets are well shown. The tall neural spine and the oblique zygapophyses at once distinguish this specimen from the cervicals of *Pariasaurus*. *Purchased from T. Bain, Esq.*, 1880.

43525 b. (*Fig.*) An imperfect anterior dorsal vertebra; from the Karoo system of Gats-Plaatz, Spreuw Fontein, Prince Albert district. Figured by Owen in his 'Catalogue,' pl. iii., and also in the 'Quart. Journ. Geol. Soc.' vol. xxxii. pl. iv., as *Tapinocephalus atherstonei*. There are no intercentral facets. The base of the neural arch remains, and the elongated rib-facet is situated entirely on the centrum. The hæmal aspect of the centrum is rounded.

Presented by W. G. Atherstone, Esq., M.D., 1872.

43525 c. A similar associated imperfect dorsal vertebra; from Gats-Plaatz. Noticed in Owen's 'Catalogue,' p. 4, no. 6.

Presented by W. G. Atherstone, Esq., M.D., 1872.

43525 d. A slightly later associated imperfect dorsal vertebra; from Gats-Plaatz. Noticed by Owen, *l. c.* (as No. 6′). The rib-facet has ascended partly on to the arch.

Presented by W. G. Atherstone, Esq., M.D., 1872.

43525 e. An imperfect associated later dorsal vertebra; from Gats-Plaatz. Noticed by Owen, *loc. cit.*, as No. 7. The rib-facet has ascended on to the arch, and the hæmal aspect of the centrum is somewhat flattened, and the sides are depressed.

Presented by W. G. Atherstone, Esq., M.D., 1872.

43525 f. An imperfect associated trunk-vertebra, still later in the series; from Gats-Plaatz. Noticed by Owen, *loc. cit.*, as No. 8. The neuro-central suture has descended upon the centrum.

Presented by W. G. Atherstone, Esq., M.D., 1872.

43525 g. A somewhat later imperfect associated trunk-vertebra; from Gats-Plaatz. Noticed by Owen, *loc. cit.*, together with the preceding specimen, as No. 8. The neuro-central suture descends on to the centrum, as in the preceding specimen; but the centrum has become shorter, and has lost the marked lateral depressions of the latter.

Presented by W. G. Atherstone, Esq., M.D., 1872.

43525 h. (*Fig.*) An imperfect associated lumbar vertebra; from Gats-Plaatz. Figured by Owen, *loc. cit.*, pl. iv. The contour of the centrum is almost identical with that of the vertebra No. R. 1706; but the present specimen is of somewhat smaller size, a difference which may be merely an individual character. There is the same notch between the arch and the centrum as in No. R. 1706.

Presented by W. G. Atherstone, Esq., M.D., 1872.

43525 i. (*Fig.*) An imperfect sacrum, associated with the preceding; from Gats-Plaatz. Figured by Owen, *op. cit.* pl. v. Four vertebræ remain; but there is no evidence to show that the 3rd and 4th are not anterior caudals, and that the true sacrals were more than two in number. In the angulated contour of the centrum the first sacral accords with the lumbar. The 3rd and 4th vertebræ have notches at the junction of the arch with the centrum, as in the lumbars.

Presented by W. G. Atherstone, Esq., M.D., 1872.

43525 k. The imperfect centrum of an associated anterior caudal vertebra; from Gats-Plaatz. Noticed by Owen, *op. cit.* p. 5, as No. 11.

Presented by W. G. Atherstone, Esq., M.D., 1872.

43525 l. (*Fig.*) The imperfect centrum of an associated later caudal vertebra, which has been bisected to exhibit the notochordal canal; from Gats-Plaatz. Figured by Owen, *op. cit.* pl. v. fig. 4. *Presented by W. G. Atherstone, Esq., M.D.*, 1872.

43525 j. The imperfect proximal portion of a dorsal rib, associated with the preceding specimens; from Gats-Plaatz. Noticed by Owen, *op. cit.* p. 5, No. 12.

Presented by W. G. Atherstone, Esq., M.D., 1872.

43525 x. An imperfect late caudal vertebra apparently belonging to this series. This specimen is labelled "Jan Willem's Farm," but it is indistinguishable in mineral characters from the preceding.

Presented by W. G. Atherstone, Esq., M.D., 1872.

The following associated series of bones of the pectoral and pelvic girdles and limbs include the types of Phocosaurus megischion, *but there is no evidence to show that they may not be referable to the present form, under which heading they are accordingly entered provisionally; they were referred by Owen to* Pariasaurus.

43525 m. The glenoidal extremities of the right scapula and coracoid; from the Karoo system of Vers Fontein (?), Prince Albert district. The coracoid approximates in contour to the corresponding bone of the Dicynodonts, and shows the straight suture for a distinct precoracoid.

Presented by W. G. Atherstone, Esq., M.D., 1872.

43525 n. The glenoidal portions of the left scapula and precoracoid; from Vers Fontein. The foramen in the precoracoid pierces the bone obliquely.

Presented by W. G. Atherstone, Esq., M.D., 1872.

43525 o. The right humerus; from Vers Fontein. Noticed by Owen in his 'Catalogue of the Fossil Reptilia of S. Africa,' p. 11, no. 23, as *Pariasaurus.*

Presented by W. G. Atherstone, Esq., M.D., 1872.

43525 p. The left humerus; from Vers Fontein. Figured in (*Fig.*) Owen's 'Catalogue,' pl. xii., as *Pariasaurus.* This and the preceding specimen show the thin flange-like expansion of the preaxial border above the condyles, characteristic of the typical forms of the suborder, and also the narrow bar in front of the entepicondylar foramen.

Presented by W. G. Atherstone, Esq., M.D., 1872.

43525 q. The right ulna; from Vers Fontein. Noticed by Owen (*Fig.*) in his 'Catalogue,' p. 12, no. 25, as *Pariasaurus*, and figured by Seeley in the 'Phil. Trans.' for 1889, pl. xxii., having been referred in the abstract to *Titanosuchus.* This specimen unquestionably belongs to the same individual as the preceding humerus, which is quite different from the corresponding bone of *Titanosuchus.*

Presented by W. G. Atherstone, Esq., M.D., 1872.

43525 r. The imperfect left side of the pelvis; from Vers Fontein. (*Fig.*) The ilium remains separate from the ischium and pubis, which are anchylosed together. The extremity of the preacetabular process of the ilium is wanting, the pubis is somewhat imperfect, and the whole of the symphysial region of the ischium is broken away. The ilium is described by Owen in his 'Catalogue,' p. 13, no. 27, and referred to *Pariasaurus*, while the ischium and pubis are noticed in the same work, p. 11, no. 22, as the right scapula and coracoid of that genus. The entire specimen is figured by Seeley in the 'Phil. Trans.' for 1889, pl. xxi., as the type of *Phocosaurus*. In woodcut fig. 17, a reversed restoration of the specimen is given, partly based on the corresponding bones of the opposite side, and partly on the Dicynodont pelvis, No. R. 1698 (p. 57). In Seeley's figure the suture between the pubis and ischium is made to pass posteriorly to the foramen; but from a comparison of the right pubis with that of No. R. 1698, the correspondence is so close that it is practically certain that the vacuity is the obturator foramen, and consequently that the ischio-pubic suture should pass through it. The right pubis seems to indicate that this is really the case, since below the sutural surface of the acetabulum there is a flat surface extending obliquely forwards to the foramen, which appears to indicate the line of suture.

Presented by W. G. Atherstone, Esq., M.D., 1872.

43525 s. The right ilium and the imperfect pubis of the same side; from Vers Fontein. The ilium is noticed by Owen in his 'Catalogue,' p. 13, no. 26, and referred to *Pariasaurus*. The symphysial portion of the pubis is broken away, but the foramen remains.

Presented by W. G. Atherstone, Esq., M.D., 1872.

43525 t. (*Fig.*) The proximal half of the right femur; said to be from Vers Fontein. Noticed by Owen in his 'Catalogue,' p. 13, no. 28, and figured by him in the 'Quart. Journ. Geol. Soc.' vol. xxxvi. pl. xvii. fig. 8. The distal surface has been cut and polished. The long and straight great (outer) trochanter found in the Dicynodont femur is wanting. The mineralogical condition is like that of the type of *Tapinocephalus*, suggesting that the specimen is from Jan Willem's Fontein.

Presented by W. G. Atherstone, Esq., M.D., 1872.

Genus **TITANOSUCHUS**, Owen[1].

Dentition of a carnivorous type, and differentiated into groups; teeth apparently $\frac{17}{16}$ in number, of which $\frac{5}{4}$ occupy the position of

[1] Quart. Journ. Geol. Soc. vol. xxxv. p. 189 (1879).

incisors. Skull probably short and deep. Coracoid fused with precoracoid. Humerus with the head placed relatively low, and consequently a short postaxial curve, the distal extremity very wide, the radial condyle extending far up on the palmar aspect, and an ectepicondylar foramen piercing the shaft in a direction ascending obliquely from the preaxial flange to the postaxial aperture of the entepicondylar foramen. Pelvis and vertebræ unknown.

The pulp-cavities of the teeth appear to have become completely closed, but (although the contrary opinion has been expressed) successional teeth were developed. The roots of the teeth terminated in a point. Unless the specimen described as *Glaridodon* (*infrà*, p. 92) prove to belong to this genus the crowns of the teeth are unknown; the cross-section of their roots is, however, of the same general type as in the teeth mentioned below under the heading of *Deuterosaurus*.

Titanosuchus ferox, Owen[1].

The type and only described species; attains dimensions nearly or perhaps quite equal to those of *Tapinocephalus atherstonei*.

Hab. South Africa.

49370. (*Fig.*) Associated fragments of the anterior portion of the cranium and mandible; from the Karoo system of Koodos-kop, Gouph district, Beaufort West. The types; described and figured by Owen in the 'Quart. Journ. Geol. Soc.' vol. xxxv. pp. 189–198, pl. xi. These specimens have been either sectionized in various planes, or their alveolar surfaces have been ground down and polished. In the fragment 49370 *b*, of which a fang of a tooth is represented in fig 4 of the plate, an inner row of successional teeth is clearly seen; a section of one of these teeth presenting the same triangular contour without a pulp-cavity observable in the anterior tooth of the type of *Tapinocephalus*. In the unfigured fragment 49370 *a*, the inner and outer rows of teeth are shown both on the worn external surface and in the section, the contour of both series being almost identical with that of the teeth of the specimen No. R. 1708, mentioned under the head of the generically undetermined specimens (p. 93). The left side of the mandibular symphysis represented in fig. 6 of the plate closely resembles the corresponding part of the jaw referred to *Deutero-*

[1] Quart. Journ. Geol. Soc. vol. xxxv. p. 189 (1879).

saurus. The incisives have less compressed roots than those which probably belong to the upper jaw.
Purchased from T. Bain, Esq., 1878.

49367. (*Fig.*) The left coracoid; from Koodos-kop. This specimen, which apparently belongs to the same individual as the following humerus, is figured by Seeley in the 'Phil. Trans.' for 1889, pl. xvi. fig. 4, as the left pubis. The bone is, however, totally unlike a pubis, and a comparison with the Dicynodont pectoral girdle, No. 36287 (p. 52), shows such a close resemblance with the coracoid and precoracoid, as to indicate that the present specimen represents those two bones conjointly. The anterior portion of the bone is incomplete, so that it cannot be determined whether there was a precoracoidal foramen. The coracoidal portion closely resembles the coracoid referred to *Tapinocephalus* (p. 88). *Purchased from T. Bain, Esq.*, 1878.

49369. (*Fig.*) The imperfect left humerus; from Koodos-kop. Figured by Seeley in the 'Phil. Trans.' for 1889, pl. xx. This specimen, together with the following one, was found lying on the surface with the types, and it is almost certain that one or the other belonged to the same individual as the latter. In the present specimen the delto-pectoral crest and the radial condyle are imperfect, and the bar in front of the entepicondylar foramen is broken away. The upward extension of the radial condyle is clearly shown; the ectepicondylar foramen is seen piercing the entire width of the shaft, its postaxial aperture opening into that of the entepicondylar foramen, while its preaxial aperture is placed on the palmar aspect of the large distal flange in the same position as in the humerus of *Brithopus* (p. 98).
Purchased from T. Bain, Esq., 1878.

49367 a. A more imperfect example of the left humerus; from Koodos-kop. Found in association with the preceding specimens. The proximal extremity is wanting, and the distal portion imperfect. The entepicondylar foramen, although crushed, is entire; and the two apertures of the ectepicondylar foramen are shown.
Purchased from T. Bain, Esq., 1878.

49368. (*Fig.*) The right femur; from Koodos-kop. Found with the preceding specimen, and, judging from its mineralogical condition, probably belonging to the same individual. Figured by Seeley, *op. cit.* pl. xix.

Purchased from T. Bain, Esq., 1878.

49367 b. (*Fig.*) An imperfect fibula; from Koodos-kop. Figured by Seeley, *op. cit.* pl. xxi. This specimen appears to have belonged to he same individual as the humerus No. 49369.

Purchased from T. Bain, Esq., 1878.

49367 c. (*Fig.*) Four phalangeal bones; from Koodos-kop. One of these is terminal; the others are figured by Seeley, *op. cit.* pl. xxiv. figs. 1, 2. *Purchased from T. Bain, Esq.*, 1878.

One or both of the two following specimens may belong to this or an allied form.

47100. A very large tibia; from the Karoo system of the Cape Colony. This specimen, although indicating an equally large animal, is longer and less massive than the tibia of *Tapinocephalus*, from which it also differs in contour.

Presented by A. G. Bain, Esq., 1853.

R. 519. A flattened tibia, approximating in size and contour to the preceding specimen; from the Karoo system of the Cape Colony. A concretion-like structure on one border of the shaft, which has been cut and polished, appears to indicate an ulceration of the bone during life.

Presented by Dr. Exton, 1884.

Generically Undetermined Specimens.

The following specimens indicate forms more or less closely allied to or identical with the preceding genera.

(a. *Glaridodon*, Seeley [1].)

49425. (*Fig.*) An imperfect incisive tooth; from the Karoo system of the Gouph district, Beaufort West. Described and figure by Seeley in the 'Proc. Roy. Soc.' vol. xliv. pp. 135, 136, fig. 2, as *Glaridodon*, of which it is the type. As already pointed out by the writer (Proc. Zool. Soc. 1889, p. 576), there are no characters by which this specimen can be generically distinguished from *Titanosuchus*. Since it was figured the root has been transversely cut, and the section

[1] Proc. Roy. Soc. vol. xliv. p. 135 (1888).

closely resembles that of the lower incisors of *Titanosuchus*. Although of the same general type, this tooth differs from the upper incisors referred to *Deuterosaurus* in the absence of the inner basal lobe, and the prominent vertical ridges on the lateral borders of the posterior aspect of the crown. *Purchased from T. Bain, Esq.*, 1878.

b. *Specimens of the dentition, some of which were referred by Owen to* Tapinocephalus.

R. 846. A very large imperfect tooth; from the Gouph district, Beaufort West. The crown has lost its summit and is otherwise imperfect, and the lower part of the root is wanting. The crown and root have the same compressed form as in the preceding specimen, although this tooth is of considerably larger size.
Purchased from T. Bain, Esq., 1880.

R. 1708. (*Fig.*) Fragment of a jaw showing on the polished surface sections of teeth; from the Karoo system of the Cape. Figured by Owen in his 'Catalogue,' pl. ii. figs. 1–3, where it is described as part of the mandibular symphysis, and referred to *Tapinocephalus*, being incorrectly stated to be one of the specimens presented by Dr. Atherstone. A thin slice, now mounted on glass, has been cut from the polished surface. The sections of the teeth appear to be very similar to those of *Titanosuchus*, an inner series of successional teeth being shown. *Presented by A. G. Bain, Esq.*, 1853.

47089. Fragment of a jaw exhibiting sections of large teeth; from the Great Karoo district. Noticed by Owen in his 'Catalogue,' p. 28, no. 53.
Presented by A. G. Bain, Esq., 1853.

c. *Specimens of vertebræ of which it is probable that a large proportion are referable to the present family, although some may belong to the* Pariasauria. *There is a probability that the associated series first mentioned may prove to belong to* Titanosuchus.

R. 1709. The centrum of an anterior dorsal vertebra; from the Karoo system of Jan Willem's Fontein [1], Prince Albert district. This specimen has been longitudinally bisected in a vertical plane. It differs from the anterior dorsals mentioned under the head of *Tapinocephalus* in its flattened hæmal and lateral surfaces; and since those vertebræ are associated with lumbars of the type of No. R. 1706 (p. 84) it would appear that the present specimen cannot belong to the same form as the latter.

Presented by W. G. Atherstone, Esq., M.D., 1872.

R. 1709 a. An imperfect anterior dorsal vertebra; from the same (*Fig.*) locality as the preceding, and apparently associated. Figured by Owen in his 'Catalogue of the Fossil Reptilia of S. Africa,' pl. xi., and referred to *Pariasaurus*, the locality being incorrectly given as Vers Fontein. This specimen, which has been longitudinally bisected in a horizontal plane, has the costal facet partly on the centrum and partly on the arch, and the centrum is more rounded than in the preceding specimen.

Presented by W. G. Atherstone, Esq., M.D., 1872.

R. 1709 b. A very imperfect dorsal centrum, showing the rib-facets; from the same locality, and probably associated with the preceding.

Presented by W. G. Atherstone, Esq., M.D., 1872.

R. 1709 c. An imperfect dorsal centrum of the same type; from the same locality.

Presented by W. G. Atherstone, Esq., M.D., 1872.

R. 1709 d. Four imperfect caudal vertebræ; from the same locality, and probably associated with the preceding.

Presented by W. G. Atherstone, Esq., M.D., 1872.

R. 852. Two imperfect centra of large trunk-vertebræ; from the Karoo system of the Gouph district, near Beaufort West. In their constricted and carinated form these specimens approximate to the later dorsals of the series noticed under the head of *Tapinocephalus*.

Purchased from T. Bain, Esq., 1880.

R. 851. Three associated imperfect dorsal centra of the same type as the preceding; from the Gouph district.

Purchased from T. Bain, Esq., 1880.

47106. The imperfect neural arches and spines of three dorsal vertebræ of a large form; from the Karoo system of the Cape Colony. The height of the neural spine and the obliquity of the zygapophyses at once distinguishes these specimens from *Pariasaurus*.

Presented by A. G. Bain, Esq., 1853.

47106 a. The greater portion of a rib, associated with the preceding.

Presented by A. G. Bain, Esq., 1853.

The following specimen not improbably belongs to this group.

R. 1663. (*Fig.*) The imperfect right quadrate of a very large form; from the Karoo system of the Cape Colony. Figured by Seeley in the 'Phil. Trans.' for 1889, pl. x. figs. 4–6, without generic determination. This specimen differs from the quadrate of the Dicynodontia by the presence of an antero-posterior perforation above the trochlea; and since there are no other forms which are sufficiently large to have had a quadrate of the size of this specimen, there is a strong presumption that it belongs to the present group.

No history.

Family Uncertain.

The incisive teeth of the type referred to *Deuterosaurus* present a remarkable approximation to the teeth of the American family *Bolosauridæ*[1], in which both the anterior and the cheek-teeth have their crowns elongated at right angles to the axis of the jaw. The American forms[2] do not have the dental series differentiated into distinct tusks and cheek-teeth, and the dentition is regarded as indicating herbivorous habits. The presence of a distinct tusk in *Diadectes* and its absence in *Empedias* show, however, that such differences are of no great importance. Whether the cheek-teeth of the skull referred to *Deuterosaurus* were transversely elongated is not apparent.

The humerus of the undermentioned forms approximates to the type of that of the *Tapinocephalidæ*, although of smaller size. The vertebræ of *Deuterosaurus* and of the undermentioned African form have a notochordal canal and the centra in the dorsal region much compressed.

[1] See Cope, 'Trans. Amer. Phil. Soc.' vol. xvi. p. 288 (1886). In a communication regarding the relations of these European forms to the *Bolosauridæ* Prof. Cope writes to the author as follows:—"Your figure of the incisor referred to *Deuterosaurus* resembles a good deal the cheek-teeth of *Bolosaurus*, and less nearly the incisors of *Empedias*, owing to the weaker basal shoulder in the latter. In the few known teeth of *Chilonyx* the basal shoulder is represented by an angle only."

[2] In *Chilonyx* the temporal fossæ are roofed. See Cope, 'Proc. Amer. Phil. Soc.' vol. xx. p. 631.

Till the form of the crowns of the incisive teeth in the *Tapinocephalidæ* is known it is impossible to say how far the present group was related to that family; but it would seem that in this group the incisive teeth were larger in proportion to the jaws than in the latter.

GENUS *non det.*

Indicated by two dorsal vertebræ, the centra of which are moderately compressed, with a subtriangular cross-section and without hæmal carina. The transverse processes appear to have been supported by flange-like buttresses, and the terminal cups are of moderate depth.

These specimens indicate a reptile of considerably larger size than the next form, but apparently allied in the structure of the vertebræ.

Hab. South Africa.

R. 847. Two imperfect adjacent dorsal vertebræ; from the Karoo system of the Gouph district, Beaufort West. The greater portion of the neural spines are wanting. The arches are fairly well preserved, and the capitular rib-facet is clearly seen on the anterior border of the upper part of the centrum. The length of the centrum is 0,070 (2·75 inches) and its vertical diameter 0·066 (2·6 inches).

Purchased from T. Bain, Esq., 1880.

GENUS *non det.*

Represented by a series of bones, among which the humerus agrees closely in size and contour with that of *Brithopus* (*infrà*), but apparently has no ectepicondylar foramen. Vertebræ with the centra much compressed and furnished with a sharp hæmal carina; the centra of the dorsals being smaller and longer than those of the lumbars. Transverse processes of dorsals supported by three prominent flange-like buttresses. No intercentra in the dorsal series.

These vertebræ closely resemble those mentioned under the head of *Deuterosaurus*.

Hab. South Africa.

The following associated series of specimens were obtained from the Karoo system at Bakkars Kraal, in the Rouxville district of the Orange Free State, and were presented by C. S. Orpen, Esq., 1884.

R. 533. Two dorsal vertebræ, cemented together by matrix. Figured (*Fig.*) by the writer in the 'Proc. Zool. Soc.' 1889, pl. liv. fig. 1.

The centrum of the anterior vertebra is imperfect, but the hinder one is well preserved. In the form of the transverse process and the sharp carination of the centrum these vertebræ resemble those of the *Clepsydropidæ*, but the terminal faces are much less deeply cupped.

R. 533 a. (*Fig.*) The left transverse process and zygapophyses of a dorsal vertebra. Figured (reversed), *op. cit.* pl. liv. fig. 2.

R. 533 b. An imperfect and crushed posterior dorsal or lumbar vertebra. The right prezygapophysis and transverse process remain, but the other parts of the arch and the spine are wanting.

R. 533 c. An imperfect vertebra belonging to the same region of the column as the preceding.

R. 533 d. (*Fig.*) The centrum and base of the arch of a lumbar vertebra. Figured, *op. cit.* pl. liv. fig. 3. The centrum has been somewhat crushed; a reflection of the posterior border on one side suggests the presence of an intercentrum, but it is more probable that this is due to crushing.

R. 533 e. Part of the arch of a trunk vertebra.

R. 533 f. Three imperfect caudal vertebræ cemented together by matrix. The neural spines and transverse processes are imperfect. The centra resemble those of the dorsals in their comparative length.

R. 533 g. (*Fig.*) The imperfect right scapula. The two extremities are preserved, but the middle is wanting. Figured, *op. cit.* pl. lv. fig. 1. In general contour and size this specimen approximates to the corresponding bone of *Platypodosaurus*, although presenting characters which indicate its generic distinctness.

R. 533 h. (*Fig.*) The imperfect distal portion of the left humerus. Figured, with a restoration of the proximal portion, *op. cit.* pl. lv. fig. 3. The region of the ulnar condyle is broken away, and the distal preaxial border is also imperfect. Sufficient of the latter remains, however, to show that an ectepicondylar foramen was probably wanting.

R. 533 i. Two portions of a femur. The upper fragment shows part of the great trochanter, while the lower one comprises the distal extremity.

R. 533 j. One extremity of an epipodial bone.

R. 533 x. Fragmentary undetermined bones.

Genus **BRITHOPUS,** Kutorga[1].

Syn. *Orthopus*, Kutorga[2].
Eurosaurus, Fischer[3].

Definitely known by the humerus, which has both ectepicondylar and entepicondylar foramina, of which the former merely pierces the supinator ridge; while its general contour resembles that of the corresponding bone of *Titanosuchus* and *Tapinocephalus*.

It has been shown by Owen that the type specimen of *Orthopus* is the proximal extremity of a left humerus, probably belonging to the same individual as the undermentioned specimen. As mentioned below there is no evidence to show that *Deuterosaurus* is really distinct from this genus. *Eurosaurus* seems to have been founded upon the evidence of the same specimen as *Orthopus*, but was subsequently confounded by Eichwald with the Labyrinthodont *Melosaurus*.

Brithopus priscus, Kutorga[4].

Syn. *Orthopus primævus*, Kutorga[5].
Eurosaurus, Fischer[6].

The type and only known species. The humerus indicates a reptile very much smaller than *Titanosuchus ferox*, although considerably larger than *Cynodraco serridens*. Trautschold[7] describes and figures part of a femur which he refers to this species.

Hab. Russia.

39358. Cast of the imperfect distal extremity of the left humerus. The original, which is the type, was obtained from the Upper Permian on the western flanks of the Urals in the Government of Perm, province of Kazan, and is preserved in the Museum of the University of Kazan. It is figured by Kutorga in pl. i. of the work above cited; and is noticed by Eichwald in his 'Lethæa Rossica,' vol. ii. p. 1626 (1860), as the coracoid of *Eurosaurus*. It is again figured

[1] Beitrage z. Kenntniss d. organischen Ueberreste d. Kupfersandsteins des Urals, p. 9 (1838).
[2] *Ibid.* p. 15.
[3] Bull. Soc. Moscou, vol. xv. p. 462 (1842).
[4] *Op. cit.* p. 9 (1838).
[5] *Ibid.* p. 15.
[6] *Loc. cit.*
[7] Mém. Soc. Moscou, vol. xv. pt. i. p. 29, pl. vi.

by Meyer in the 'Palæontographica,' vol. xv. art. 3, pl. xv. figs. 5–7 (1866). Another figure is given by Owen in the 'Quart. Journ. Geol. Soc.' vol. xxxii. p. 353, fig. 1, in conjunction with the proximal extremity of the humerus on which *Orthopus* was founded. The specimen is also figured by the present writer in the 'Proc. Zool. Soc.' 1889, pl. lv. fig. 4, with a restoration of the imperfect postaxial border. *Purchased*, 1865.

Genus **DEUTEROSAURUS**, Eichwald [1].

Founded upon a portion of the vertebral column [2], which apparently closely resembles that of the African form mentioned above. The skull referred to this genus by its describer indicates a considerably larger individual than the one to which the vertebræ belonged; and since the humerus of *Brithopus* is proportionately somewhat larger than those vertebræ, there would be more reason for referring the skull to that genus. In the absence of any decisive evidence one way or other and bearing in mind that *Deuterosaurus* may be generically identical with *Brithopus*, the teeth of the above-mentioned type are noticed under the present heading.

The above-mentioned skull has the incisive teeth $\frac{5}{4}$ in number, and comparative small tusks; the number and contour of the cheek-teeth being unknown. In the incisives (fig. 19) the crowns have a claw-like outer cusp, with a well-developed basal inner ledge; their roots are laterally compressed, with a pulp-cavity of moderate size.

Deuterosaurus biarmicus, Eichwald [3].

The type and only described species. The typical vertebral column indicates a smaller individual than the African form mentioned on p. 96; but the teeth and skull which have been referred to this species are of relatively larger dimensions.

Hab. Russia.

R. 303. (*Fig.*) A somewhat imperfect incisive tooth of the same type as the incisives of the skull referred by Eichwald to this form; from the Upper Permian of the Kargalinsk copper-mines, Government of Orenburg. Described and figured by Twelvetrees in the 'Geol. Mag.' dec. ii. vol. ix. p. 338, pl. viii. fig. 2 (1882), and again by Seeley in the 'Proc. Roy. Soc.' vol. xliv. p. 136, fig. 1 (1888); the specimen

[1] Bull Soc Moscou, vol. xxi No. iii p. 151 (1848).

[2] Figured in 'Lethæa Rossica,' pl. lix.

[3] *Loc. cit.*

being in both instances referred without any proviso to *Deuterosaurus*. Since the specimen (fig. 19) was figured the lower surface of the root has been cut and polished; and thus exhibits the compressed pulp-cavity, and the

Fig. 19.

? *Deuterosaurus biarmicus*.—Lateral aspect of an incisive tooth, wanting the lower portion of the root; from the Upper Permian of Kargalinsk, Orenburg. ½.

concentric layers of dentine noticed by Eichwald. Similar teeth are figured by Eichwald in his 'Lethæa Rossica,' pl. lviii. fig. 3, and also by Trautschold in the 'Mém. Ac. Imp. Nat. Moscou,' vol. xv. pt. i. pl. vii. figs. 3–5.

Presented by W. H. Twelvetrees, Esq., 1882.

GENUS *non det.*

R. 304. (*Fig.*) The crown of a tusk; from the Upper Permian of Kargalinsk, near Orenburg, Russia. Described and figured by Twelvetrees in the 'Geol. Mag.' dec. ii. vol. ix. p. 338, pl. viii. fig. 3, as the tooth of a Dinosaur. This specimen is laterally compressed and has serrated edges, and thus resembles the tusks of *Lycosaurus*; it is smaller and more slender than the tusks of the skull referred by Eichwald to *Deuterosaurus*.

Presented by W. H. Twelvetrees, Esq., 1882.

Family DIADECTIDÆ.

Dentition of an herbivorous type, the cheek-teeth (fig. 20) being transversely elongated, with oval crowns, divided into flattened and subequal moieties by a median vertical ridge at right angles to the longer diameter; alveoli not separated. Teeth on the vomer. No secondary posterior nares. Vertebræ notochordal, with intercentra; neural spines of moderate height, zygosphenal (hyposphenal) articulations present, and the pre- and postzygapophyses approximated to one another, with horizontal articular surfaces. A tusk may or may not be present, and there may be either one or two rows of cheek-teeth.

This family has been regarded by its founder, Cope [1], as representing a distinct subordinal group—Cotylosauria—on account of an apparent peculiarity in the articulation of the cranium with the vertebral column. Later on, however, it was suggested that this peculiarity might be due to the loss of the basioccipital [2]; and in a subsequent memoir [3] this family is placed immediately after the *Bolosauridæ*, which apparently implies that its subordinal distinction had been withdrawn.

Fig. 20.

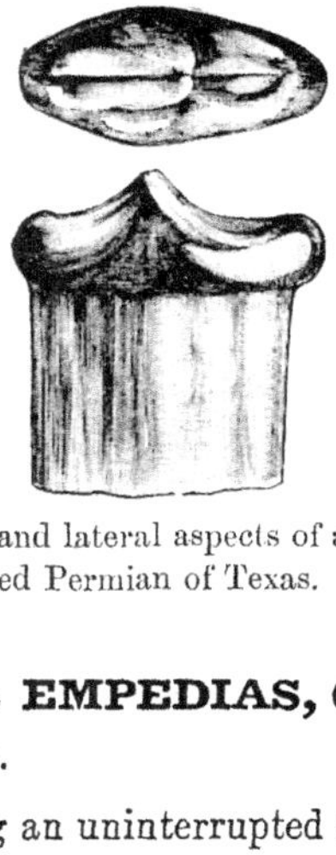

Empedias molaris.—Oral and lateral aspects of a cheek-tooth; from the reputed Permian of Texas. 1/1.

Genus **EMPEDIAS**, Cope [4].

Syn. *Empedocles*, Cope [5].

The dentition forming an uninterrupted series without a distinct tusk, and the incisives only distinguished from the cheek-teeth by their form, having more or less distinct transverse edges [6]. Number of upper teeth varying from 14 to 16; cheek-teeth in a single row.

[1] Amer. Nat. vol. xiv. p. 304 (1880).
[2] Proc. Amer. Phil. Soc. vol xix. p. 47 (1880).
[3] Trans. Amer. Phil. Soc. vol. xvi. p. 288 (1886).
[4] Proc. Amer. Phil. Soc. vol. xix. pl. v. (1881). This plate, although referring to p 56, was not issued till the next No. (108) in June 1881.
[5] *Ibid.* vol. xvii. p 516 (1878).—Preoccupied by Stål, 1867.
[6] See Cope, Amer. Nat. vol xiv. p. 304 (1880).

Dorsal vertebræ with the capitular facet rising on to the arch, which is coossified with the centrum; terminal faces of latter deeply cupped.

Fig. 21.

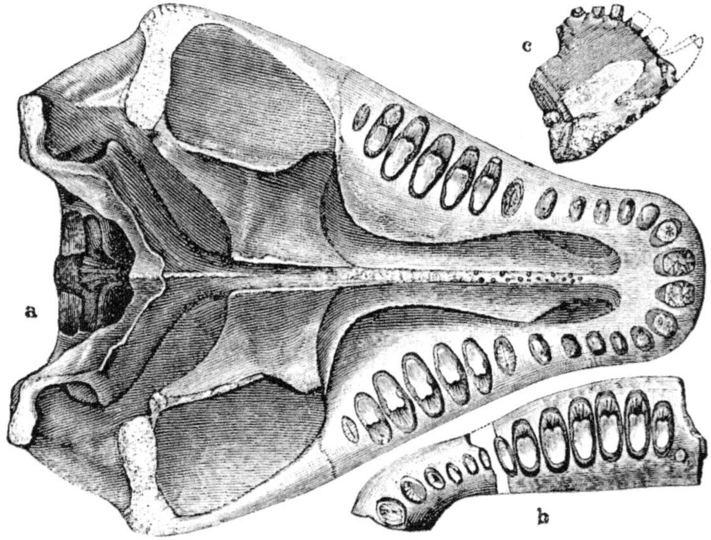

Empedias molaris.—*a*, palatal aspect of cranium; *b*, do. of right ramus of mandible; *c*, lateral aspect of mandibular symphysis. From the reputed Permian of Texas. $\frac{1}{2}$. (*After Cope.*)

Empedias molaris, Cope[1].

Syn. *Diadectes molaris*, Cope[2].
Empedocles molaris, Cope[3].
Empedias molaris, Cope[4].

Teeth $\frac{14}{14}$; upper ones with very wide crowns (fig. 21), the last one being small; inner and outer triturating surfaces of cheek-teeth large and flat. Presphenoid with a median keel; pterygoids wide.

Hab. North America (Texas).

R. 613. (*Fig.*) Part of the left maxilla, containing five posterior cheek-teeth, mostly imperfect; from the reputed Permian of Texas. The one entire tooth (fig. 20) is figured by Seeley in the 'Proc. Roy. Soc.' vol. xliv. p. 137, fig. 4

[1] Amer. Nat. vol. xii. p. 565 (1878).—*Diadectes*. [2] *Loc. cit.*
[3] Proc. Amer. Phil. Soc. vol xix. p. 47 (1880).
[4] *Ibid.* pl. v. (1881), see note 4 on preceding page.

(1888). The specimen agrees with the corresponding portion of the skull figured by Cope in the 'Proc. Amer. Phil. Soc.' vol. xix. pl. v. (reproduced in fig. 21).

Presented by R. Lydekker, Esq., 1885.

Fig. 22.

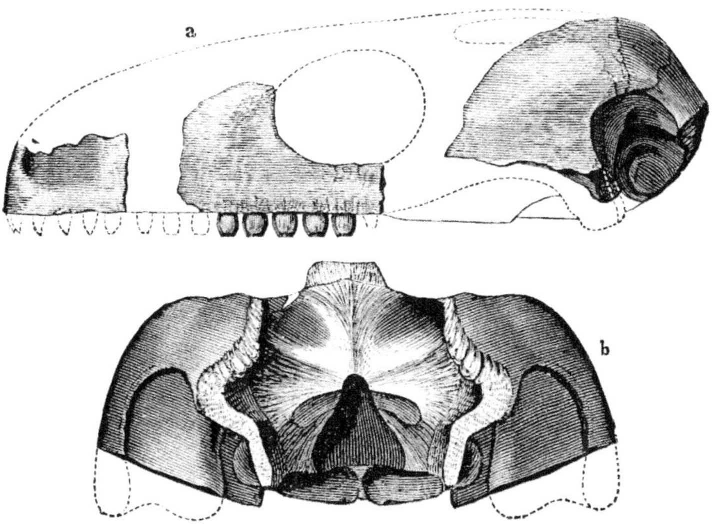

Empedias molaris.—Left lateral (*a*) and occipital (*b*) aspects of the cranium; from the reputed Permian of Texas. ½. (*After Cope.*)

R. 572. A somewhat imperfect dorsal vertebra, which from its comparatively large size may be provisionally referred to this species; from Texas. The zygosphenes (hyposphenes) are well shown, and the characters of the specimen agree closely with those of the smaller type vertebræ described by Cope in the 'Proc. Amer. Phil. Soc.' vol. xvii. p. 517. Thus the zygapophyses are much elevated and laterally expanded, and are connected together back and front. From the junction the transverse process (diapophysis) descends as a nearly vertical septum; the inferior part of this septum appears to have a distinct capitular facet, which is thus placed on the side of the arch. The neural spine is entire, and exhibits the stoutness and quadrangular cross-section characteristic of the types. The total height of this specimen is 0,130. *Purchased*, 1885.

Empedias phaseolinus, Cope[1].

Syn. *Diadectes phaseolinus*, Cope[2].
Empedias phaseolinus, Cope[3].

Upper cheek-teeth 16 in number, with narrower crowns than in *E. molaris*, the last one small. Presphenoid flat; pterygoids narrow.
Hab. Texas.

R. 575. Fragment of a left maxilla probably belonging to this species; from the reputed Permian of Texas. This specimen shows the broken bases of three cheek-teeth, the largest of which has a transverse diameter of 0,0145, or nearly the same as in the type specimen.
Purchased, 1885.

Genus **DIADECTES,** Cope[4].

The type genus. Distinguished from *Empedias* by the presence of a distinct tusk separating the incisive from the cheek-teeth, and of a fossa in the alveolar border of the jaw at the inner extremity of each cheek-tooth. The outer alveolar border of the maxilla diverges from the line of the tooth-series backwardly and outwardly, which is not generally the case in *Empedias*.

Diadectes sideropelicus, Cope[5].

The type and only described species.
Hab. North America (Texas).

R. 575 a. Part of the left maxilla; from the reputed Permian of Texas. The broken bases of four cheek-teeth are shown, with the fossa in the jaw at the inner extremity of each. The backward divergence of the outer side of the alveolar border is also shown. The largest tooth has a transverse diameter of 0,012, as in the type. *Purchased*, 1885.

R. 575 b. A smaller fragment of a jaw showing the bases of two teeth; from Texas. *Purchased*, 1885.

Family CLEPSYDROPIDÆ.

Dentition of a carnivorous type, and teeth present on the palatines and pterygoids; marginal teeth, although frequently having one or more enlarged tusks, not differentiated into the three regular

[1] Proc. Amer. Phil. Soc. vol. xix. p. 46 (1880).—*Diadectes*.
[2] *Loc. cit.*
[3] *Ibid.* vol. xx. p. 635 (1883).
[4] *Ibid.* vol. xvii. p. 505 (1878).
[5] *Loc. cit.*

groups found in the *Galesauridæ*. Vertebræ notochordal, with intercentra, and frequently with an extraordinary development of the neural spines (fig. 24); no zygosphenes, and the pre- and post-zygapophyses widely separated from one another, with oblique articular surfaces. Humerus typically with the shaft much constricted, the distal expansion wider than the proximal, and the proximal articular surface at right angles to the sides of the bone.

Fig. 23.

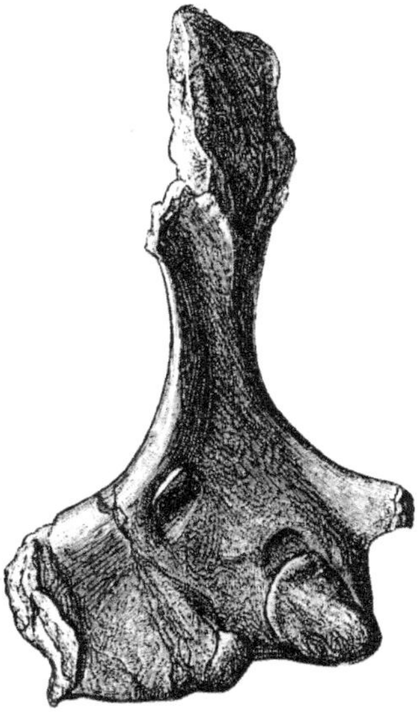

Stereorhachis dominans.—Palmar aspect of the left humerus, imperfect proximally; from the Lower Permian of France. $\frac{1}{2}$. (*After Gaudry.*)

In the type genus *Clepsydrops*, Cope [1], the three elements of the pectoral girdle are coossified into a single bone [2]; while in *Dimetrodon* [3] the pelvic bones form an innominate.

The humerus of *Stereorhachis*, Gaudry [4] (fig. 23), from the Upper Permian of France, corresponds with the description of that of *Clepsydrops*.

1 Proc. Ac. Nat. Sci. Philad. 1875, p. 407.

2 See Rep. Amer. Assoc. vol. xxxiii. pl. facing p. 481.

3 See Proc. Amer. Phil. Soc. vol. xvii pp. 513, 514 (1878).

4 Les Enchaînements du Monde Animal, etc.—Fossiles Primaires, p. 279 (1883).

Genus **NAOSAURUS**, Cope[1].

The neural spines of the dorsal and lumbar vertebræ enormously elongated, with paired horizontal processes at intervals. Transverse processes of dorsals large and prominent, no distinct capitular facet on centrum, although all the ribs have double heads. In *Dimetrodon*, Cope[2], the horizontal processes of the neural spines of the vertebræ are absent.

Naosaurus cruciger, Cope[3].

The type species. The distal portions of the neural spines of the vertebræ cylindrical, and the distal horizontal processes represented by tubercles.

Fig. 24.

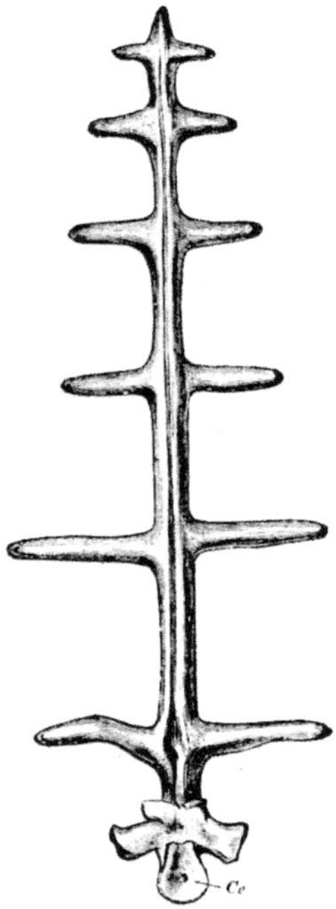

Naosaurus claviger.—Anterior aspect of a dorsal vertebra (partially restored); from the reputed Permian of Texas. $\frac{1}{6}$. *Ce*, centrum. (See Trans. Amer. Phil. Soc. vol. xvi. pl. iii. fig. 1, 1886.)

Hab. North America (Texas).

[1] Amer. Nat. vol. xx. p. 545 (1886).
[2] Proc. Amer. Phil. Soc. vol. xvii. p. 512 (1878).
[3] Amer. Nat. *loc. cit.*

R. 570*. Fragment of a maxilla with broken teeth belonging either to a species of *Naosaurus* or *Dimetrodon*; from the reputed Permian of Texas. The broken bases of six teeth remain, and the crown of one germ-tooth is seen *in alveolo*. The teeth accord in size and character with the hinder teeth of *Naosaurus claviger*, figured by Cope in the 'Trans. Amer. Phil. Soc.' vol. xvi. pl. ii. fig. 1. *Purchased*, 1885.

R. 564. Two portions of the neural spine of a vertebra; from the reputed Permian of Texas. The smaller specimen, which is evidently from near the distal extremity, exhibits the subcylindrical section and the lateral tubercles, and thus resembles the vertebra figured by Cope in the 'Trans. Amer. Phil Soc.' vol. xvi. pl. iii. fig. 3. *Purchased*, 1885.

R. 564 a. Part of the neural arch and spine of a dorsal vertebra apparently referable to this species; from Texas. In the contour of the transverse processes and the lower part of the spine, this specimen approximates to the vertebra of this species figured by Cope, *loc. cit.*, rather than to the vertebra of *N. claviger* represented in figs. 1, 2 of the same plate. In the latter figure, the lower part of the spine is represented as inclining backwards, and the lower horizontal processes curving forwards, but in the present specimen both these directions are reversed. *Purchased*, 1885.

Naosaurus claviger, Cope[1].

Larger than the preceding, with the distal extremities of the neural spines of the vertebræ dilated and compressed, and the processes large throughout the extent of the spine.

It does not appear possible to distinguish the spines of this species from those of *N. microdus*, Cope[2], which is characterized by its smaller teeth.

Hab. North America (Texas).

R. 564 b. A considerable portion of the distal half of the neural spine of a dorsal vertebra; from the reputed Permian of Texas. Resembles the spine of the vertebra figured by Cope in the 'Trans. Amer. Phil. Soc.' vol. xvi. pl. iii. figs. 1, 2. *Purchased*, 1885.

[1] Amer. Nat vol. xx. p. 545 (1886).

[2] See Trans Amer. Phil. Soc vol. xvi. p. 294.

R. 564 c. Fragment of the spine of a vertebra, showing the base of a horizontal process; from Texas. *Purchased*, 1885.

Specifically undetermined specimens.

R. 576 a. An imperfect dorsal vertebra; from the reputed Permian of Texas. Only the proximal portion of the neural spine remains, the zygapophyses are imperfect, the transverse processes broken off, and the margins of the terminal cups of the centrum damaged. As in No. R. 564 *a*, the neural spine is inclined backwards, instead of forwards as in Cope's figures. *Purchased*, 1885.

R. 576 b. Portion of the neural spine of a vertebra; from Texas. *Purchased*, 1885.

R. 564 d. Fragment of a neural spine; from Texas. *Purchased*, 1885.

R. 564 e. Part of a neural spine, with the bases of the horizontal processes; from Texas. *Purchased*, 1885.

Genus **EMBOLOPHORUS**, Cope[1].

Known by vertebræ, in which the neural arches are coossified with the centra, and each dorsal intercentrum carries a facet on either side for the articulation of the capitulum of the rib. The dorsal centra have cylindroidal and deeply cupped terminal faces, and a compressed median portion with a sharp hæmal carina; the transverse processes being short. The height of the neural spine appears to be unknown.

The intercentra project beyond the edges of the centra.

Embolophorus, sp.

The following specimens indicate a species of much larger size than the typical *E. fritillus*, Cope[2], in which the length of the vertebral centrum is given as 0,004. In the present specimens the length of the centrum reaches 0,025; and these vertebræ agree in absolute size with the half-sized figures of *E. dolloverianus* given by Cope in the 'Proc. Amer. Phil. Soc.' vol. xxii. pl. i. fig. 4.

Hab. North America (Texas).

R. 569 x. Two somewhat imperfect dorsal vertebræ and an intervening intercentrum; from the reputed Permian of Texas.

[1] Proc. Amer. Phil. Soc. vol. xvii. p. 518 (1878). [2] *Loc. cit.*

The length of the larger centrum is 0,025, and the diameter of its terminal face 0,020. The capitular facets are well shown on the posterior border of the hæmal surface of the intercentrum. *Purchased*, 1885.

Embolophorus dolloverianus, Cope [1].

The vertebræ are still larger than those of the preceding form.

Hab. North America (Texas).

R. 569. (*Fig.*) Two imperfect dorsal vertebræ and an intercentrum, apparently associated; from the reputed Permian of Texas. These specimens (fig. 25) agree closely with the types described and figured by Cope, of half the natural size, in the 'Proc. Amer. Phil. Soc.' vol. xxii. p. 43, pl. i. fig. 4 (1884), without specific name. One specimen exhibits the centrum and a part of the neural arch, with the zygapophyses. The intercentra are of large size and exhibit the facet for the ribs on the posterior border of the hæmal surface. *Purchased*, 1885.

Fig. 25.

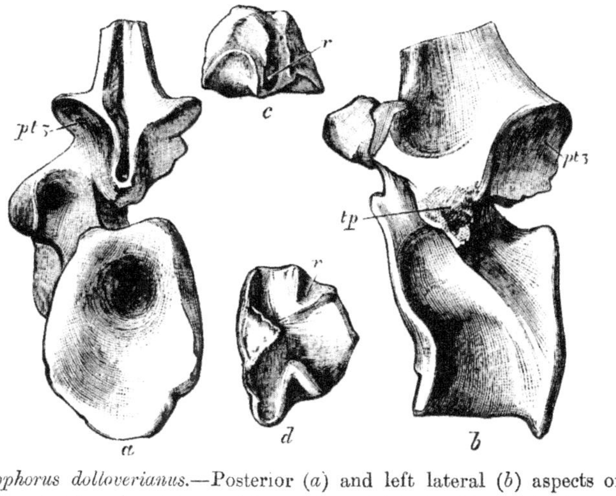

Embolophorus dolloverianus.—Posterior (*a*) and left lateral (*b*) aspects of an imperfect dorsal vertebra, and lateral (*c*) and inferior (*d*) aspects of an intercentrum; from the reputed Permian of Texas. $\frac{2}{3}$. *pt.z*, postzygapophysis; *tp*, broken transverse process; *r*, capitular facet of intercentrum.

[1] Trans. Amer. Phil. Soc. vol. xvi. p. 287 (1886).

R. 571. The imperfect centrum and base of the neural arch of a dorsal vertebra; from Texas. *Purchased*, 1885.

R. 576. The centrum and base of the neural arch of a dorsal vertebra; from Texas. *Purchased*, 1885.

Generically Undetermined Specimens.

It is probable that a large proportion of the following specimens are referable to various genera of the present family, but in the absence of figures of so many of the typical forms it is impossible to refer them to their proper position.

All the following specimens were obtained from the reputed Permian of Texas, and were purchased in 1885.

R. 576 c. An imperfect right quadrate. This specimen approximates, both in size and contour, to the corresponding bone of the skull of *Naosaurus claviger* figured by Cope in the 'Trans. Amer. Phil. Soc.' vol. xvi. pl. ii. fig. 1, although it does not appear to belong to that species. The squamous articular surface for the pterygoid is seen on the lower part of the inner surface.

R. 566. An imperfect cervical (?) vertebra, wanting the greater portion of the neural spine. The terminal faces of the centrum are deeply cupped; the arch, which is distinct from the centrum, has no transverse process, showing that the ribs articulated solely with the centrum.

R. 566 a. The centrum of a vertebra probably referable to the same form as the preceding. There is a prominent capitular rib-facet on the anterior border.

R. 568. An imperfect caudal (?) vertebra. The terminal faces are but slightly cupped, and there is a large double-headed rib anchylosed to the sides of the centrum, as in the caudals of *Dimetrodon*.

R. 573. An imperfect dorsal vertebra of a large form. The transverse process is large and prominent, and the terminal faces of the centrum are nearly flat.

R. 576 d. Part of the neural arch of a dorsal vertebra, showing the entire transverse process.

INCERTÆ SEDIS.

Family GORGONOPIDÆ.

In the one known representative of this family the temporal fossæ were roofed over, but the cranial bones are not sculptured; the dentition being of a carnivorous type, and differentiated into incisives, tusks, and cheek-teeth. Apparently no secondary posterior nares. No teeth on vomer.

In the roofing of the temporal fossa *Gorgonops* agrees with *Chilonyx* (*suprà*, p. 95, note 2), and also with the *Pariasauridæ* (*infrà*), although differing from the latter and also from the American *Pariotichidæ* in the absence of sculpturing on the cranial bones. This genus probably forms a connecting-link between the typical Theriodontia and the Pariasauria.

Genus **GORGONOPS**, Owen[1].

The type and only known genus; described upon the evidence of the cranium. Skull depressed, long and narrow, with the orbits small and lateral, approximated to the occiput, and far removed from the muzzle. Nares forming small horizontal slits, overhung by the large nasals. Upper incisives five in number.

Gorgonops torvus, Owen[2].

The type and only described species. Of considerable size, the length of the cranium being 0,205 (8·1 inches).

Hab. South Africa.

R. 1647. (*Fig.*) The imperfect cranium; from the Beaufort beds of the Karoo system at Mildenhalls near Fort Beaufort. The type specimen; figured by Owen in his 'Catalogue of the Fossil Reptilia of S. Africa,' pls. xxi., xxii. The preorbital portion of the specimen is nearly entire, but the hinder region is imperfect, portions of the roof of both temporal fossæ being broken away. The broken bases of the incisives are shown on both sides, behind which the alveolar margins of the jaws have been ground down to exhibit the alveoli of the tusks; no traces of cheek-teeth are visible. The vomer is well displayed and is seen to have no teeth. *Presented by A. G. Bain, Esq.*, 1853.

[1] Cat. Foss. Rept. S. Africa, p. 27 (1876). [2] *Loc. cit.*

Suborder *PARIASAURIA.*

Skull with the temporal fossæ roofed over, two temporal arcades, cranial bones typically sculptured, and apparently no secondary posterior nares. Dentition fully developed. Vertebræ notochordal, frequently with intercentra, and not more than two in sacrum. Humerus with the inferior aperture of the entepicondylar foramen situated on the distal aspect. Ilium (fig. 26) with its supra-acetabular portion forming an elongated triangular plate lying obliquely to the axis of the sacrum, to which it is attached by a single rib; ischium and pubis uniting without the intervention of an obturator foramen; but a small vascular foramen in the pubis, and the anterior border of the latter everted, with a broad roughened surface which may have supported an epipubic ossification. No median bar to interclavicle.

The typical members of this group were regarded by Owen as referable to the Dinosauria, under the name of Tretospondyli [1], but their Anomodont character was first pointed out by Cope [2], who included them in his Pelycosauria.

Family PARIASAURIDÆ.

Dentition forming an uninterrupted series of similar teeth, the crowns of which have more or less distinct vertical flutings or serrations [3]. Typically the vertebræ with very low and wide arches, extremely short neural spines, and horizontal zygapophyses.

It is at present difficult to determine whether some of the characters given under the head of the genus should not rather be given as family ones.

Genus **ANTHODON,** Owen [4].

Imperfectly known, cranium apparently more vaulted and narrower than in *Pariasaurus.* Teeth with short, compressed, and antero-posteriorly expanded crowns, having a convex subtrenchant free edge, marked by numerous crenulations; roots long, and narrower than the crowns. Centra of dorsal vertebræ with deeply cupped terminal faces, and no evidence of the presence of intercentra.

[1] *Tapinocephalus* was included in the same group. The name Tretospondyli is not adopted for the reason that it is also applicable to most of the Theriodontia.

[2] Proc. Amer. Phil. Soc. vol. xvii. pp. 529, 530 (1878).

[3] The *Pariotichidæ,* from the reputed Permian of North America, are distinguished by their dentition being of a carnivorous type (see Nicholson and Lydekker, 'Manual of Palæontology,' 3rd ed. vol. ii. p. 1056, 1889).

[4] Cat. Foss. Rept. S. Africa, p. 14 (1876).

This genus is provisionally included in the *Pariasauridæ*, being apparently closely allied to the type genus. The teeth present a marked resemblance to those of the Dinosaurian family *Stegosauridæ*[1], but there is no doubt that the present form is a true Anomodont.

Anthodon serrarius, Owen[2].

The type species. Approximately half the dimensions of *Pariasaurus bombidens*.

Hab. South Africa.

47337. (*Fig.*) The imperfect skull; from the Karoo system near Styl-Krantz, Sneewberg range[3]. The type; figured by Owen in his 'Catalogue of the Fossil Reptilia of S. Africa,' pl. xiii. fig. 1. The muzzle is wanting, and posteriorly the specimen is broken off through the line of the parietal foramen, and thence to the quadrate. The outer surface of the bones of the skull is mostly chipped away, but traces of sculpture apparently remain in the middle line. The postorbital flange is distinctly seen in the quadratic region. The teeth are mostly shown in section. In the description in Owen's 'Catalogue' the locality is incorrectly given.

Presented by A. G. Bain, Esq., 1853.

47337 a. (*Fig.*) Portion of the dorsal region of the vertebral column, associated with the preceding. This specimen, which contains three imperfect vertebræ, and has been vertically and longitudinally bisected, is described by Owen, *op. cit.* p. 71, no. 123, with a representation of a section of one of the vertebræ in pl. lxx. fig. 2. The vertebræ approximate very closely in structure to those of *Pariasaurus*, having similar widely extended and horizontal zygapophyses, and apparently extremely short neural spines. The transverse process occupies a similar position between the zygapophyses, and the rib-facets were more or less pedunculated. The width between the outer extremities of the postzygapophyses is 0,067 (2·63 inches), and the length from the posterior extremity of the zygapophysis

[1] See Marsh, Geol. Mag. dec. iii. vol. vi. p. 207 (1889).

[2] *Loc. cit.*

[3] It is uncertain whether the beds from which this specimen was obtained are the Beaufort or Stormberg beds; see Quart. Journ. Geol. Soc. vol. xxiii. p. 143, where this species is described as "having the characters of the Blinkwater Monster."

of one vertebra to that of another 0,042 (1·65 inches); well illustrating the Amphibian feature of the extreme width of the neural arches.

Presented by A. G. Bain, Esq., 1853.

47338. (*Fig.*) Fragment of rock showing in a split surface a portion of the left mandibular ramus, with teeth; from the Karoo system on the Bushman's river, below Graham's Town, Albany. Figured by Owen in his 'Catalogue,' pl. xiii. figs. 2, 3.

Presented by A. G. Bain, Esq., 1853.

Anthodon (?), sp.

The following specimen, which presents no characters by which it can be distinguished from *Anthodon*, indicates a much smaller reptile than adult individuals of the type. Its higher geological horizon tends to confirm its specific distinctness from *A. serrarius.*

Hab. South Africa.

R. 502. Cast of a fragment of rock showing one lateral aspect of an imperfect tooth. The original was obtained from the Wood-bed stage of the Uitenhage system on the Sunday's river, Uitenhage, and is preserved in the Museum of the Geological Society. It is noticed in the 'Quart. Journ. Geol. Soc.' vol. xxiii. p. 164.

Presented by Prof. T. Rupert Jones, 1884.

Genus **PARIASAURUS**, Owen[1].

The type genus. Cranium very broad and much depressed, with deeply descending postorbital flanges. Marginal teeth with moderately tall, swollen, and narrow crowns, with a few deeply marked flutings descending from the cutting-edge; roots moderately long and nearly as wide as the crowns; numerous conical teeth on palate. Vertebræ with comparatively short centra, which apparently have flattened terminal faces; the dorsal ribs articulating with a pedunculate rib-facet on the centrum, and a long transverse process on the arch; intercentra in dorsal and caudal regions. When known, some 29 vertebræ, of which 18 are presacral. Ischia much produced backwardly. It is suggested that there may have been a dermal armour.

Pariasaurus serridens, Owen[2].

The type species; known only by the skull. Teeth apparently some $\frac{14}{14}$ in number. If the type specimen retains an approximation

[1] Cat. Foss. Rept. S. Africa, p. 6 (1876).—*Pareiasaurus.*

[2] *Loc. cit.*

to its original contour it would appear that the mandible was comparatively narrow and pointed at the symphysis, and the cranium somewhat less depressed, and with a deeper postorbital flange than in the next species.

Hab. South Africa.

R. 1710. Fragment of the right mandibular ramus; from the (*Fig.*) Beaufort beds of the Karoo system on the Blinkwater river, south of the Winterberg peak [1], north of Fort Beaufort. This specimen, which appears to be all that now remains of the type skull, has been vertically cut and polished, and exhibits three teeth in use, and a germ-tooth in its alveolus. Described and figured by Owen in his 'Catalogue of the Fossil Reptilia of S. Africa,' p. 8, no. 15, pl. viii. figs. 1, 2, and also by Seeley in the 'Phil. Trans.' for 1888, p. 74, pl. xvi. fig. 2.

Presented by A. G. Bain, Esq., 1853.

R. 1710 a. Cast of the imperfect and distorted skull. The original, (*Fig.*) of which the preceding specimen is a fragment, was the type. The cast is figured by Owen in his 'Catalogue,' pls. vi., vii., and also (from the palatal aspect) by Seeley in the 'Phil. Trans.' for 1888, pl. xvi. fig. 1.

Presented by Sir R. Owen, K.C.B.

Pariasaurus bombidens, Owen [2].

This species was originally said to differ from the type by the teeth having broader serrations, and a more convex contour of the outer surface, but subsequent observations throw doubt on the validity of these differences [3]. It is suggested [4] that the cranium was more depressed than in the type, with a deeper postorbital flange, and a wider and more rounded mandible. It is, however, difficult to say whether the latter differences are not at least partially due to the crushed and imperfect condition of the type skull. The length of the entire skeleton is approximately about 8 feet. The ilium has a prominence in advance of the acetabulum.

Hab. South Africa.

[1] See Trans. Geol. Soc. ser. 2, vol. vii. p. 56, and Quart. Journ. Geol. Soc. vol. xxiii. p. 143, where the original skull is alluded to as the "Blinkwater Monster."

[2] Cat. Foss. Rept. S. Africa, p. 9 (1876).

[3] Seeley, Phil. Trans. for 1888, p 75.

[4] Seeley, *loc. cit.*

R. 1714. (*Fig.*) The imperfect anterior portion of the skull; from the Karoo system of Vers Fontein, in the Cape Colony. The type; figured by Owen in his 'Catalogue of the Fossil Reptilia of S. Africa,' pl. viii. fig. 3, pl. ix.; and noticed by Seeley in the 'Phil. Trans.' for 1888, pp. 70, 75 (as No. 43525). The nearly entire left ramus of the mandible, with its teeth, is preserved; and the outer aspect of the dental series of the upper jaw is also shown.

Presented by W. G. Atherstone, Esq., M.D., 1872.

49426. (*Fig.*) The nearly entire skeleton; from the Karoo system of Palimeet[1] Fontein in the Fraserburg district. This specimen, which is provisionally referred to the present species, is described and figured by Seeley in the 'Phil. Trans.' for 1888, pp. 59–109, pls. xii.–xv. and xvii.–xx. The axial skeleton is fairly well preserved as far back as the hinder part of the caudal region, but the whole of the bones of the limbs are wanting. The frontal aspect of the skull is imperfect, many of the bones being broken; but there seems to be no evidence in support of the existence of the so-called "infranasal" bone introduced in pl. xiii. of the memoir cited between the maxilla and the premaxilla. In the pectoral girdle a considerable portion of the large interclavicle and clavicles is preserved; and parts of the left scapula and coracoid remain, although not sufficiently well preserved as to show whether the precoracoid was a separate bone. In the pelvic girdle a considerable portion of the left innominate remains, which shows that the ilium was almost identical in general contour with that of the undermentioned specimen No. 36251 (p. 117), having the same prominence in advance of the acetabulum. In the restoration of a cervical vertebra given in pl. xvii. of the memoir, both the upper and lower facets are represented as situated on the centrum, considerably behind its anterior border; but there does not appear to be any means of determining whether the upper costal facet was really so placed, and it is clear that the lower one was situated on the anterior border of the centrum. The prominent (parapophysial) rib-facets of the dorsal vertebræ are very clearly shown. There is no

[1] Incorrectly given by Owen as Palinut.

evidence to show whether the terminal faces of the dorsal vertebræ were flattened or cupped.

Purchased from T. Bain, Esq., 1878.

49374 a. An imperfect middle caudal vertebra, apparently referable to a species of this genus; from the Karoo system of the Cape. This specimen can only be distinguished from the middle caudals of the skeleton of *P. bombidens* by its somewhat superior size.

Purchased from T. Bain, Esq., 1878.

SPECIFICALLY UNDETERMINED SPECIMENS.

The following specimens are provisionally referred to *Pariasaurus.*

36248. The proximal extremity of a left humerus; from the Karoo system of the Brak river, near Fort Beaufort. This specimen is of the same type as the one mentioned below (p. 119) under the name of *Propappus*, but is of larger size, and is therefore provisionally referred to the present genus. The upper extremity of the delto-pectoral crest is shown, and the postaxial border is entire.

Presented by A. G. Bain, Esq., 1853.

36247. (*Fig.*) A specimen which is apparently the imperfect distal extremity of a left femur, probably referable to this or an allied form; from the Brak river. Figured by Owen in the 'Trans. Geol. Soc.' ser. 2, vol. vii. pl. xxxiv. fig. 4, where it is suggested that it may be either a humerus or a femur; and noticed in his 'Catalogue of the Fossil Reptilia of South Africa,' p. 57, no. 96, where it is regarded as the distal extremity of the left humerus of *Udenodon.* The specimen is certainly not Dicynodont, and from its aberrant type probably belongs to the present group; it agrees fairly in size with the proximal half of the femur, and may have been associated with No. 36248. The inner condyle is much larger than the outer, thus presenting a marked resemblance to the femur of *Echidna.*

Presented by A. G. Bain, Esq., 1853.

36251. (*Fig.*) The imperfect pelvic and sacral region; from the Karoo system of the Brak river. Figured by Owen in the 'Phil. Trans.' for 1862, pls. xxiii., xxiv., and also in his 'Catalogue of the Fossil Reptilia of S. Africa,' pls. xxxvi., xxxvii., as *Dicynodon tigriceps*; a small figure being given by Seeley in the 'Phil. Trans.' for 1888, p. 107, under the same name. As pointed out on p. 102 of the memoir last cited, the ilium is supported by only a single sacral rib, which is of very large size, and the whole structure of the specimen is essentially the same as in *Pariasaurus*. The ischium and pubis accord closely with the corresponding bones of the innominate No. 47090 mentioned below under the heading of *Propappus*, and the position of the pubic foramen is the same in both. The ischia must, however, when entire, have been more produced backwardly than in the latter. The centra of the sacral and adjacent vertebræ are longer and more compressed laterally than in the skeleton referred to *Pariasaurus bombidens*; but the ilium agrees with that of the latter in the presence of a protuberance immediately in advance of the acetabulum, and in the long and narrow form of the preacetabular portion, which has a prominent ridge on the anterior border of the dorsal aspect. This specimen may belong to *P. serridens*.

Presented by A. G. Bain, Esq., 1853.

Genus **PROPAPPUS**, Seeley[1].

Founded upon the humerus, which, as being of a totally different type from that of the Theriodontia and Dicynodontia, may be regarded as referable to this group. If the under-mentioned pelvis belong to this form there will be evidence that *Propappus* is generically distinct from *Pariasaurus*.

The under-mentioned pelvis has the ilium wider and more upright than in *Pariasaurus*, and without such a prominent ridge on the anterior border of the dorsal surface, and no prominence immediately in advance of the acetabulum; while the ischium was less produced posteriorly. The humerus (like that referred to *Pariasaurus*) is characterized by the comparatively slight expansion of its extremities, the stoppage of the delto-pectoral crest below the head, the inferior position of the entepicondylar foramen, which pierces the bone obliquely from the posterior to the distal aspect, and the large single condyle for the radius and ulna, above which there is no expanded flange.

[1] Proc. Roy. Soc. vol. xliv. p. 142 (1888).

This form of humerus—especially in the position of the entepicondylar foramen—approximates more closely to a mammalian type than is the case in the Theriodontia. *Propappus* was at first regarded as allied to *Stereorhachis* (fig. 23), both forms being referred to a separate group under the name of Gennetotheria [1]. The undermentioned innominate presents characters approximating both to a Labyrinthodont and to a Mammalian type.

Propappus omocratus, Seeley [2].

The type species. Of smaller dimensions than *Pariasaurus bombidens.*

Hab. South Africa.

36250. The somewhat imperfect right humerus; from the Karoo system of the Brak river, near Fort Beaufort. The type; noticed by Seeley in the 'Proc. Roy. Soc.' vol. xliv. p. 142. This specimen has a length of 0,320 (19·7 inches). The delto-pectoral crest and the opposite border are imperfect, and the distal condyle has been somewhat flattened.

Presented by A. G. Bain, Esq., 1853.

R. 848. The proximal portion of the left femur of this or a closely allied form; from the Karoo system of Vinderaars Fontein, near Beaufort West, on the southern flank of the Nieuwveldt range. This specimen agrees so well in relative size and structure with the humerus as to leave little doubt that it indicates an allied or identical form. It differs from the femur of the Dicynodontia in the absence of the long outer trochanter, and approximates to a monotrematous type.

Purchased from T. Bain, Esq., 1880.

36246. A left innominate, probably referable to this form; from the (*Fig.*) Karoo system on the road to Block Drift, near Fort Beaufort. The mineral condition of this specimen is identical with that of No. 36250, so that it might well have belonged to the same individual. Figured by Owen in the 'Trans. Geol. Soc.' ser. 2, vol. vii. pl. xxxiv. fig. 1, and also in his

[1] When the term Gennetotheria was first proposed it was used in an ordinal sense, although subsequently regarded as a suborder of the Anomodontia (Proc. Roy. Soc vol xliv p. 383) Still later (Phil Trans for 1889, p. 292) *Lycosaurus* is given as the typical representative of the Gennetotheria, although no mention of that genus was made when the group was proposed.

[2] Proc. Roy. Soc. vol. xliv. p. 142 (1888).

Catalogue of the Fossil Reptilia of S. Africa,' pl. xxviii., and referred to *Dicynodon leoniceps*. A small figure is also given by Seeley in the 'Phil. Trans.' for 1888, p. 103, where this reference is provisionally accepted. This specimen (fig. 26) differs totally from the innominates of Dicynodonts (as shown by specimens both from Africa and India), and agrees so closely with that of *Pariasaurus* (as pointed out by Seeley), that there can be but little hesitation in referring it to the same group. This is confirmed

Fig. 26.

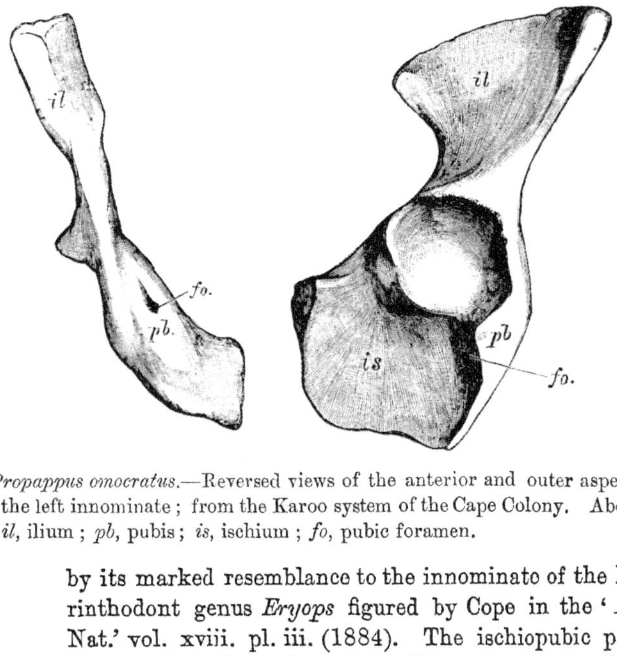

(?) *Propappus omocratus*.—Reversed views of the anterior and outer aspects of the left innominate; from the Karoo system of the Cape Colony. About $\frac{1}{6}$. *il*, ilium; *pb*, pubis; *is*, ischium; *fo*, pubic foramen.

by its marked resemblance to the innominate of the Labyrinthodont genus *Eryops* figured by Cope in the 'Amer. Nat.' vol. xviii. pl. iii. (1884). The ischiopubic portion differs widely from that of the Dicynodontia (fig. 3), the postacetabular free border of the ischium being very short and more nearly horizontal. The foramen in the pubis differs totally in position from the obturator foramen of the Dicynodontia, piercing the bone in an oblique direction, and evidently corresponding to the pubic foramen of the Labyrinthodont pelvis. The thickened everted preaxial border of the pubis has a roughened surface suggestive of the articulation of an epipubic bone.

Presented by A. G. Bain, Esq., 1853.

Class **AMPHIBIA.**

Order ECAUDATA.

Body of adult short, destitute of a tail, and furnished with four limbs, of which the pelvic pair are the larger and adapted for leaping. No gills in adult. Skull short and wide, with the parietals confluent with the frontals, and the orbits usually undefined; a "girdle-bone" in the ethmoidal region, and a predentary ossification. Presacral vertebræ few in number, and generally procœlous; vertebral column terminating in a long urostyle; dorsal ribs usually absent. Ilia prolonged backwards, so as to throw the acetabulum far behind the sacrum; radius and ulna, and tibia and fibula respectively fused together; and the calcaneum and astragalus greatly elongated. Four digits in the manus and five in the pes, with an additional ossicle in the pes which apparently represents a prehallux.

Section **FIRMISTERNIA.**

Coracoids firmly united by a single epicoracoid cartilage; precoracoids, when present, resting with their distal extremity upon the coracoids, or connected with the latter by the epicoracoid cartilage. A tongue.

Family RANIDÆ.

Teeth in the upper jaw; sacral processes with slight or no terminal expansion; vertebræ procœlous; no dorsal ribs; urostyle articulating with the sacrum by two condyles, and without transverse processes.

Genus **OXYGLOSSUS,** Tschudi [1].

No teeth on the vomer. Omosternum with a bony style; sternum forming a cartilaginous plate. Terminal phalangeals simple; digits of manus free, those of pes webbed. Head relatively large.

The species of this Oriental genus are of small size, usually not exceeding $1\frac{1}{4}$ inch in length.

[1] Mém. Soc. Neuchâtel, vol. ii. art. iii. p. 85 (1839).

Oxyglossus pusillus (Owen [1]).

Syn. *Rana pusilla*, Owen [2].
Oxyglossus pusillus, Stoliczka [3].

Head large, triangular, about one third broader than long behind. Radius and ulna relatively long, and calcaneum and astragalus comparatively short.

The association of a great number of specimens in the intertrappean clays of Bombay renders it probable that this species does not belong to *Ixalus* (in which vomerine teeth are also wanting), since the members of that genus are of arboreal habits.

Hab. India (Bombay).

35107. Slab of rock showing several more or less nearly perfect skeletons; from the intertrappean Eocene beds of Bombay. Similar to the specimens figured by Stoliczka, *op. cit.* pl. ix.
Presented by the Rev. Stephen Hislop, 1861.

39485. Three fragments of rock showing imperfectly preserved skeletons; from Bombay.
Presented by A. B. Wynne, Esq., 1864.

Genus **RANA**, Linn.[4]

Teeth on the vomer; tympanum distinct or hidden. Omosternum and sternum with a bony style. Terminal phalangeals acute, transversely dilated or T-shaped; digits of manus free, those of pes webbed. Vertebral centra elongated. Metacarpals short and stout. In the type species and allied forms the length of the humerus is somewhat more than half that of the femur, but in the 'Bull-frogs' the humerus is relatively longer.

For synonymy see Boulenger, 'Catalogue of Batrachia Salientia,' pp. 6, 7 (1882).

Rana meriani, Meyer [5].

Apparently allied to *R. temporaria*, but with the head somewhat larger and more pointed, the transverse processes of the 2nd vertebra directed more anteriorly, and the digits shorter.

Hab. Europe (Germany).

[1] Quart. Journ. Geol. Soc. vol. iii. p. 224 (1847).—*Rana.* [2] *Loc. cit.*
[3] Mem. Geol. Surv. Ind. vol. vi. p. 387 (1869).
[4] Syst. Nat. ed. 12, vol. i. p. 354 (1766).
[5] Neues Jahrb. 1853, p. 163.

33064 a. Slab of lignite showing the impression of the skeleton; (*Fig.*) from the Lower Miocene of Rott, near Bonn, Siebengebirge. Described and figured by Meyer in the 'Palæontographica,' vol. vii. p. 130, pl. xvi. fig. 2. The impression is of the ventral aspect of the skeleton. This specimen may be regarded as one of the types. *Purchased. About* 1860.

41086. A split slab of lignite showing the impression of the nearly entire skeleton; from Rott. This specimen is slightly larger than the preceding. *Purchased,* 1868.

R. 60. A split slab of lignite with the impression and some of the bones of the imperfect skeleton; from Rott. The skull is very incomplete.
Transferred from the Museum of Practical Geology, 1880.

35657. Slab of lignite with the impression and some of the bones of a rather smaller skeleton; from Rott. One humerus is entire. This specimen agrees very closely in size with the skeleton figured by Meyer, *op. cit.* pl. xvi. fig. 3. The contour of the soft parts is exhibited. *Purchased,* 1859.

33064. A split slab of lignite showing the impression and some of the bones of a nearly entire skeleton, together with the contour of the soft parts; from Rott. This specimen agrees in size with the preceding.
Purchased. About 1860.

30267. Slab of lignite, showing the impression of the nearly entire skeleton of a somewhat smaller frog not improbably belonging to this species. The sacrum is not shown. The contour of the soft parts of the hind limbs is preserved. *Purchased,* 1855.

Rana noeggerathi, Meyer [1].

A small species, of less than half the size of *R. temporaria.* Body slender, the hind limbs very long, and the skull longer than broad. Humerus scarcely more than half the length of the femur.

The form of the sacral processes and the terminal phalangeals distinguishes this species from *Hyla*; while the shortness of the humerus at once shows its distinctness from *Palæobatrachus*, apart from the characters of the sacrum.

Hab. Europe (Germany).

[1] Neues Jahrb. 1852, p. 466.

36379–80. A split slab of lignite showing the impression and traces of the bones of the skeleton, with some indications of the contour of the soft parts; from the Lower Miocene of Rott, near Bonn, Siebengebirge. This specimen agrees in all respects with the type skeleton from the same deposits, figured by Meyer in the 'Palæontographica,' vol. vii. pl. xviii. fig. 9. *Purchased*, 1862.

Section ARCIFERA.

Coracoids and precoracoids connected by an arched epicoracoid cartilage, of which one side overlaps the other. Tongue present or absent [1].

Family LEPTODACTYLIDÆ [2].

Upper jaw toothed; sacral processes with slight or no terminal expansion; terminal phalangeals never claw-like; vertebræ procœlous; no dorsal ribs; urostyle articulating with the sacrum by two condyles, and without transverse processes. Tongue present.

In the type genus the humerus of the male is expanded into a more or less largely developed flange-like plate extending from the posterior surface.

Genus **LEPTODACTYLUS**, Fitzinger [3].

Syn. *Cystignathus*, Wagler [4].

The type species. Skull smooth and without dermal ossifications; teeth on the vomer placed behind the posterior nares; tympanum very distinct. Outer metatarsals united. Sacral processes without any distinct expansion.

Leptodactylus pentadactylus (Laurenti [5]).

Syn. *Rana pentadactyla*, Laurenti [6].
Rana labyrinthica, Spix [7].
Cystignathus labyrinthicus, Duméril & Bibron [8].

---

[1] The *Palæobatrachidæ* seem to connect the Phaneroglossa and Aglossa of Boulenger's classification so closely, that such divisions are not advisable in a palæontological classification.

[2] This family is usually known as *Cystignathidæ*, but since *Cystignathus* is a synonym the above name is substituted.

[3] Class. Reptilium, p. 38 (1826).

[4] Syst. Amphib. p. 202 (1830).

[5] Synopsis Reptilium, p. 32 (1768).—*Rana*.

[6] *Loc. cit.*

[7] Nov. Sp. Ranar. Brasil. p. 31 (1824).

[8] Erpétologie Générale, vol. viii. p. 407 (1841).

Of large size. The conjoint radius and ulna short and broad; and the humerus of the male only moderately expanded.

Hab. South America.

18895. The left humerus of a male, imperfect proximally; from the (*Fig.*) Pleistocene cavern-deposits of Lagoa Santa, Minas Geraes, Brazil. Described and figured by Günther in the 'Ann. Mag. Nat. Hist.' ser. 3, vol. iii. p. 384, pl. xv. fig. D as *Cystignathus labyrinthicus*.
Claussen Collection. Purchased, 1845.

18895 a. The right radius and ulna; from Lagoa Santa. Agrees in relative size with the preceding, and precisely resembles the corresponding bone of a recent male skeleton, from Bahia (Zool. Dep. No. *r*). *Claussen Collection.*

18895 b. A right femur, imperfect proximally, apparently belonging to this species; from Lagoa Santa. *Claussen Collection.*

18895 c. A conjoint tibia and fibula agreeing in relative size with the preceding; from Lagoa Santa. Noticed by Günther, *op. cit.* p. 384. This and the preceding specimen are smaller than the corresponding bones of the above mentioned recent skeleton. *Claussen Collection.*

18895 d. A similar specimen; from Lagoa Santa.
Claussen Collection.

The following larger specimens may probably be referred to female individals of this species: all are from the cavern-deposits of Lagoa Santa, and belong to the Claussen Collection.

18895 e. The left femur.

18895 f. A conjoint tibia and fibula, wanting the terminal epiphyses, and agreeing in relative size with the preceding specimen.

18895 g. The conjoint left radius and ulna.

Leptodactylus ocellatus (Linn.[1]).

Syn. *Rana ocellata*, Linn.[2]
Cystignathus ocellatus, Duméril & Bibron[3].

The conjoint radius and ulna short and wide; humerus of male with the flange enormously expanded, and having a highly convex posterior border.

Hab. South America (Eastern side).

[1] Syst. Nat. ed. 12. vol. i. p. 356 (1766).—*Rana*
[2] *Loc. cit.*
[3] Erpétologie Générale, vol. viii. p. 396 (1841).

The following specimens were obtained from the Pleistocene cavern-deposits of Lagoa Santa, Minas Geraes, Brazil; and belong to the Claussen Collection. Purchased, 1845.

18895 h. A left humerus belonging to a small male individual of this species, or to a closely allied form. Except for its smaller size, this specimen closely resembles the humerus of a recent male skeleton of *L. ocellatus*, figured by Günther in the 'Ann. Mag. Nat. Hist.' ser. 3, vol. iii. pl. xv. fig. E, although the radial ridge does not extend quite so high up. The head is wanting.

18895 i. A similar specimen.

18895 j. A left radius and ulna agreeing in relative size with the humerus 18895 *h*.

18895 j'. A similar specimen.

Leptodactylus ?, sp.

The radius and ulna longer and more slender than in the preceding species.

Hab. South America (Brazil).

18395 k. The left radius and ulna, imperfect distally; from the Pleistocene cavern-deposits of Lagoa Santa, Minas Geraes, Brazil. This specimen is larger and yet much more slender than the corresponding bone of the male of *L. pentadactylus*, No. 18895 *a*. *Claussen Collection. Purchased,* 1845.

SPECIFICALLY UNDETERMINED SPECIMENS.

The following were obtained from the Pleistocene cavern-deposits of Lagoa Santa, Minas Geraes, Brazil, and belong to the Claussen Collection. Purchased, 1845.

18895 l. The left humerus of a female, imperfect proximally. This specimen is of the narrow type characteristic of *L. pentadactylus*.

18895 m. A femur agreeing in relative size with the preceding.

18895 n. A calcaneum and astragalus.

18895 o. An imperfect left ilium of the size of that of *L. ocellatus*.

18895 p. A vertebra.

18895 q. A parasphenoid.

18895 r. The imperfect left humerus of a male. This specimen appears to be of the type of *L. pentadactylus*, although much smaller than No. 18895.

Genus **CERATOPHRYS,** Wied.[1]

Typically the skull with extensive dermal ossifications, which completely surround the orbit and have a pitted sculpture; teeth on the vomer; tympanum more or less distinct or hidden. Outer metatarsals united.

For synonymy see Boulenger, 'Catalogue of Batrachia Salientia,' p. 221 (1882).

Ceratophrys cornuta (Linn.[2]).

Syn. *Rana cornuta*, Linn.[3]

Skull very large, vaulted, and fully ossified; the vomerine teeth in two small patches between the posterior nares; a broad ridge from the orbit to above the tympanum; orbit very small; interorbital bar concave; and tympanum moderately distinct.

For synonymy see Boulenger, *op. cit.* p. 224.

Hab. South America (Surinam and N. Brazil).

18896. The cranium of a female, imperfect anteriorly; from the (*Fig.*) Pleistocene cavern-deposits of Lagoa Santa, Minas Geraes, Brazil. Described and figured by Günther in the 'Ann. Mag. Nat. Hist.' ser. 3, vol. iii. pp. 380–384, pl. xv. figs. B, C. *Claussen Collection. Purchased*, 1845.

18896 a. The cranium of a male, imperfect anteriorly; from the (*Fig.*) cavern-deposits of Lagoa Santa. Described and figured by Günther, *op. cit.* pp. 377–380, pl. xv. fig. A. *Claussen Collection.*

18896 b. Portions of the upper jaw of a female; from Lagoa Santa. *Claussen Collection.*

18896 c. The occipital region of the cranium of a female; from Lagoa Santa. *Claussen Collection.*

1 Beitrage z. Naturgeschichte v. Brasiliens, vol. i. p. 569 (1825).
2 Syst. Nat. ed. 12, vol. i. p. 356 (1766).—*Rana.*
3 *Loc. cit.*

Family BUFONIDÆ.

Upper jaw toothless; sacral processes somewhat expanded at their extremities; vertebræ procœlous; no dorsal ribs; urostyle articulating with the sacrum by two condyles, and without transverse processes. Tongue present.

Genus **BUFO**, Laurenti[1].

The type genus. No teeth on vomer.

Bufo melanostictus, Schneider[2].

Hab. India and China.

Fig. 27.

(?) *Bufo melanostictus.*—An imperfect humerus; from the Pleistocene of Madras. ½. (From the 'Palæontologia Indica.')

R. 1732. Two imperfect humeri, provisionally referred to this species; from the Pleistocene cave-deposits of the Karnul district, Madras (see 'Palæontologia Indica,' ser. 10, vol. iv. pt. i.). In their great distal expansion these bones (fig. 27) agree with the humerus of *Bufo* and differ from *Rana*. *Presented by the Director of the Geological Survey of India.*

Family DISCOGLOSSIDÆ.

Upper jaw toothed; sacral processes expanded terminally; vertebræ opisthocœlous; small ribs attached to the transverse processes of the anterior dorsal vertebræ; urostyle articulating with the sacrum either by one or two condyles, and with a pair of diverging transverse processes at its anterior extremity. Tongue present.

The opisthocœlian character of the vertebræ and the presence of ribs are features in which this family resembles the Caudata.

[1] Synopsis Reptilium, p. 25 (1768).
[2] Hist. Amphib. p. 216 (1791-1801).

Genus **LATONIA**, Meyer[1].

Skull with sculptured ossifications, and the parieto-frontal bones completely ossified [as in *Ceratophrys*]. Ribs with short, posteriorly directed processes; urostyle articulating with sacrum by two condyles.

This genus was regarded by its founder as closely allied to *Ceratophrys*, the vertebræ being described as procœlous[2]. Cope[3], however, pointed out that it should be referred to the present family, the transverse processes to the urostyle and the expanded sacral processes being distinctly shown in Meyer's figure of the type. The second vertebra in advance of the sacrum in this figure seems to be distinctly opisthocœlous; and vertebræ of this type are found in association with the Sansan species.

Latonia seyfriedi, Meyer[4].

Syn. *Latonia (Ceratophrys) seyfriedi*, Meyer[5].

The type species. Of the approximate dimensions of *Leptodactylus pentadactylus*.

Hab. Europe (Switzerland).

42737. Slab showing the ventral aspect of the imperfect skeleton; from the Upper Miocene of Œningen, Switzerland. This specimen is considerably smaller than the type skeleton figured by Meyer in his 'Fauna der Vorwelt—Säugethiere etc. aus dem Molasse,' pl. iv., and may therefore indicate a male individual. The character of the vertebræ cannot be determined. *Van Breda Collection. Purchased,* 1871.

Latonia gigantea, Lartet[6].

Syn. *Rana gigantea*, Lartet[7].
Rana rugosa, Lartet[8].
Latonia rugosa, Cope[9].

Very imperfectly known, and therefore only provisionally separated from the type species. There is every probability that *Rana rugosa* was founded upon small (? male) individuals of this species;

[1] Neues Jahrb. 1843, p. 580.

[2] Meyer, Fauna der Vorwelt—Säugeth. etc. aus dem Molasse, p. 19 (1845).

[3] Nat. Hist. Rev. ser. 2, vol. v. p. 105 (1865), and Journ. Ac. Nat. Sci. Philad. ser. 2, vol. vi. p. 75 (1867).

[4] Neues Jahrb. 1843, p. 580. [5] *Loc. cit.*

[6] Notice de la Colline du Sansan, p. 41 (1851).—*Rana.* [7] *Loc. cit.*

[8] *Loc. cit.* Preoccupied by Schlegel, 'Fauna Japonica—Rept.' pl. iii. fig 3, (1838).

[9] Nat. Hist. Rev. ser. 2, vol. v. p. 105 (1865).

the opisthocœlous charecter of the vertebræ of the latter is noticed by Cope in the 'Nat. Hist. Rev.' vol. v. p. 105.

Hab. Europe (France).

33271. Part of the right maxilla; from the Middle Miocene of Sansan (Gers). Agrees with the type fragment figured by Gervais in his 'Zool. et Pal. Françaises,' pl. lxiv. fig. 24.
Presented by Monsieur Ed. Lartet, 1857.

Family PALÆOBATRACHIDÆ.

Upper jaw toothed; sacral processes much expanded at the termination; vertebræ procœlous; no dorsal ribs; urostyle articulating with the sacrum by two condyles, and without transverse processes. Tongue unknown.

Wolterstorff [1] regards this family as allied on the one hand to the *Pelobatidæ* and on the other to the *Xenopodidæ* [2].

Genus **PALÆOBATRACHUS,** Tschudi [3].

Syn. *Probatrachus,* Peters [4].

The type genus. Skull very large, being longer than the whole of the vertebral column in advance of the urostyle, with the frontoparietals anchylosed in the middle line, 1st and 2nd vertebræ anchylosed together, and the sacral and two vertebræ immediately in advance also anchylosed; vertebral centra very short. No trace of pollex; metacarpals long and slender.

The humerus is frequently nearly as long as the femur. The transverse process of the 8th vertebra, and more rarely that of the 7th, may join with the sacral process in supporting the ilium.

Palæobatrachus diluvianus (Goldfuss [5]).

Syn. *Rana diluviana,* Goldfuss [6].
Palæobatrachus goldfussi, Tschudi [7].

The type species. Smaller than *Rana temporaria.* Sacral process anchylosing with the transverse processes of both the 7th and 8th vertebræ.

[1] Jahrb. nat. Ver. Magdeburg for 1886, p. 156 (1887).

[2] =*Dactylethridæ.* *Dactylethra* being a synonym of *Xenopus,* the latter should be taken as the base of the family name.

[3] Mém. Soc. Neuchâtel, vol. ii. art. 3, p. 81 (1839).

[4] Monatsb. k. Ak. Wiss. Berlin, 1877, p. 678.

[5] Nov. Acta Ac. Cæs. Leop.-Car. vol. xv. p. 119 (1831).—*Rana.*

[6] *Loc. cit.*

[7] Mém. Soc. Neuchâtel, vol. ii. art. 3, p. 81 (1839).

According to Wolterstorff[1], the skeleton figured by Meyer in the 'Palæontographica,' vol. vii. pl. xviii. fig. 1, may be regarded as a characteristic specimen. The variety *elegans*, Wolterstorff[2], is of small size.

Hab. Europe (Germany).

35549. Fragment of lignite showing the impression of a young larva; from the Lower Miocene of Rott, near Bonn, Siebengebirge. This specimen accords with the one figured by Meyer in the 'Palæontographica,' vol. vii. pl. xx. figs. 1, 3, which is referred by Wolterstorff, in the 'Jahrb. nat. Ver. Magdeburg' for 1885, p. 51, to this species.
Presented by W. J. Neville, Esq., 1859.

49464. Fragment of lignite with the impressions of two similar larvæ; from Rott. *Purchased*, 1877.

30271. Four fragments of lignite showing impressions of similar larvæ; from Rott. *Purchased*, 1855.

Palæobatrachus gigas, Meyer[3].

The largest species; nearly the dimensions of *Rana adspersa.* The transverse process of the 8th vertebra joining the sacral process.

Hab. Europe (Germany).

R. 61. Fragment of lignite showing the impression of a larva which may belong to this species; from the Lower Miocene of Rott, near Bonn, Siebengebirge. This specimen corresponds very nearly in size with the one figured by Wolterstorff in the 'Jarhb. nat. Ver. Magdeburg' for 1886, pl. x. fig. 3, and provisionally referred to this species.
Transferred from the Museum of Practical Geology, 1880.

Palæobatrachus meyeri, Troschel[4].

Somewhat more slender and smaller than *P. diluvianus,* with a peculiarity in the structure of the coracoid. The structure of the sacrum is not fully known in the type, but in a specimen provisionally referred to this species by Wolterstorff it seems that the transverse process of the 3rd vertebra alone joined the sacral process,

[1] Jahrb. nat. Ver. Magdeburg for 1885, p. 47 (1886).
[2] *Ibid* for 1886, p 95 (1887).
[3] Palæontographica, vol. vii. p. 169 (1859).
[4] In Dechen's Geognost. Beschreib. d. Siebengebirges, 2nd ed. p. 328 (1861).

and that there was a long interval between the compound iliac bar and the transverse process of the 6th vertebra.

Hab. Europe (Germany).

41087. A split slab of lignite showing the impression of the skeleton and the contour of the soft parts of a small frog provisionally referred to this species; from the Lower Miocene of Rott, near Bonn, Siebengebirge. This specimen is slightly smaller than the type specimen figured by Wolterstorff in the 'Jahrb. Ver. nat. Magdeburg' for 1886, pl. viii. fig. 4; and the sacrum appears to be of the same character as in the larger skeleton represented in fig. 1 of the same plate, and provisionally referred to this species. The sacrum is altogether different from that of the small frog represented in fig. 2 of the same plate as *P. diluvianus*, var. *elegans*. All the above specimens are from Rott. *Purchased*, 1868.

Palæobatrachus bohemicus, Meyer[1].

Larger than *P. diluvianus*, and apparently allied to the Prussian *P. grandipes*, Giebel[2]. Only the transverse process of the 8th vertebra joining the sacral process; the former process being comparatively small. The vertebral column is stated to be shorter than in *P. grandipes*.

Hab. Europe (Bohemia).

35808. Slab of lignite showing the impression and some of the bones of an imperfect skeleton apparently belonging to this species; from the Lower Miocene of Markersdorf, Bohemia. This specimen accords in size with the imperfect skeleton from the same locality figured by Wolterstorff in the 'Jahrb. nat. Ver. Magdeburg,' for 1886, pl. ix. fig. 1, and also with that of *P. grandipes* represented in fig. 2 of the same plate. *Purchased*, 1859.

Palæobatrachus luedeckei, Wolterstorff[3].

Smaller and slighter than *P. diluvianus*, with very slender limbs; the transverse process of the 7th vertebra remaining entirely distinct from the sacral process; vacuity between 8th transverse process and sacral process much smaller than the one between 7th and 8th transverse processes, which remains open.

Hab. Europe (Bohemia).

[1] Palæontographica, vol. vii. p. 180 (1859).
[2] Jahresber. nat. Ver. Halle, 1850, p. 44—*Palæophrynus*.
[3] Jahrb. nat. Ver. Magdeburg for 1885, p. 67 (1886).

37232. Fragment of lignite, showing the impression of one nearly entire skeleton and part of that of another; from the Lower Miocene of Kostálova (? Kostelitz), Bohemia. This specimen accords with the skeleton from Markersdorf figured by Wolterstorff in the 'Jahrb. nat. Ver. Magdeburg' for 1885, pl. vi. fig. 4. *Purchased*, 1863.

35809. Fragment of lignite showing the ventral aspect of the skeleton of a rather smaller individual; from the Miocene of Markersdorf, Bohemia. Resembles the skeleton from the same locality figured by Wolterstorff, *op. cit.* pl. vi. fig. 6. The impression of the left manus is well preserved. *Purchased*, 1859.

It is probable that all or the majority of the following specimens are referable to this species, as being the common Bohemian form.

37231. Fragment of lignite showing the impression and a few fragments of bone of the nearly entire skeleton; from Kostálova. Resembles the specimen figured by Wolterstorff, *op. cit.* pl. vi. fig. 1, but the characters of the sacrum are not shown. *Purchased*, 1863.

36586. Fragment of lignite containing the skeleton of a small individual, with the bones in an imperfect condition; from Markersdorf. The sacral characters are not apparent. *Purchased*, 1862.

35810. Fragment of lignite showing the imperfect skeleton; from Markersdorf. Agrees in size with the preceding. The left metacarpals are finely preserved, and show the slender elongated form characteristic of the genus. *Purchased*, 1860.

35811. Fragment of lignite with the imperfect anterior portion of the skeleton; from Markersdorf. One pectoral limb is well preserved with the exception of the phalangeals. *Purchased*, 1860.

35812. The imperfect skeleton of a small individual, in lignite; from Markersdorf. Resembles the specimen figured by Wolterstorff, *op. cit.* pl. vi. fig. 6. *Purchased*, 1860.

R. 1731. The impression of the nearly entire skeleton of a small individual, in lignite; from Markersdorf. *Purchased.*

35814. Slab of lignite showing the impression and fragments of the bones of the imperfect skeleton of a large individual; from

Markersdorf. The ilia and pelvic limbs are wanting, but the characters of the sacrum are well shown.
Purchased, 1860.

35815. Fragment of lignite showing portions of skeleton of two individuals; from Markersdorf. The bones are much decomposed. *Purchased*, 1860.

37233. The impression of a distorted skeleton, in lignite; from Kostálova. *Purchased*, 1863.

35813. A split fragment of lignite showing the impression of the pes; from Markersdorf. *Purchased*, 1860.

35817. Slab of lignite showing the impressions of the skeletons of a number of larvæ of this species; from Markersdorf. This specimen resembles the larva figured by Wolterstorff, *op. cit.* pl. iv. fig. 10, and are smaller and narrower than the larvæ of *P. diluvianus*. *Purchased*, 1860.

36587. Slab of lignite showing impressions of similar larvæ; from Markersdorf. *Purchased*, 1862.

Palæobatrachus, sp.

The following specimen agrees in size with small individuals of *P. luedeckei*, but does not show characters which admit of its specific determination.

Hab. Europe (Germany).

35867. Slab of lignite showing the impressions of the bones of the skeleton and the contour of the soft parts; from the Lower Miocene of Osberg, Rhenish Prussia. *Purchased*, 1860.

Specimens of which the position is uncertain.

30971. The ilia of a comparatively large species; from the Lower Miocene (Upper Oligocene) of Allier, France.
Bravard Collection. Purchased, 1852.

30972. A right radius and ulna agreeing in relative size with the preceding; from Allier. *Bravard Collection.*

30974. The distal extremity of the tibia and fibula; from Allier.
Bravard Collection.

26654. Several imperfect bones in matrix; from the Lower Miocene of Issoire, Puy-de-Dôme.
Pomel Collection. Purchased, 1851.

Order CAUDATA.

Body long, and either lacertiform or anguiform, with a tail; pectoral, and generally the pelvic, limbs present. Skull without postorbital, supratemporal, and supraoccipital bones, and no parietal foramen. Ribs short, and the vertebræ either amphi- or opisthocœlous. No thoracic buckler or ventral scutes. Gills persistent or caducous.

The palatines, when present, are approximated in the middle line, and placed internally to the vomers and pterygoids.

Family SALAMANDRIDÆ.

Adult without gills; maxillæ present; teeth in both jaws; vertebræ opistho- or amphicœlous. Palatines present.

Subfamily SALAMANDRINÆ.

Palatine teeth in two longitudinal series, diverging posteriorly; no teeth on parasphenoid; vertebræ opisthocœlous and fully ossified.

Genus **MEGALOTRITON,** Zittel[1].

Known by detached vertebræ and limb-bones. Vertebræ much depressed, and strongly opisthocœlous the arch expanding into a broad plate overhanging the centrum, and the zygapophyses very wide.

Megalotriton filholi, Zittel[2].

The type species; typically of large size, the vertebræ measuring 15 millim. in length. The type specimens are from the Quercy Phosphorites.

Hab. Europe (France, and ? Germany).

30484. Twelve vertebræ belonging either to the present or an allied species; from the Lower Miocene (Upper Oligocene) of Mayence. Although of smaller size, these specimens closely resemble the type vertebræ figured by Zittel on page 420 of the work cited. They are doubtless specifically identical with the smaller vertebræ mentioned on page 421 of the same work.

Hastings Collection. Purchased, 1855.

[1] Handbuch der Palæontologie, Abth iii. Band 2 Lief. p. 420 (1888).

[2] *Loc. cit.*

Genus **HELIARCHON**, Meyer[1].

Tail long; 13 or 14 presacral vertebræ, of which the characters are unknown; ribs long, with a spine-like posterior projection near the proximal end; carpus and tarsus unossified.

Heliarchon furcillatus, Meyer.

The type and only known species. Length about 0,110; skull as long as broad, and pointed anteriorly; orbits ovoid.

Hab. Europe (Germany).

42731. (*Fig.*) A split fragment of lignite, showing the impression of the skeleton; from the Lower Miocene of Rott, near Bonn, Siebengebirge. The type; figured by Meyer in the 'Palæontographica,' vol. x. pl. ii. figs. 5, 6.
Van Breda Collection. Purchased, 1871.

Genus **MOLGE**, Merrem[2].

Syn. *Triton*, Laurenti[3] (*non* Linn.).

Skull with a ligamentous or bony fronto-squamosal arch; tail compressed; carpus and tarsus ossified; ribs without proximal processes; 5 digits in pes.

Molge noachica (Goldfuss[4]).

Syn. *Triton noachicus*, Goldfuss[5].

A small and imperfectly known species; length about 0,064.

Hab. Europe (Germany).

30268. Fragment of lignite, showing the impression of the skeleton; from the Lower Miocene of Rott, near Bonn, Siebengebirge. Resembles the specimen figured by Meyer in the 'Palæontographica,' vol. vii. pl. viii. fig. 7.
Purchased, 1855.

42759. Fragment of lignite, showing the imperfect impression of the skeleton; from Rott. *Purchased*, 1871.

[1] Palæontographica, vol. x. p. 292 (1863).
[2] Tentamen Systematis Amphib p 185 (1820).
[3] Synopsis Reptilium, p. 37 (1768).
[4] Nova Acta Ac. Cæs. Leop.-Car. vol. xv. p. 126 (1831).—*Triton.*
[5] *Loc. cit.*

GENUS *non det.*

The following specimens may belong either to *Molge* or *Salamandra.*

33273. Two imperfect presacral vertebræ; from the Middle Miocene of Sansan (Gers), France.

Presented by Monsieur Edouard Lartet.

Family AMPHIUMIDÆ.

No gills in the adult; maxillæ present; teeth in both jaws; palatines aborted; vertebræ amphicœlous.

Genus **CRYPTOBRANCHUS,** Leuckart[1].

Syn. *Proteocordylus*, Eichwald[2].
Andrias, Tschudi[3].
Megalobatrachus, Tschudi[4].
Hydrosalamandra, Leuckart[5].

A series of vomerine teeth parallel with and approximated to those of the margins of the jaws. Two or four branchial arches. Digits 4–5.

The reduction of the branchial arches to two and the abortion of the spiracles in *C.* (*Megalobatrachus*) *maximus* do not seem to be characters worthy of generic distinction.

Cryptobranchus scheuchzeri (Holl[6]).

Syn. *Protée gigantesque, Salamandre gigantesque*, Cuvier[7].
Salamandra scheuchzeri, Holl[8].
Proteocordylus diluvii, Eichwald[9].
Salamandra gigantea, Meyer[10].
Andrias scheuchzeri, Tschudi[11].
Cryptobranchus primigenius, Van der Hoeven[12].

---

[1] Isis, 1821, p. 260.
[2] Zool Specialis, vol. iii. p 165 (1831).
[3] Neues Jahrb. 1837, p. 545. [4] *Ibid.* p. 547.
[5] Froriep's Neue Notizen, vol. xiii. p. 19 (1840).
[6] Handbuch d. Petrefactenkunde, p 95 (1830).—*Salamandra.*
[7] Ann. d. Muséum, vol. xiii. p. 411 (1809). Clashes with *Salamandra gigantea*, Barton. [8] *Loc. cit.*
[9] Zool. Specialis, vol iii p. 165 (1831).
[10] Palæologica, p 117 (1832).
[11] Neues Jahrb. 1837, p 545. [12] *Ibid.* 1838, p. 165.

Cryptobranchus diluvii-testis, Bronn [1].
Hydrosalamandra prisca et *primigenia*, Leuckart [2].
Megalobatrachus scheuchzeri, Lydekker [3].

Larger than *C. maximus* of Japan, with a shorter and wider skull, which approximates to that of *C. alleghaniensis*. Manus larger than in *C. maximus*.

It is probable that in most cases the width of the skull is exaggerated by pressure.

Fig. 28.

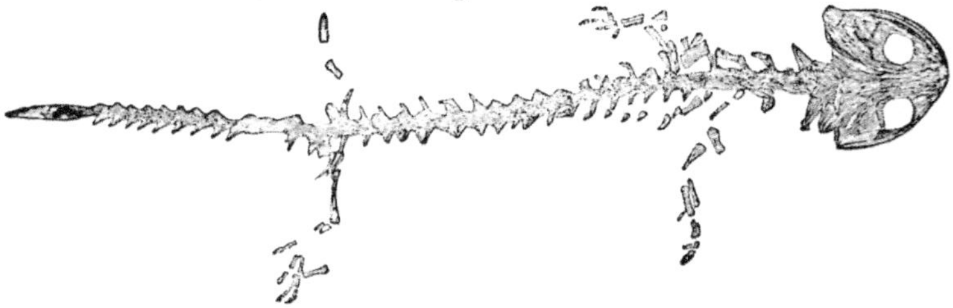

Cryptobranchus scheuchzeri.—The imperfect skeleton; from the Upper Miocene of Œningen. Greatly reduced.

Hab. Europe (Switzerland).

The following specimens are from the Upper Miocene of Œningen, Switzerland.

R. 963. (*Fig.*) Slab of rock showing the greater part of an adult skeleton. Figured by Karg in the 'Denkschr. Nat. Schwabens,' vol. i. pl. ii. fig. 3, as *Lacerta*; and also by Cuvier in the 'Ossemens Fossiles,' 2nd ed. vol. v. pt. 2, pl. xxv. fig. 5, and pl. xxvi. fig. 1. Noticed by Meyer in his 'Fauna der Vorwelt—Säugethiere etc. aus dem Molasse,' p. 30. This specimen was originally in the collection of Dr. Ammon of Schaffhausen. *No history.*

42740. Slab showing the dorsal aspect of the skeleton of a half-grown individual, with the limbs imperfectly preserved. The skull is less expanded than in the preceding specimen, and is thus more like that of the existing species.
Van Breda Collection. Purchased, 1871.

42741. Slab showing the ventral aspect of the nearly entire skeleton of a young individual. The skull is extremely expanded by pressure. *Van Breda Collection.*

[1] Lethæa Geognostica, vol. ii. p. 1166 (1838).
[2] Froriep's Neue Notizen, vol. xiii. p. 19 (1840).
[3] In Nicholson and Lydekker's 'Manual of Palæontology,' 3rd ed. vol. ii. p. 1040 (1889).

42742. A split slab showing the imperfect skeleton of a still younger individual. *Van Breda Collection.*

2761. Fragment of rock showing the imperfectly preserved skull of a young individual. *Van Breda Collection.*

Cryptobranchus tschudii (Meyer[1]).

Syn. *Andrias tschudii*, Meyer[2].

Smaller than *C. maximus*, with a skull of nearly the same form, and a similar short manus.

Hab. Europe (Germany).

42730. (*Fig.*) Slab showing the impression of the greater part of the skeleton; from the Lower Miocene of Rott, near Bonn, Siebengebirge. The type specimen; figured by Meyer in the 'Palæontographica,' vol. vii. pl. ix. fig. 1. The animal lay on its back; most of the bones have perished, and the impressions of the bones of the manus have disappeared since the specimen was figured. A comparison of Meyer's figure with that of *C. maximus* given by Moesch in the 'Neujahrsblatt nat. Ges. Zurich,' 1887, will show the resemblance between the two. *Van Breda Collection. Purchased*, 1871.

Order LABYRINTHODONTIA.

Body long, and usually lacertiform (occasionally anguiform), with a tail; pectoral limbs (when present) shorter than the pelvic limbs; the latter, and usually the former, being pentadactylate. Skull (fig. 29) with the temporal region completely roofed over by postorbital and supratemporal bones, and with paired supraoccipitals and distinct epiotics[3], and a parietal foramen. Teeth pointed, with a large pulp-cavity, and the dentine either simple or more or less plicated. Vertebræ either amphicœlous and fully ossified, or with a notochordal canal, or with large intercentra and the centra represented by paired lateral pieces (pleurocentra). A bony thoracic buckler on the ventral aspect composed of a median (interclavicle) and two lateral (clavicles) plates (fig. 44). Bony scutes frequently present on the ventral aspect of the body.

Teeth are very generally present on the palatines and vomers, and more rarely on the pterygoids. There is very generally an ossified sclerotic ring.

[1] Palæontographica, vol. vii. p. 49 (1860). [2] *Loc. cit.*

[3] Baur regards the bone here termed supratemporal as the squamosal, and *vice versâ*. He also regards the bone generally termed epiotic as the episthotic.

Fig. 29.

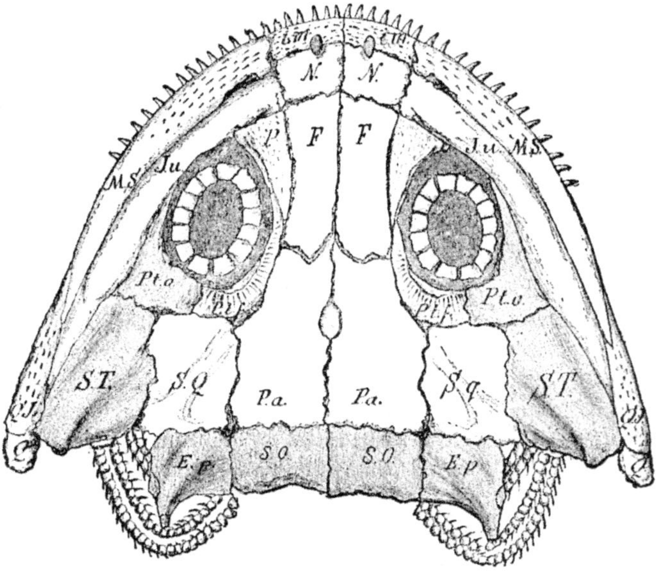

Protriton salamandroides.—Greatly enlarged view of the frontal aspect of the skull of a larva, with the characteristic bones shaded; from the Gaskohle (Lowest Permian) of Bohemia. *N*, nasal; *F*, frontal; *P.a*, parietal; *S.O*, supraoccipital; *E.p*, epiotic; *S.T*, supratemporal; *S.Q*, squamosal *P.t.f*, postfrontal; *Ju*, jugal; *P.t.o*, postorbital; *im*, premaxilla; *M.S*, maxilla; *Q.J*, quadratojugal; *Q*, quadrate. (*After Fritsch.*)

Suborder *LABYRINTHODONTIA VERA.*

Crocodiliform Labyrinthodonts, frequently of large size, in which the bodies of the vertebræ in the adult may be either fully ossified, or of a rhachitomous or embolomerous type[1]; the vertebræ in the young being apparently always of one of the two latter types. Dentine of teeth usually more or less plicated. Cranial bones deeply sculptured, and usually with mucous canals forming a "lyra." No internal gills developed.

[1] In certain genera like *Diplospondylus* and *Cricotus* each caudal vertebra consists of a centrum mainly carrying the neural arch, and an equally large and discoidal posterior intercentrum to which the chevrons are united. These intercentra, according to the views of Cope, correspond with the chevron-bearing intercentra of *Clepsydrops* among the Anomodont Reptilia, and

Some writers divide this group into the two suborders of Stereospondyli and Temnospondyli, according as to whether the vertebræ are fully or imperfectly ossified, but the presence of rhachitomous vertebræ in the young of *Mastodonsaurus* seems to render such division inadvisable.

Family MASTODONSAURIDÆ.

Skull triangular, and more or less elongated, with the cranial bones very strongly sculptured, the occipital condyles ossified, and large palatal vacuities (fig. 32); dentine of teeth with very complex plications; no bony ring in sclerotic; and no ventral scutes. Bodies of vertebræ fully ossified in the adult.

There are large palato-vomerine tusks on the inner side of the maxillary teeth; and the palatines run parallel to the maxilla. The mandible has a large postarticular process; and there is a small inner series of mandibular teeth. In the type genus the pubes are separate from the ischia, and do not enter into the formation of the acetabulum; and the sacral ribs form kidney-like disks.

---

the wedge-bones of *Sphenodon*; this type of structure being known as the *embolomerous*. In the trunk-vertebræ of other genera like *Trimerorhachis* (fig. 46) and *Archegosaurus* each vertebra (fig. 30) consists of four portions—viz., a basal intercentrum (hypocentrum), a pair of pleurocentra, and a neural arch. In this *rhachitomous* type Cope regards the pleurocentra as representing the centrum of the embolomerous type, since they both carry the arch; and as he finds that the functional centra in other forms, like *Chelyosaurus*, apparently correspond to the intercentra of *Archegosaurus*, while the pleurocentra are small and apparently about to disappear, it is argued that in other Amphibia the real centra are totally wanting, and the vertebral bodies, which in the caudal region have the chevrons united to them, are really intercentra, to which the neural arches have been shifted.

Fig. 30.

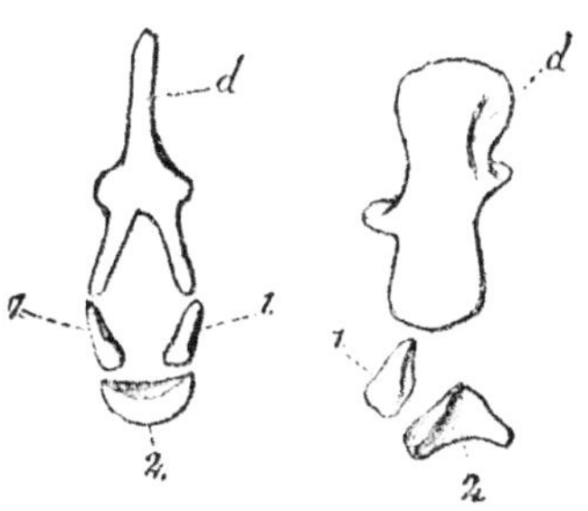

Diagram of a rhachitomous vertebra; from the front and left side. 1, pleurocentra; 2, intercentrum; *d*, neural spine. (*After Fritsch.*)

Genus **MASTODONSAURUS**, Jæger[1].

Syn. *Salamandroides*, Jæger[2].

The type genus. Skull (figs. 31, 32) broad, with slightly concave lateral borders and an obtuse muzzle; palatal vacuities broadest near the middle and far removed from the muzzle; premaxillary vacuities double and somewhat approximated; two small perforations for the mandibular tusks in advance of the nares; nares oval and widely separated; orbits oval, narrowing in front, and situated some distance in advance of the parietal foramen; lyra distinct, and enclosing an elongated diamond-shaped space between the orbits and nares; frontal long, pointed in front, and entering extensively into the formation of the inner border of the orbit; squamosal, postorbital, and postfrontal short. Usually from three to five enlarged premaxillary teeth, lateral teeth smaller; anterior vomerine teeth parallel to those of premaxillæ; two or more enlarged palato-vomerine tusks, the palatines and vomer anchylosing together in this neighbourhood. Cranial sculpture either radiating ridges and grooves, or pitted. Median plate[3] (interclavicle) of thoracic buckler diamond-shaped, with a long posterior spine, and no distinct lateral wings; lateral plates (clavicles) not meeting in a long suture in advance of the middle one.

Mastodonsaurus giganteus, Jæger[4].

Syn. *Salamandroides giganteus*, Jæger[5].
Mastodonsaurus giganteus, Jæger[6].
Mastodonsaurus jægeri, Alberti[7].
Labyrinthodon salamandroides, Owen[8].
Mastodonsaurus salamandroides, Plieninger[9].
Labyrinthodon jægeri, Owen[10].

The type species. Length of skull varying from 0,700 to 1,000. Cranial and thoracic sculpture coarsely radiate.

Hab. Europe (Germany[11]).

[1] Foss. Rept. Württemberg, p. 35 (1828). [2] *Ibid.* p. 38.

[3] See Meyer and Plieninger, 'Beiträge Pal. Wurttembergs,' pl. iii.; and E. Fraas, Palæontographica, vol. xxxvi. p. 85, fig. 5.

[4] Foss. Rept. Wurttemberg, p. 39 (1828).—*Salamandroides.* [5] *Loc. cit.*

[6] Bull. Soc. Géol. France, sér. 1, vol. iii. p. 86 (1833).

[7] Beitrag z. Monograph. d. Bunten Sandsteins, etc. p. 236 (1834).

[8] Trans. Geol. Soc. ser. 2, vol. vi. p. 512 (1842).

[9] Ebenda, p 57 (1844).

[10] Odontography, pt. ii. p. 198 (1841).

[11] There is no sufficient evidence in support of the alleged occurrence of this species in Britain.

Fig. 31.

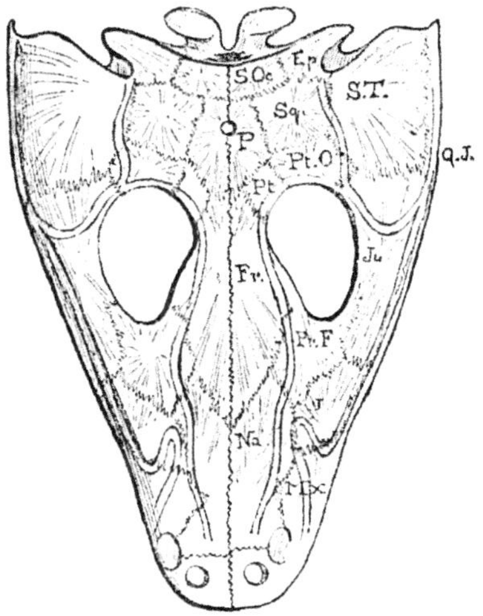

Mastodonsaurus giganteus.—Frontal aspect of the cranium, with the sculpture omitted; from the Lettenkohle (Lowest Keuper) of Wurtemberg. About $\frac{1}{8}$. *SOc*, supraoccipital; *Ep*, epiotic; *P*, parietal; *Sq*, squamosal; *S.T*, supratemporal; *Q J*, quadratojugal; *Ju*, jugal; *Pt*, postfrontal; *Pt.O*, postorbital; *Fr*, frontal; *Pr.F*, prefrontal; *L*, lachrymal; *Na*, nasal; *Mx*, maxilla; the premaxilla has no letter. (*After Fraas.*)

Fig. 32.

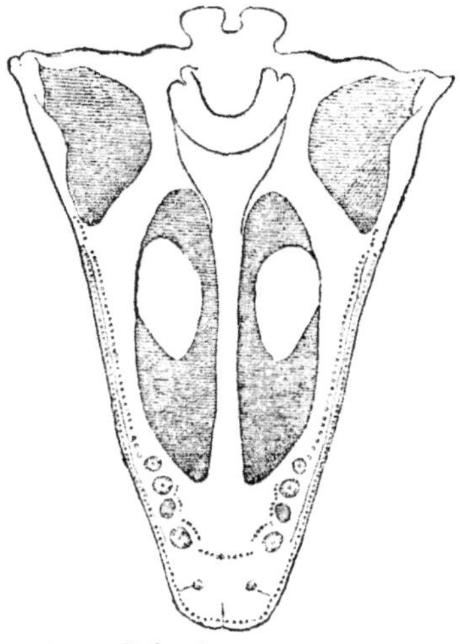

Mastodonsaurus giganteus.—Palatal aspect of the cranium; from the Lettenkohle of Würtemberg. About $\frac{1}{8}$. (*After Miall.*)

11959. Cast of one of the large anterior tusk-like teeth. The original, which is the type of the genus, was obtained from the Alum-shales at the base of the Lettenkohle (Lowest Keuper [1]) at Gaildorf, Würtemberg. The specimen is described and figured by Jæger in his 'Foss. Rept. Württemberg,' p. 35, pl. iv. fig. 4; and is preserved in the Museum at Stuttgart. It is also figured in Owen's 'Odontography,' pl. 63. fig. 1.
Mantell Collection. Purchased, 1838.

R. 84. The base of a somewhat smaller tooth: from the bone-bed at the base of the Lettenkohle at Gaildorf. This specimen has been cut and polished to exhibit the labyrinthic structure.
Transferred from the Museum of Practical Geology, 1888.

43637. A smaller tooth; from the Lettenkohle of Gaildorf. The summit is wanting. *Purchased*, 1859.

33091. A still smaller imperfect tooth, in matrix, from the bone-bed of the Lettehkohle of Gaildorf. *Purchased*, 1848.

11959 a. Cast of the exoccipitals and basioccipital. The original was obtained from the Lettenkohle of Gaildorf, and is described and figured by Jæger in his 'Foss. Rept. Wurttemberg,' p. 38, pl. v., as the type of *Salamandroides*.
Mantell Collection. Purchased, 1838.

48206. Cast of the cranium. The original, which is preserved in the Museum at Stuttgart, was obtained in 1833 from the Alum-shale of the Lettenkohle at Gaildorf. It is described and figured by Meyer and Plieninger in their 'Beitrag. Pal. Württembergs,' p. 64, pl. vi. fig. 1, and pl. vii. fig. 1; and also by E. Fraas in the 'Palæontographica, vol. xxxvi. pl. 32, pls. i. and ii. *Purchased*, 1877.

33092. The imperfect centrum of a dorsal vertebra, in a somewhat crushed condition; from the Lettenkohle of Gaildorf. This specimen resembles the vertebræ figured by Meyer and Pleininger, *op. cit.* pl. iv.; of which the characters are noticed by E. Fraas in the 'Palæontographica,' vol. xxxvi.

[1] The Lettenkohle is regarded by E. Fraas as a primary division of the Trias, ranking with the Keuper and Muschelkalk.

pp. 78, 79. The characteristic diminution in the thickness of the dorsal half of the centrum is well exhibited.
Purchased, 1848.

33093. Fragments of ribs; from the Lettenkohle of Gaildorf.
Purchased, 1848.

Mastodonsaurus granulosus, E. Fraas[1].

Imperfectly known, but readily distinguished by the finely pitted sculpture of the cranial bones and thoracic buckler.

Hab. Europe (Germany).

33075. Five imperfect sculptured bones; from the Upper Muschelkalk bone-bed (Middle Trias) of Crailsheim, Würtemberg. The sculpture agrees precisely with that of the type specimens from the same locality figured by Fraas in the 'Palæontographica,' vol. xxxvi. pl. vi. figs. 1, 2.
Purchased, 1848.

Mastodonsaurus keuperinus, E. Fraas[2].

Imperfectly known. Nearly or quite as large as *M. giganteus*, from which it is readily distinguished by the larger size and greater relative width of the posterior nares (choanæ).

Hab. Europe (Germany).

15270. Slab of sandstone exhibiting the external surface of the left lateral thoracic plate probably belonging to this species; from the Middle Keuper Sandstone (Upper Trias) of Feuerbach-Haide, near Stuttgart, Würtemberg. In its pitted structure this specimen differs from the thoracic plates referred to *Capitosaurus*, and resembles the larger left lateral plate of *M. giganteus* figured by Meyer and Plieninger in their 'Beitr. Pal. Württembergs,' pl. iv. figs. 1, 2.
Purchased.

Mastodonsaurus indicus, Lydekker (n. sp.).

Known only by a fragment of the thoracic buckler, which indicates a form fully as large as *M. giganteus*. Definite specific characters cannot at present be given.

Hab. India.

R. 596. Cast of a considerable portion of the right lateral plate

[1] Palæontographica, vol. xxxvi. art i. p. 94 (1889).
[2] *Ibid.* p 116.

(clavicle) of the thoracic buckler. The original, which is the type, was obtained from the Denwa group of the Gondwanas, on the Denwa river, Satpura district, Central Provinces, and is preserved in the Indian Museum, Calcutta. It is figured by the writer in the 'Palæontologia Indica,' ser. 4, vol. i. pt. 3, pl. vi. fig. 8, as a Crocodilian scuta, and in pt. 5, p. 30, fig. 2 of the same as a supratemporal of *Mastodonsaurus*. Since the date of the latter figure a cast has been submitted to Dr. E. Fraas, of Munich, who has compared it with the thoracic buckler of *M. giganteus*, and states that it agrees so closely as to leave no doubt that it indicates an allied species. The specimen

Fig. 33.

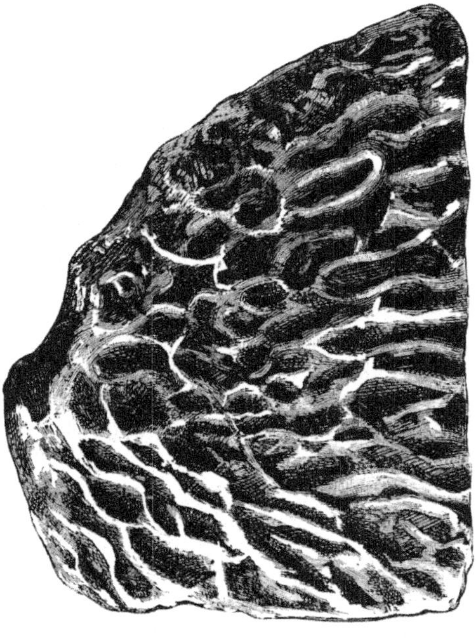

Mastodonsaurus indicus.—Part of the right lateral thoracic plate; from the Denwa stage of the Gondwanas of Central India. $\frac{1}{2}$. (From the 'Palæontologia Indica.')

is represented in fig. 33; the upper part of the border sloping away from the superior angle is part of the external natural border of the plate; the postero-external angle of the specimen being part of the region from which the posterior clavicular process projected.

Made in the Museum, 1885.

Genus CAPITOSAURUS, Münster [1].

Including *Cyclotosaurus*, Fraas [2].

Skull (fig. 34) with the muzzle often broad and obtuse; palatal vacuities expanded in front and approximated to muzzle; a large single premaxillary vacuity in the middle line; orbits elliptical, and approximated to the parietal foramen, generally small; lyra indistinct and external to orbits and nares; nares large, rounded, and widely separated; frontal short, pointed in front, and forming part of inner border of orbit; squamosal, postorbital, and postfrontal very short. Premaxillary and maxillary teeth uniform; a number of small vomerine teeth parallel to those of the premaxillæ; one or two large tusks in advance of and behind the posterior nares. Each cranial bone strongly pitted at the centre, and in most cases radiately grooved towards the periphery, but such groovings absent in the supraoccipital, parietal, squamosal, postorbital, and postfrontal. The auditory notch may be either open or converted into a foramen (fig. 34). Median plate (interclavide) of thoracic buckler intermediate between that of *Mastodonsaurus* and that of *Metoposaurus* [3]. The fully adult teeth have no central pulp-cavity.

Capitosaurus nasutus, Meyer [4].

Average length of cranium 0,330. Auditory slit open. Cranium somewhat elongated, and narrowing anteriorly, with a sharp muzzle and comparatively large orbits, of which the axis is somewhat oblique to that of the cranium. Interclavicle with rounded posterior extremity like that of *Metoposaurus*.

Hab. Europe (Germany).

The following specimens are from the Bunter (Lower Trias) of Bernburg, Northern Germany.

40041. The cranium, in a somewhat imperfect condition. Both frontal and palatal aspects are shown, but the extremity of the muzzle is concealed by matrix. This specimen accords with the imperfect type skull figured by Meyer

[1] Neues Jahrb 1836, p 580.

[2] Palæontographica, vol. xxxvi. art 1, p 121 (1889).—Quoted from MS. notes by Zittel in 1888

[3] See E Fraas, Palæontographica, vol xxxvi p 135.

[4] Palæontographica. vol vi. art. 6 p 222 (1856).

in the 'Palæontographica,' vol. vi. pls. xxv. & xxvi. The somewhat narrow contour is well shown; and the open auditory notch (Meyer, *op. cit.* pl. xxiv.) is very distinctly displayed. *Purchased*, 1866.

42751. Mass of sandstone showing the inner laminæ of the bones of the cranial roof, with the exception of those of the muzzle. This specimen agrees in size with the preceding. *Purchased*, 1871.

36346. The imperfect cranium of a smaller individual. The frontal aspect is represented by a sandstone cast, or portions of the inner lamina of bone, of the cranial roof. The upper portion of the premaxillæ is wanting, so that the large premaxillary vacuity is shown. *Purchased*, 1862.

36344. Slab of sandstone exhibiting the palatal aspect of a cranium agreeing in size with the preceding. The contour precisely resembles that of the larger specimen figured by Meyer, *op. cit.* pl. xxvi. *Purchased*, 1862.

36347. Slab of sandstone showing the inner lamina of the bones of the cranial roof, imperfect posteriorly. This specimen is rather larger than No. 36346. *Purchased*, 1862.

36348. Slab of sandstone showing the upper surface of the inferior lamina of the bones of the anterior extremity of the palate of a small individual. *Purchased*, 1862.

36350. Fragment of rock showing the inner layer of the bones of the posterior portion of the cranial roof. This specimen indicates a small individual, and may be referable to the next species. *Purchased*, 1862.

36352. Mass of sandstone showing the greater portion of a mandibular ramus. *Purchased*, 1862.

30270. Slab of sandstone showing the impression of the ventral surface of the median place (interclavicle) of the thoracic buckler, together with a plaster cast taken from the same. This specimen indicates a small individual. Its contour is very similar to that of the corresponding bone of *Metoposaurus* figured by E. Fraas in the 'Palæontographica,' vol. xxxvi. pl. xv., and less like that of *C. robustus* figured by Quenstedt in his 'Mastodonsaurier Württembergs,' pl. iv. fig. 1. *Purchased*, 1855.

Capitosaurus fronto, Meyer[1].

Smaller than *C. nasutus*, with the facial portion of the cranium shorter, and much wider at the muzzle; orbits smaller, with the axis less oblique to that of the cranium. Auditory slit apparently open.

Hab. Europe (Germany).

36345. Slab of sandstone showing the frontal aspect of the imperfect cranium: from the Bunter (Lower Trias) of Bernburg, Northern Germany. The right orbit is preserved, but the left orbital region is crushed in and concealed by matrix. This specimen appears to agree very closely with the imperfect posterior portion of the cranium from the same locality figured by Meyer in the 'Palæontographica,' vol. vi. pl. xxviii. fig. 2, as the type of this species. The difference in the length of the muzzle from that of *C. nasutus* is apparent by comparison with the small specimen No. 36346; thus in the latter the interval between the orbit and extremity of the muzzle is 0,175, against 0,150 in the present specimen; the two skulls being approximately equal in size. *Purchased*, 1862.

Capitosaurus robustus, Meyer[2].

Syn. *Mastodonsaurus robustus*, Quenstedt[3].
Cyclotosaurus robustus, E. Fraas[4].

The type of *Cyclotosaurus*. Of very large size, the length of the cranium being about 0,610. Auditory slit converted into a foramen by the junction of the epiotic with the supratemporal; cranium comparatively short and broad, with a blunt, rounded muzzle, and relatively small orbits. Epiotics extending behind line of condyles. Interclavicle long and diamond-shaped, without distinct lateral wings.

A nearly entire skull is figured by E. Fraas in the 'Palæontographica,' vol. xxxvi. pls. ix., x., from which the woodcut on the next page is drawn.

Hab. Europe (Germany).

[1] Palæontographica, vol. vi. art. 6, p. 229 (1856).
[2] In Meyer and Plieninger's 'Paläontologie Württembergs,' p. 11 (1844).
[3] Mastodonsaurier im grünen Keupersandsteine Württembergs (1850).
[4] Palæontographica, vol. xxxvi art. 1, p. 121 (1889).

The originals of the following specimens were obtained from the Middle Keuper (Upper Trias) of Feuerbach-Haide, near Stuttgart, Würtemberg.

37993. Cast of a slab of rock showing the inner lamina of the bones of the cranial roof. Original figured in Quenstedt's 'Mastodonsaurier Württembergs,' pl. i. fig. 1.
Purchased, 1864.

37992. Cast of slab showing the palatal aspect of the cranium. Original figured by Quenstedt, *op. cit.* pl. ii. fig. 1.
Purchased, 1864.

Fig. 34.

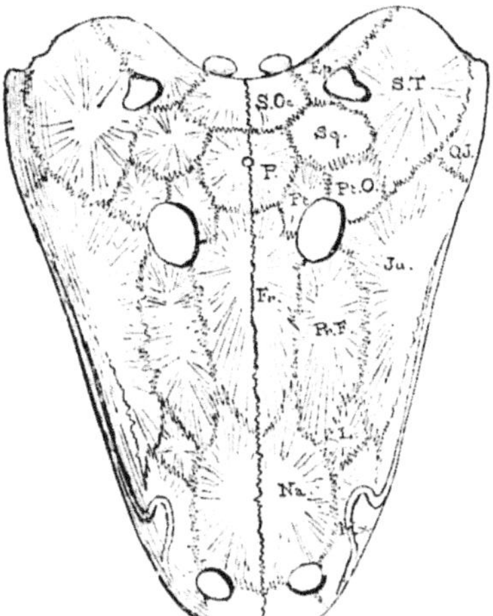

Capitosaurus robustus.—Frontal aspect of the cranium, with the sculpture omitted; from the Middle Keuper (Upper Trias) near Stuttgart. About $\frac{1}{8}$. Letters as in figure 31. (*After E. Fraas.*)

R. 1439. Cast of slab showing the frontal aspect of the muzzle. Original in the Museum at Munich.
Presented by Prof. K. A. von Zittel, 1888.

R. 1439 a. Cast of slab showing palatal aspect of the anterior portion of the cranium, with the large vomerine tusks. Original in the Museum at Munich.
Presented by Prof. K. A. von Zittel, 1888.

R. 1439 b. Cast of slab showing the oral surface of the mandibular

symphysis, with its tusks. The original is in the Munich Museum, and is figured by E. Fraas in the 'Palæontographica,' vol. xxxvi. pl. xi. fig. 2.
Presented by Prof. K. A. von Zittel, 1888.

R. 1439 c Cast of slab with the ventral surface of part of the mandible exposed. The original is in the Museum at Munich. *Presented by Prof. K. A. von Zittel*, 1888.

Of the following specimens from the Middle Keuper Sandstone of Feuerbach-Haide at least the majority may be referred to this species.

33074. Slab of sandstone exhibiting the external surface of the median thoracic plate. This specimen, which seems to include the imperfect anterior and left lateral moiety of the bone, accords well with the specimen of which the left half is figured by Quenstedt in his 'Mastodonsaurier Württembergs,' pl. iv. fig. 1; and is unlike the corresponding bone of *Metoposaurus*. *Purchased*, 1848.

14675. Slab of sandstone showing the impression of the external surface of the right lateral thoracic plate. This specimen, which has a simple radiate sculpture, accords with the imperfect lateral plate figured by Quenstedt, *op. cit.* fig. 12, and is unlike the corresponding bone of *Metoposaurus*.
Purchased.

14676. Fragment of sandstone showing the impression of an imperfect lateral plate agreeing with the preceding. The proximal portion is wanting. *Purchased.*

15271. Mass of sandstone showing the outer aspect of the greater part of the left thoracic plate, and the impression of part of the inner surface of the median plate. *Purchased.*

R. 281. Fragment of sandstone showing the outer surface of an imperfect and smaller lateral thoracic plate.
Egerton Collection. Purchased, 1882.

R. 84. Fragment of sandstone showing the outer surface of a sculptured bone. The sculpture differs from that of the preceding specimen by its pitted character, but it is not certain that the specimen belongs to the thoracic buckler.
Transferred from the Museum of Practical Geology, 1884.

43636. Slab of sandstone exhibiting the outer surface of an imperfect sculptured bone. The sculpture is pitted.
Purchased, 1859.

14675. Fragment of sandstone exhibiting the impression of the outer surface of a bone with pitted sculpture. *Purchased.*

33073 Slab of sandstone with portions of a sculptured bone.
Purchased, 1848.

Genus **METOPOSAURUS**, Lydekker (*n. nom.*).

Syn. *Metopias*, Meyer [1].

Skull (fig. 35) broad, with obtuse muzzle; palatal vacuities expanded in front; premaxillary vacuities large and double; orbits large, oval, situated in the anterior half of the skull, and widely

Fig. 35.

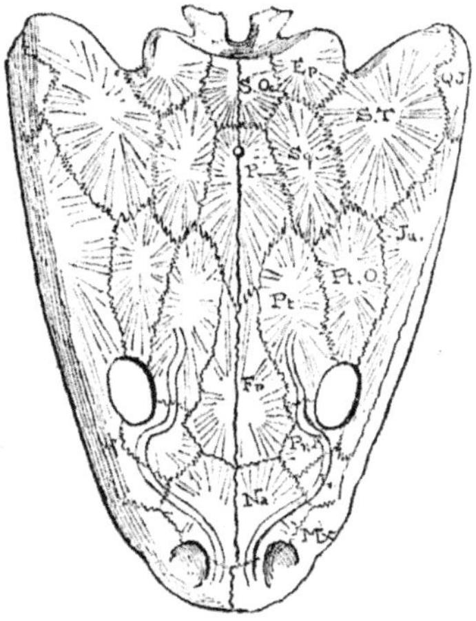

Metoposaurus diagnosticus.—Frontal aspect of the cranium, with the sculpture omitted; from the Keuper (Upper Trias), near Stuttgart. ⅛. Letters as in fig. 31. (*After E. Fraas.*)

separated from one another; lyra enclosing an oblong space between orbits and nares; nares terminal, large, rounded and approximated; frontals pointed behind, and excluded from the border of the orbit by the junction of the postfrontal with the prefrontal; squamosal, postorbital, and postfrontal elongated. Teeth apparently arranged

[1] Neuer Jahrb. 1842, p. 302. Preoccupied by Gory in 1832 for a genus of Coleoptera.

like those of *Capitosaurus*. Each cranial bone strongly pitted at the centre, and radiately grooved at the periphery. Median plate (interclavicle) of thoracic buckler with a rounded posterior extremity and distinct lateral wings; lateral plates (clavicles) meeting in a long suture in advance of the median plate[1]. Dentition weak; teeth strongly fluted externally, especially at the base, with but slight internal foldings[2].

Metoposaurus diagnosticus (Meyer[3]).

Syn. *Metopias diagnosticus* (Meyer[4]).
Labyrinthodon diagnosticus, Owen[5].

The type and only described species. Skull attaining a length of 0,450.

Hab. Europe (Germany).

37938. The imperfect cranium; from the Middle Keuper (Upper Trias) near Stuttgart, Würtemberg. The occiput is imperfect, and the extremity of the muzzle is wanting. Both the frontal and palatal aspects are shown. This specimen, which is mentioned in the 'Rep. Brit. Assoc.' for 1874, p. 157, is considerably smaller than the entire skull figured by Fraas in the 'Palæontographica,' vol. xxxvi. pls. xii., xiii. The vomers are very clearly shown, and have the teeth arranged as in *Capitosaurus*.
Häberlein Collection. Purchased, 1862.

GENUS *non det.*

The Labyrinthodont represented by the undermentioned species appears to be allied to *Metoposaurus* and *Capitosaurus*; the associated teeth represented in fig. 36 are, however, larger than those of

Fig. 36.

Labyrinthodont Jaw.—From the Maleri stage of the Gondwanas of Central India. ½. (From the 'Palæontologia Indica.')

[1] See E. Fraas, Palæontographica, vol. xxxvi. pl. xv.
[2] *Ibid.* p. 148. [3] Neuer Jahrb. 1842, p. 302 —*Metopias*. [4] *Loc. cit.*
[5] Palæontology, 2nd ed p. 216 (1861).

Metoposaurus, and have a distinct pulp-cavity. Fully ossified vertebræ are found with the other remains. The bone figured by the writer in the 'Palæontologia Indica,' ser. iv. vol. i. pt. 5, pl. vi. fig. 3, as a squamosal, appears to be a supratemporal; and apparently makes a close approximation to the corresponding bone of *Metoposaurus*.

Hab. India.

R. 580. Five fragments of sculptured cranial bones; from the Maleri stage of the Upper Gondwanas at Maleri, 32 miles north-west of Sironcha, Central Provinces. The sculpture of some of these specimens closely resembles that of *Metoposaurus diagnosticus*.

Presented by the Director of the Geological Survey of India, 1885.

The following specimen may indicate an allied form.

Hab. Africa.

R. 513. Fragment of a sculptured bone; from the Karoo system of the Orange Free State. The sculpture is unlike that of *Rhitidosteus*.

By exchange with the Blomfontein Museum, 1888.

Genus **TREMATOSAURUS**, Braun[1].

Skull long, and gradually narrowing in front, with a rounded muzzle; palatal vacuities pointed in front, and far removed from extremity of muzzle; premaxillary vacuities small and widely separated; orbits oval, and situated far in advance of parietal foramen, although not in anterior half of skull; nares some distance behind muzzle, large, approximated, and elongated; lyra distinct, and enclosing an ovoid space between the orbits and nares: frontal long, pointed at both extremities, and excluded from border of orbit by junction of postfrontal with prefrontal; postorbital and postfrontal elongated. Premaxillary and maxillary teeth small and uniform; 4 small vomerine teeth parallel to those of the maxilla; two large vomerine tusks in advance of posterior nares, and three or four palatine ones behind the same. Cranial bones pitted at the centre, and radiately grooved at the periphery. Posterior spine of median plate of thoracic buckler larger and wider than in *Mastodonsaurus*, but the general shape of the bone similar.

[1] Bericht Naturf. Aerzte, 1841, p. 74.

Trematosaurus brauni, Burmeister[1].

Syn. *Labyrinthodon brauni,* Owen[2].

The type species. Skull with an average length of 0,240, and a width posteriorly of 0,130; orbits situated midway between the muzzle and the occiput.

Hab. Europe (Germany).

The following specimens are from the Bunter (Lower Trias) of Bernburg, Northern Germany.

40042. The imperfect cranium of a small individual. The muzzle is wanting, but the orbital and postorbital regions are well preserved, and exhibit the sculpture and lyra very distinctly. *Purchased,* 1866.

36354. The imperfect cranium of an adult individual. The muzzle is wanting, and only the inner lamina of the cranial bones remains. This specimen agrees in size with the one figured by Burmeister, *op. cit.* pl. i. *Purchased,* 1862.

36369. An imperfect adult cranium. The muzzle is broken off, and only portions of the cranial bones remain. *Purchased,* 1862.

36360. A smaller imperfect cranium. The muzzle is wanting, but most of the cranial bones remain, with the loss of the sculpture. The palate is concealed. *Purchased,* 1862.

42810 a. An imperfect adult cranium, with the counterpart in matrix. The muzzle is wanting, and the greater portion of the cranial bones is adherent to the matrix of the counterpart. *Purchased,* 1871.

36374. A small imperfect cranium. Portions of the sculpture remain, and the contour of the right epiotic cornu and auditory notch is well shown. *Purchased,* 1862.

36356. The parieto-occipital region of an adult cranium, showing the sculpture. *Purchased,* 1862.

36357. A nearly similar specimen, without the sculpture. *Purchased,* 1862.

[1] Labyrinthodonten aus dem bunten Sandstein, pt. i. p. 69 (1849).

[2] Palæontology, 2nd ed p. 215 (1861).

36349. The posterior portion of a small cranium, with the counterpart in matrix. The counterpart shows the under surface of the anterior region of the cranial roof. *Purchased*, 1862.

R. 1750. A split slab of sandstone showing the imperfect skull. One piece shows the under surface of the cranial bones of the parietal and occipital regions. A middle portion contains the hinder region of the cranium. The lower slab contains the pterygoids, the summits of the maxillary and palatine teeth, and traces of the mandible. *Purchased.*

R. 1751. Slab of sandstone showing the under surface of the anterior half of the cranium. In the region of the muzzle the bones of the palate, with the premaxillary vacuities and posterior nares, are exhibited; while behind the anterior border of the palatal vacuities only the under surface of the cranial roof remains. *Purchased.*

42810 b. Slab of sandstone showing the dorsal aspect of the bones of the palate. The muzzle is wanting, and the bones remaining are imperfect. *Purchased*, 1871.

30269. Slab of sandstone with the dorsal aspect of the bones of the hinder half of the palate of a smaller cranium. *Purchased*, 1855.

33068. Fragment of sandstone exhibiting the dorsal surface of the hinder half of a nearly similar palate. The bone of part of the right maxilla is broken away, and thus exhibits the bases of the teeth. *Purchased*, 1858.

36363. Part of a small cranium, exhibiting both the frontal and palatal aspects. *Purchased*, 1862.

36375. Fragment of sandstone showing the anterior portion of the palate of a small individual. The specimen is broken off near the anterior extremity of the palatal vacuities. *Purchased*, 1862.

42810. Mass of sandstone showing a portion of a mandibular ramus. *Purchased*, 1871.

36369. Slab of sandstone showing the median plate (interclavicle) of the thoracic buckler. Nearly all the sculpture is lost; in contour this specimen agrees with the larger example figured by Burmeister, *op. cit.* pl. iv. fig. 2. *Purchased*, 1862.

42753. Fragment of rock with the greater part of the right lateral (clavicle) and a fragment of the median plate of the thoracic girdle. Part of the sculpture is preserved.
Purchased, 1871.

GENUS *non det.*

Known by fragmentary bones, characterized by a pustular sculpture (like that of *Micropholis*), and indicating a large form.
Hab. Europe (Germany).

21530 x. Three imperfect sculptured bones; from the Muschelkalk bone-bed (Middle Trias) of Crailsheim, Würtemberg. The sculpture accords with that of the specimens from the same locality figured by E. Fraas in the 'Palæontographica,' vol. xxxvi. pl. vi. figs. 15, 16, as *Labyrinthodon* sp.; that term being apparently used as equivalent to Labyrinthodont. *Purchased*, 1847.

GENUS *non det.*

The undermentioned specimens indicate a large Labyrinthodont in which the teeth have a large pulp-cavity, and much simpler foldings than in *Mastodonsaurus*. The specimens No. 48833 are said to be some of those referred, in the 'Rep. Brit. Assoc.' for 1874, p. 157, to *Metoposaurus diagnosticus*, but the teeth are very much larger than in that species.
Hab. Europe (England).

44833. Fragments of jaws with teeth, in matrix; from the Rhætic of Aust Cliff, near Bristol. Some of the teeth have been cut to show the internal structure.
Presented by Benjamin Bright, Esq., 1873.

R. 394. Other fragments of similar teeth and jaws, in matrix; from Aust Cliff. *Enniskillen Collection. Purchased*, 1882.

Family ANTHRACOSAURIDÆ.

Skull usually triangular and more or less angulated, with the cranial sculpture well marked, the occipital condyles ossified, and the palatine foramina very small and placed far back; dentine of the teeth more or less complexly plicated. A ventral armour of elongated dermal scutes; and probably a sclerotic ring. Bodies of vertebræ fully ossified in the adult; intercentra absent or present.

According to Atthey's figure of the skull of the type genus, the palatines bear teeth and are situated immediately on the inner side of the maxillæ, as in the *Mastodonsauridæ*. In the typical forms there is no postarticular process to the mandible.

Genus **ANTHRACOSAURUS**, Huxley[1].

The type genus. Skull broadly triangular, with large postero-lateral expansions and small epiotic cornua; orbits very small, subtriangular, approximated, and situated in the hinder third of the skull; nares small, round, and widely separated; lyra indistinct; frontal excluded from orbit by junction of prefrontal with postfrontal; anterior border of orbit formed by prefrontal; vomer small and toothless. Premaxillary and maxillary teeth few, unequal, and forming an irregular series; one large palatine tusk near the posterior nares, and others further back; mandibular teeth irregular; pterygoids apparently carrying a number of denticules. Crowns of teeth ridged, conical, with a transversely oval section at the base, and laterally compressed and curved near the summit[2], where the marginal ones are carinated; plications of dentine very complex. Mandibular ramus short and highly arcuated inferiorly[3]. Cranial sculpture pitted and very sparsely distributed. Intercentra apparently absent in vertebral column. *Anthracosaurus raniceps*, Goldenberg, does not belong to this genus.

Anthracosaurus russelli, Huxley[4].

The type and only described species. Of large size, the cranium having a length of about 0,360 (14·2 inches), and a width of 0,330 (11·8 inches). Muzzle very broad and rounded. Figures of the skull &c. are given by Atthey in the 'Ann. Mag. Nat. Hist.' ser. 4, vol. xviii. pls. viii.–xi. (1876). The type specimen is from the Coal-Measures (Upper Carboniferous) of Lanarkshire[5].

Hab. Europe (Britain).

30532. Slab of rock showing a large portion of the palate in a very imperfect condition; probably from the Lower Carboniferous of Burdie House, near Edinburgh. This specimen extends as far back as the posterior extremity of the

[1] Quart. Journ. Geol. Soc. vol. xix. p. 56 (1863).

[2] Atthey, Ann. Mag. Nat. Hist. ser. 4, vol. xviii. p. 160 (1876).

[3] See Ann. Mag. Nat. Hist. ser. 4, vol. xviii. pl. x.; it was stated in the Rep. Brit. Assoc. for 1874, p. 159, that the mandible had a postarticular process, but its absence is clearly shown in this figure.

[4] *Loc. cit.*

[5] See Quart. Journ. Geol. Soc. *loc. cit.*

dental series. The anterior palatine tusk is seen immediately behind the posterior nares on the left side: while other large tusks are shown further back on the palate in the same position as in Huxley's figure of the type specimen in the 'Quart. Journ. Geol. Soc.' vol. xix. p. 59, fig. 1. The contour of the palatine foramina is obscurely indicated. One of the palatine tusks is beautifully preserved. *Purchased*, 1856.

45859. Slab of shale showing the inner aspect of a portion of the right ramus of the mandible: from the Coal-Measures (Upper Carboniferous) of Newcastle. This specimen apparently accords with the entire ramus figured by Atthey in the 'Ann. Mag. Nat. Hist.' ser. 4, vol. xviii. pl. x. figs. 1, 2. Several of the teeth remain, and show the oval contour of the cross-section, with its larger diameter transverse to the axis of the jaw. *Purchased*, 1874.

R. 1740. The articular region of the right ramus of the mandible of a small individual of this species or of an allied smaller form; from the Coal-Measures of Northumberland. Resembles the corresponding part of the larger ramus figured by Atthey, *op. cit.* pl. x. fig. 1, and shows very clearly the absence of a postarticular process.
Presented by J. Wood Mason, Esq., 1880.

37324. Slab showing three vertebræ and ribs; from the Coal-
(*Fig.*) Measures of Airdrie, Lanarkshire. The one vertebra in which the greater portion of the arch remains and a rib are figured by Huxley in the 'Quart. Journ. Geol. Soc.' vol. xix. p. 63, fig. 2. The figured vertebra resembles the entire one figured by Atthey, pl. x. fig. 4.
Purchased, 1863.

45860. Slab showing six imperfect vertebral centra; from Newcastle. *Purchased*, 1874.

45861. The centrum of a larger vertebra; from Newcastle.
Purchased, 1874.

45862. Slab of shale showing the greater portion of a rib; from Newcastle. *Purchased*, 1874.

41851 x. Slab showing portion of the vertebral column with ribs; from the Coal-Measures of Jarrow Colliery, Kilkenny. The vertebræ are badly preserved; one rib shows its double head. *Purchased*, 1870.

41851 y. Slab showing the terminal faces of eight vertebral centra; from Jarrow. *Purchased*, 1870.

R. 1740 a. Two small imperfect vertebræ apparently belonging to the same individual as the mandible No. R. 1740; from Northumberland.
Presented by J. Wood Mason, Esq., 1880.

45863. Slab showing a number of dermal scutes of the ventral armour; from Newcastle. These scutes resemble the one figured by Atthey, *op. cit.* pl. viii. figs. 2, 3.
Purchased, 1874.

R. 43. Fragment of shale showing a similar scute; from Newcastle.
Presented by T. P. Barkas, Esq., 1880.

Genus **MACROMERIUM,** Fritsch[1].

Imperfectly known. Chiefly characterized by the peculiar features of the pelvis, which approximates to that of *Sphenodon*, the ilium being slender, and the ischium having a very short symphysis, and thereby differing widely from that of *Mastodonsaurus*. Teeth large, with subconical crowns, grooved inferiorly, but towards the summit smooth, flattened on the inner side, and having distinct fore-and-aft carinæ; a distinct pulp-cavity, and the dentine much infolded at the base. Cranial sculpture pitted and irregularly distributed. Vertebral centra perforated. Ventral scutes very large.

Macromerium schwarzenbergi, Fritsch[2].

Syn. *Macromerion (Labyrinthodon) schwarzenbergi*, Fritsch[3].

The type species. Of large size, the skull being probably fully as large as that of *Loxomma allmani*. Ventral scutes three times as long as wide.

Hab. Europe (Bohemia).

The originals of the following type specimens were obtained from the Rothliegendes (Lower Permian) of Bohemia, and are described and figured by Fraas in his 'Fauna der Gaskohle.' The casts were purchased in 1888.

R. 1465. Cast of a fragment of the cranial roof, embedded in matrix. Original figured, *op. cit.* pl. lxv. fig. 1; from Kounova.

[1] Fauna der Gaskohle, vol. i. pt. i. p. 29 (1879).—Amended from *Macromerion*.

[2] Sitzungsb. k. bohm. Ges. Wiss. for 1875, p. 71 (1876).—*Labyrinthodon*.

[3] *Loc. cit.*

R. 1466. Cast of a slab showing one side of part of the upper jaw with teeth. Original figured, *op. cit.* pl. lxv. figs. 6–9; from Kounova.

R. 1468. Cast of a slab exhibiting parts of the pelvis and an imperfect vertebral centrum. Original figured, *op. cit.* pl. lxix. fig. 1; from Kounova. Portions of the right ilium and ischium are seen in apposition, while the left pubis is detached.

R. 1460. Cast of the left ilium and ischium. Original figured, *op. cit.* pl. lxvi. figs. 1, 2; from Kounova.

R. 1461. Cast of a limb-bone, probably the humerus. Original figured, *op. cit.* pl. lxvi. fig. 4; from Kounova.

R. 1462. Cast of a limb-bone, probably the femur. Original figured, *op. cit.* pl. lxix. fig. 6; from Kounova.

R. 1463. Cast of an undetermined limb-bone. Original figured, *op. cit.* pl. lxviii. fig. 12; from Kounova.

Macromerium bayeri, Fritsch[1].

Cranial sculpture more strongly marked than in the type species; ilium smaller, but with a larger process than in the latter; and the scutes of the ventral buckler shorter and wider.

Hab. Europe (Bohemia).

R. 1467. Cast of a slab showing a sacral rib, ilium, and ischium. The original, which is one of the types, was obtained from the Rothliegendes (Lower Permian) of Kounova, and is figured by Fritsch in his 'Fauna der Gaskohle,' vol. ii. pl. lxiv. fig. 3. There is a discrepancy between the cast and the figure as to the position of the boundary between the ischium and the sacral rib. *Purchased*, 1888.

Macromerium (?) simplex, Fritsch[2].

A smaller form provisionally referred to this genus, mainly known by the pelvis.

Hab. Europe (Bohemia).

[1] Fauna der Gaskohle, vol. i. pt. i. p. 29 (1879).
[2] *Ibid.* vol. ii. pt. ii. p. 41 (1885).

R. 1464. Cast of an imperfect innominate. The original, which is the type, was obtained from the Rothliegendes (Lower Permian) of Kounova; and is figured by Fritsch in his 'Fauna der Gaskohle,' vol. ii. pl. lxvii. figs. 1, 2.
Purchased, 1888.

Macromerium scoticum, Lydekker[1].

Much smaller than the type species, and agreeing in the size of the teeth with *M. bicolor*, Fritsch[2], of the Bohemian Permian. The teeth appear to be very similar to those of the latter[3], showing the same large pulp-cavity, but are generally somewhat less curved and more slender. This feature is at present the chief evidence for regarding this form as specifically distinct from *M. bicolor*, although its lower geological horizon is probably alone sufficient. There is no evidence to show in what respects *M. bicolor*, which was founded upon the evidence of the teeth, differs from some of the above-mentioned small species based upon the pelvis.

Hab. Europe (Scotland).

R. 310. (*Fig.*) Slab of rock showing the outer surface of the greater portion of the dentary element of the right ramus of the mandible; from the Lower Carboniferous of Gilmerton, near Edinburgh. The type specimen; figured by the writer in the 'Quart. Journ. Geol. Soc.' vol. xlvi. pl. xii. fig. 1. One large tusk-like tooth is preserved a short distance behind the extremity of the symphysis, while the base of another remains at the extremity; traces of two larger tusks are shown at a distance of about 1½ inch behind the second tusk, posteriorly to which all the teeth are small. The crowns of the teeth accord generally with those of *M. bicolor*, figured in pl. lxx. of Fritsch's memoir, showing the same distinct carinæ, and strongly marked grooves on the smooth surface of the crown; while the broken teeth exhibit a similar large pulp-cavity. In their less marked backward curvature the teeth are more like those of the type species.
Enniskillen Collection. Purchased, 1882.

[1] Quart. Journ. Geol. Soc. vol. xlvi. p. 290 (1890).
[2] Fauna der Gaskohle, vol. ii. pt. ii. p. 41, pl. lxvii. fig. 15 (1885).
[3] See Fritsch, *op. cit.* pl. lxx. figs. 1–5.

Genus **LOXOMMA**, Huxley[1].

Syn. *Orthosaurus*, Barkas[2].
Macrosaurus, Barkas[3].
Megalocephalus, Barkas[4].

Skull (fig. 37) forming an isosceles triangle, with large postero-external expansions, small epiotic cornua, and very deep auditory slits; orbits very large, of irregular oval contour, with a projecting

Fig. 37.

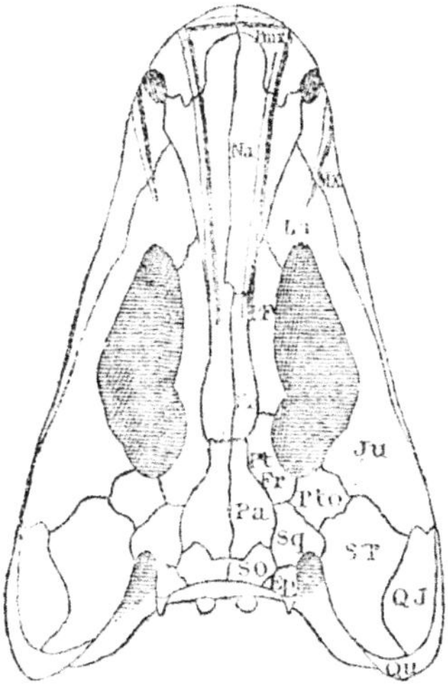

Loxomma allmani.—Frontal aspect of cranium, with the sculpture omitted; from the Carboniferous of Northumberland. About ⅛. *PF*, prefrontal. Other letters as in fig. 31. (*After Miall.*)

process from the inner and outer borders a short distance behind the middle; nares small, oval, and widely separated; a lyra situated on two preorbital ridges; frontal excluded from orbit by junction of prefrontal and postfrontal; anterior border of orbit formed by prefrontal and lachrymal. Premaxillary teeth three or four in number, and larger than those of maxilla; large palato-vomerine tusks before and behind posterior nares; mandibular teeth few, unequal, and many of them very large. Crowns of teeth

[1] Quart Journ. Geol. Soc. vol xviii. p. 293 (1862)
[2] Coal-Measure Palæontology, p. 61, pl viii. (1873).
[3] *Ibid.* p. 58, pl vii.
[4] *Ibid.* p. 69, pl. ix. fig. 189. In description incorrectly given as pl. viii.

much compressed laterally, straight, smooth, with very prominent fore-and-aft carinæ [1]; a large pulp-cavity, and the dentine much folded. Cranial sculpture pitted, and extending over all the bones. Mandibular rami slender. Vertebral column with very large inter-centra, which carry the chevrons in the caudal region.

Orthosaurus and *Megalocephalus* were founded on the evidence of imperfect crania, and *Macrosaurus* on a portion of the vertebral column; all being apparently referable to the type species. The vertebral column on which *Macrosaurus* was based is described and figured by Embleton in 'Nat. Hist. Trans. Northumb. & Durham,' vol. viii. p. 349, pl. vi., and referred to *Loxomma*. The large inter-centra are well shown, and approximate to those of the embolomerous type.

Loxomma allmani, Huxley [2].

Syn. *Orthosaurus pachycephalus*, Barkas [3].
Macrosaurus polyspondylus, Barkas [4].
Megalocephalus macromma, Barkas [5].

The type species. Skull vaulted, with a broad and somewhat spatulate muzzle; total length usually about 0,333 (13·5 inches), length in middle line 0,291 (11·5 inches); greatest width 0,227 (9 inches). Crowns of teeth not serrated.

A skull is figured by Embleton and Atthey in the 'Ann. Mag. Nat. Hist.' ser. 4, vol. xiv. pls. iv.–viii. (1874); the skulls figured by Barkas are crushed flat, and therefore appear larger. The type specimen is from the Lower Carboniferous ironstone of Gilmerton [6], near Edinburgh; an horizon probably corresponding to the upper part of the Mountain Limestone of England.

Hab. Europe (Britain).

R. 585. (*Fig.*) The cranium, wanting the greater part of the inferior moiety of the mandible; from an ironstone band in the Coal-Measures (Upper Carboniferous) of Dawley, near Coalbrookdale, Shropshire. This specimen is the only known European Labyrinthodont in which the contour of the skull is completely preserved; the restored figure

[1] In the Permian *L. bohemicum*, Fritsch (Fauna der Gaskohle, vol. ii. pt. i. p. 16, pl. lviii., 1885), the carinæ are serrated.

[2] Quart. Journ. Geol. Soc. vol. xviii. p. 293 (1862)

[3] Coal-Measure Palæontology, p. 61, pl. viii. (1873).

[4] *Ibid.* p. 58, pl. vii.

[5] *Ibid.* p. 69, pl. ix. fig. 189. In description incorrectly given as pl. viii.

[6] See Quart. Journ. Geol. Soc. vol. xix. p. 56, note.

given in the 'Rep. Brit. Assoc.' for 1873, pls. i., ii., is chiefly taken from it. *Presented by George Maw, Esq.*

R. 306. Slab of shale showing the external surface of the left (*Fig.*) dentary element of the mandible; from the Coal-Measures (Upper Carboniferous) near Hamilton, Lanarkshire. Five large tusk-like teeth are shown, one of which is figured in the accompanying woodcut. This specimen accords with the imperfect mandible figured by Barkas in his 'Coal-Measure Palæontology,' pl. ix. fig. 195, and referred to *Pteroplax*. *Egerton Collection*. *Purchased*, 1882.

R. 57. Fragment of ironstone showing the imperfect crown of one of the large mandibular tusks; from the Coal-Measures of Shelton, North Staffordshire. *Purchased*, 1881.

Fig. 38.

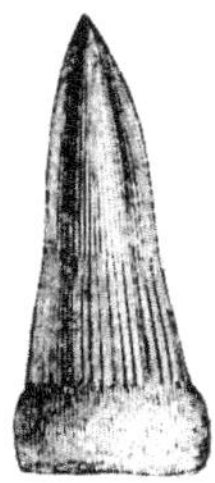

Loxomma allmani.—Outer aspect of a lower tooth; from the Coal-Measures of Lanarkshire. $\frac{1}{1}$.

R. 1738. Fragment of shale showing the lateral aspect of the crown of a tusk; from the Coal-Measures of (?) Northumberland. *Egerton Collection*. *Purchased*, 1882.

R. 1739. Fragment of shale showing one side of a somewhat smaller tusk; from the Coal-Measures of Northumberland. *Presented by T. P. Barkas, Esq.*, 1871.

41126. Fragment of shale showing a median thoracic plate (interclavicle) not improbably belonging to this form; from the Lower Carboniferous of Burdie House, near Edinburgh. Resembles the specimen figured by Huxley in the 'Quart. Journ. Geol. Soc.' vol. xviii. pl. xi. fig. 2, and provisionally referred to *Loxomma*. Nearly the whole of the unsculptured posterior spine is wanting in this specimen. If rightly referred this specimen will indicate that the thoracic plates were small in comparison with the skull. *Purchased*, 1868.

Genus **EOSAURUS**, Marsh[1].

Known only by vertebræ, which from their fully ossified centra may indicate a member of this family.

Eosaurus acadianus, Marsh[2].

The type species; of large size, the vertebral centra having a diameter of about 0,063 (2·5 inches).

Hab. North America.

36820. Casts of two vertebral centra. The originals, which are the types, were obtained from the Upper Carboniferous of South Joggins, Nova Scotia, and are described and figured by Marsh in the 'Amer. Journ.' ser. 2, vol. xxxiv. p. 1, pl. i. figs. 1, 2 (1862), as belonging to an Ichthyopterygian Reptile. *Presented by Prof. O. C. Marsh*, 1862.

Family NYRANIIDÆ.

Skull (figs. 39, 40) with the palatines situated near the middle line, internally to the vomers and pterygoids, and the palatine vacuities small and placed far back. Vertebræ (*Ichthyerpetum*) discoidal. Teeth less complex than in the *Anthracosauridæ*. A ventral armour.

The type genus was placed by Fritsch with the *Archegosauridæ*, although its resemblance to *Anthracosaurus* was pointed out; it was subsequently made the type of a family by the present writer[3], and placed next the *Archegosauridæ*.

Genus **NYRANIA**, Fritsch[4].

The type genus. Skull (figs. 38, 39) short and wide, with the epiotic cornua directed outwardly, the orbits situated in the hinder half, with a wide interorbital bar; position of nares unknown; palate covered with minute denticules; a tusk-like tooth in each vomer. Known only by the skull.

Nyrania trachystoma, Fritsch[5].

The type and only described species. Length of skull about 0,125 (5 inches).

Hab. Europe (Bohemia).

1 Amer. Journ. ser. 2, vol. xxxiv. p. 1 (1862). 2 *Loc. cit.*

3 In Nicholson and Lydekker's 'Manual of Palæontology,' 3rd ed. vol. ii. p. 1032 (1889).

4 Fauna der Gaskohle, vol. ii. pt. ii. p. 33 (1885). 5 *Loc. cit.*

R. 1451. Electrotype of a slab of shale showing the frontal aspect of the imperfect cranium, wanting the greater part of the external sculptured layer. The original, which is one of the types, was obtained from the Gaskohle (Lowest Permian) of Nyřan (Pilsen); and is figured by Fritsch in his 'Fauna der Gaskohle,' vol. ii. pl. lxii. fig. 1.
Purchased, 1888.

Fig. 39.

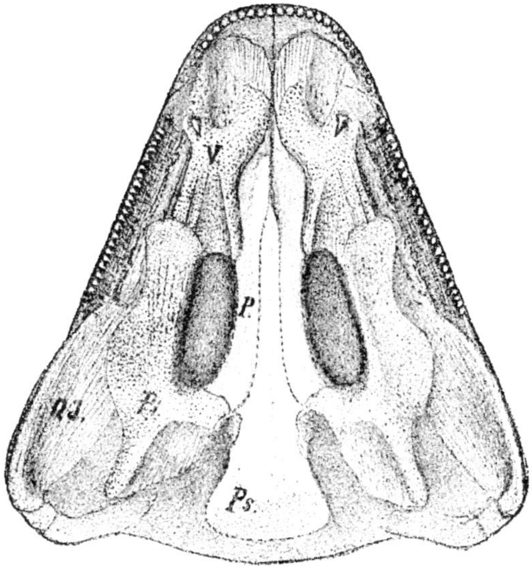

Nyrania trachystoma.—Palatal aspect of the cranium; from the Gaskohle (Lowest Permian) of Nyřan, Bohemia. $\frac{1}{2}$. *V*, vomer; *P*, palatine; *Ps*, parasphenoid; *Pt*, pterygoid; *QJ*, quadratojugal. (*After Fritsch.*)

R. 1452. Electrotype showing the left half of the oral aspect of the palate. The original is one of the types, and was obtained from Nyřan; it is figured by Fritsch, *op. cit.* pl. lxiii. fig. 2. The denticules on the pterygoids are very distinct; the rough surface of these bones resembling that of *Anthracosaurus* as figured by Atthey in the 'Ann. Mag. Nat. Hist.' ser. 4, vol. xviii. pl. ix. *Purchased*, 1888.

R. 1453. Electrotype exhibiting the bones of the palate. Original from Nyřan; figured by Fritsch, *op. cit.* pl. lxii. fig. 5.
Purchased, 1888.

R. 1454. Electrotype showing one of the bones of the palate (apparently a vomer) exhibiting denticules and two tusks. Original from Nyřan; figured by Fritsch, *op. cit.* pl. lxii. figs. 3, 4. *Purchased*, 1888.

Fig. 40.

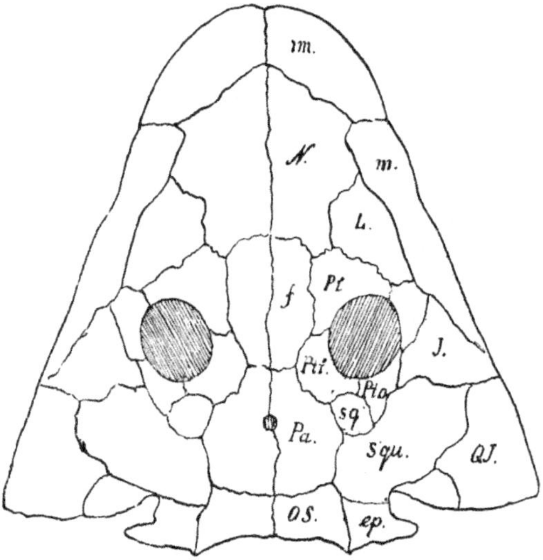

Nyrania trachystoma.—Frontal aspect of the cranium; from the Gaskohle of Bohemia. $\frac{1}{2}$. *im*, premaxilla; *m*, maxilla; *N*, nasal; *L*, lachrymal; *f*, frontal; *Pf*, prefrontal; *Ptf*, postfrontal; *Pto*, postorbital; *Pa*, parietal; *OS*, supraoccipital; *sq*, squamosal; *Squ*, supratemporal; *ep*, epiotic; *QJ*, quadratojugal; *J*, jugal. (*After Fritsch.*)

Genus **ICHTHYERPETUM,** Huxley[1].

Syn. *Erpetocephalus*, Huxley[2].

Apparently very closely allied to *Nyrania*, but the epiotic cornua less produced laterally, and the orbits relatively larger. Palate unknown.

Whether these slight differences are sufficient to justify the generic distinctness of *Nyrania* may be doubtful. The present genus was founded upon the evidence of the vertebral column, while *Eupetocephalus* was based upon that of the skull, but it was suggested in the original description that the two might prove to be identical.

[1] Trans. Roy. Irish Academy, vol. xxiv. p. 367 (1867).—Amended from *Ichthyerpeton.*

[2] *Ibid.* p. 368.

Ichthyerpetum bradleyæ, Huxley[1].

Syn. *Erpetocephalus rugosus*, Huxley[2].

The type and only described species. Length of skull about 0,080 (3·15 inches).

Hab. Europe (Ireland).

41851 g. Slab of shale showing the frontal aspect of the cranium and the lateral aspects of the mandibular rami; from the Coal-Measures (Upper Carboniferous) of Jarrow Colliery, Kilkenny. The contour of the cranium is fairly well preserved, and agrees with that of the imperfect specimen figured by Huxley in the 'Trans. Roy. Irish Academy,' vol. xxiv. pl. xxxiii. fig. 2, as *Erpetocephalus*, showing the same slight outward inclination of the extremities of the epiotic cornua, and the concave posterior border of the supraoccipitals. Compared with *Nyrania* (fig. 40) the resemblance is so close as to leave no reasonable doubt of the close alliance of the two forms. *Purchased*, 1870.

41851 h. Slab showing the imperfect skull and a portion of the vertebral column; from Jarrow. The frontal aspect of the posterior portion of the cranium is preserved, on the right side of which is seen the hinder part of a mandibular ramus. The supraoccipital and epiotic region corresponds precisely with that of the preceding specimen, and the position of the orbits is also similar. The vertebral column agrees with the following examples. *Purchased*, 1870.

41851 i. Slab of shale showing traces of the precaudal portion of the skeleton; from Jarrow. The posterior half of the cranium is very imperfectly preserved, with the loss of nearly the whole of the sculpture. The left lateral thoracic plate and the left pelvic limb are indistinctly indicated; but the vertebral column is badly preserved. *Purchased*, 1870.

41851 j. Fragment of shale showing a portion of the vertebral column, bent upon itself; from Jarrow. The vertebræ agree with those of the type specimen figured by Huxley

[1] Trans. Roy Irish Academy, vol. xxiv p. 367 (1867). [2] *Ibid.* p. 368.

in the 'Trans. Roy. Irish Academy,' vol. xxiv. pl. xxviii. fig. 1, and do not appear to be of an embolomerous type. *Purchased*, 1870.

41851 k. Fragment of shale containing traces of the anterior part of a skeleton apparently referable to this form; from Jarrow. *Purchased*, 1870.

Family DENDRERPETIDÆ[1].

Imperfectly known. Skull large and tending more or less to a parabolic form; the parasphenoid with a short stem, and rough and expanded body. Teeth smooth at the summit, but deeply grooved at the base, with simple and irregular plications of the dentine. According to Dawson the vertebral centra are discoidal.

Genus **DENDRERPETUM,** Owen[2].

The type genus. Skull of an ovoid or subparabolic contour, with the orbits placed in the middle of the length or posteriorly, the nares very minute, the nasals large and expanded anteriorly, and the muzzle rounded.

Dendrerpetum pyriticum, Fritsch[3].

Length of skull about 0,065 in middle line, its length exceeding the width; orbits in middle of length.

Hab. Europe (Bohemia).

R. 1448. Electrotype from a fragment of shale, showing the ventral aspect of the cranium. The original, which is the type, was obtained from the Gaskohle (Lowest Permian) of Nyřan (Pilsen); and is figured by Fritsch in his 'Fauna der Gaskohle,' vol. ii. pl. xlix. fig. 1. *Purchased*, 1888.

Dendrerpetum (?) deprivatum, Fritsch[4].

Length of skull about 0,045 in middle line; skull longer and

[1] =*Dendrerpetontidæ*, Fritsch.

[2] Quart. Journ. Geol. Soc. vol. ix. p. 64 (1853).—Amended from *Dendrerpeton*.

[3] Fauna der Gaskohle, vol. i. pt. i. p. 28 (1879).

[4] *Ibid.* vol. ii. pt. i. p. 9 (1885).

narrower than in the preceding species, with nearly straight lateral margins. The generic reference is provisional.

Hab. Europe (Bohemia).

R. 1449. Electrotype from a fragment of shale, showing the frontal aspect of the cranium. The original is the type, and was obtained from the Gaskohle (Lowest Permian) of Nyřan (Pilsen); it is figured by Fritsch in his 'Fauna der Gaskohle,' vol. ii. pl. li. fig. 1. *Purchased*, 1888.

The three following genera are very imperfectly known, but approximate in the more or less parabolic figure of the skull to Dendrerpetum [1]. *The imperfect young skeleton noticed below* (p. 172), *under the name of* Platyceps, *appears to have rhachitomous vertebræ, but this may be merely a feature of immaturity. If the vertebræ of the adult were fully ossified, it is not improbable that these genera should be included in the* Dendrerpetidæ.

Genus **BRACHYOPS**, Owen [2].

Skull very broad; orbits situated far forwards, large, oval, converging anteriorly, and separated by a very wide interorbital bar; nares apparently small. Cranial sculpture faintly radiate; an indistinct lyra present.

Brachyops laticeps, Owen [3].

The type and only known species. Of considerable size; skull rather broader than long, with a rounded muzzle, its total length being 0,107 (4·25 inches), and its greatest width 0,121 (4·75 inches.

Hab. India (Central Provinces).

R. 1737. Cast of the imperfect cranium. The original, which is preserved in the Museum of the Geological Society, was obtained from the Mangli stage of the Lower Gondwanas near Mangli, a small deserted village lying at the northern extremity of the Wardha Gondwana basin, about 50 miles south of Nagpur, and 35 north-west of Chanda. It is figured by Owen in the 'Quart. Journ. Geol. Soc.' vol. xi.

[1] Compare Owen's figure of the skull of *Brachyops* with that of *Dendrerpetum foveolatum*, given by Fritsch in his 'Fauna der Gaskohle,' vol. ii. p. 9, fig. 126.

[2] Quart. Journ. Geol. Soc. vol x. p. 473 (1854).

[3] *Loc. cit.*

pl. ii., and also by Medlicott and Blanford in their 'Manual of the Geology of India,' pt. i. p. 131.

Made in the Museum.

Genus **BOTHRICEPS,** Huxley [1].

Syn. (?) *Platyceps,* Stevens [2].

Skull more elongated than in the preceding genus; orbits placed near the middle of the skull, and the width of the interorbital bar only slightly exceeding that of the orbit; nares large. Epiotic cornua long and narrow. Cranial sculpture closely and irregularly pitted.

It is highly probable that the flattened skeleton of a small Labyrinthodont described and figured as *Platyceps wilkinsoni,* Stephens [3], belongs to a very young individual of a species of this genus, in which the skull has its width somewhat increased by pressure, although (as in *Archegosaurus*) it was relatively wider than in the adult.

Bothriceps australis, Huxley [4].

The type species. Rather smaller than *Brachyops laticeps,* the extreme length of the skull being 0,095 (3·7 inches), and its greatest width about 0,095 (3·7 inches). Skull relatively wide, with a blunted muzzle.

Hab. Australia.

23110. (*Fig.*) The skull, wanting nearly all the cranial bones; from Australia. The type specimen; described and figured by Huxley in the 'Quart. Journ. Geol. Soc.' vol. xv. p. 647, pl. xxii. fig. 1. This specimen was doubtless obtained from the Hawkesbury beds of New South Wales. The contour of the epiotic cornua is lost. *Purchased,* 1848.

Bothriceps huxleyi, Lydekker [5].

Smaller than the preceding species, with a narrower and more pointed skull (fig. 41). In typical specimens the extreme length of the skull is about 0,065 (2·6 inches), and its greatest width 0,062 (2·45 inches).

Hab. South Africa (Orange Free State).

[1] Quart. Journ. Geol. Soc. vol. xv. p. 649 (1859).
[2] Proc. Linn. Soc. N. S. Wales, ser. 2, vol. i. p. 1175 (1887).—Preoccupied.
[3] *Loc. cit.* [4] *Loc. cit.*
[5] Ann. Mag. Nat. Hist. ser. 6, vol. iv. p. 476 (1889).

R. 506. The skull, with a mass of matrix attached which contains portions of the thoracic buckler and vertebral column; from the Karoo system of the Orange Free State. In those portions of the cranium which are free from matrix the characteristic pitted sculpture is well shown.

By exchange with the Blomfontein Museum, 1884.

R. 507. (*Fig.*) The skull; from the Orange Free State. The type; figured in woodcut fig. 41. The palate is concealed by matrix. A portion of the thoracic buckler is shown.

By exchange with the Blomfontein Museum, 1884.

Fig. 41.

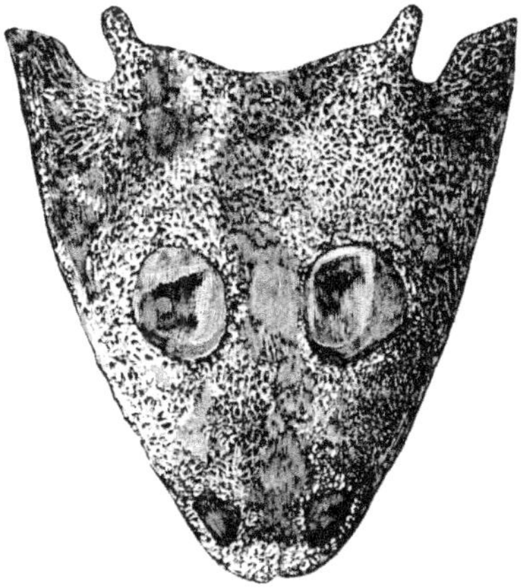

Bothriceps huxleyi.—Frontal aspect of the skull; from the Karoo system of the Orange Free State. $\frac{4}{5}$.

R. 508. The skull: from the Orange Free State. This specimen is much obscured by matrix, but the sculpture is very clearly displayed near the margins of the upper jaw.

By exchange with the Blomfontein Museum, 1884.

R. 504. Mass of matrix containing two entire skulls and the anterior portion of a third one, together with imperfect limb-bones and vertebræ; from the Orange Free State. The skulls are rather larger than the preceding ones; in the best preserved example the orbits appear to have been enlarged by fracture. The vertebræ are too imperfectly preserved to admit of the determination of their characters.

By exchange with the Blomfontein Museum, 1884.

R. 505. A larger skull belonging either to the present or a closely allied species; from the Orange Free State. This specimen is much obscured by matrix, but the sculpture is shown on the mandibular rami.

By exchange with the Blomfontein Museum, 1884.

Bothriceps (?) major (Owen[1]).

Syn. *Petrophryne* (?) *major*, Owen[2].

Very imperfectly known. Of the approximate size of *B. australis*, but with the skull apparently resembling that of *B. huxleyi* in contour.

Hab. South Africa.

47340. Cast of a fragment of sandstone, containing impressions of the dental series, with small portions of the contiguous alveolar plate of the upper jaw. The original, which is the type, was obtained from (? the Stormberg beds of) the Karoo system of the Stormberg range, and is preserved in the Museum of the Geological Society. It is described by Owen in his 'Catalogue of the Fossil Reptilia of South Africa,' pp. 68–69. *Made in the Museum.*

Genus MICROPHOLIS, Huxley[3].

Syn. *Petrophryne*, Owen[4].

Skull approximating in contour to that of *Bothriceps*, but with a pustular sculpture, and the interorbital bar frequently or always narrower than the transverse diameter of the orbit, and the epiotic cornua less produced.

Although the sculpture of the upper cranial bones is not shown in the type of *M. stowei*, Huxley[5], there can be no doubt of the generic identity of *Petrophryne* with *Micropholis*, as was first pointed out by Zittel.

Micropholis granulata (Owen[6]).

Syn. *Petrophryne granulata*, Owen[7].
Micropholis granulata, Zittel[8].

The type of *Petrophryne*. Distinguished from *M. stowei* by its

[1] Cat. Foss. Rept. S. Africa, p. 68 (1876).—*Petrophryne*. [2] *Loc. cit.*

[3] Quart. Journ. Geol. Soc. vol. xv. p. 649 (1859).

[4] Cat. Foss. Rept. S. Africa, p. 67 (1876). [5] *Loc. cit.*

[6] *Loc. cit.*—*Petrophryne.* [7] *Loc. cit.*

[8] Handbuch der Palæontologie, 1 Abth. iii. Band, 2 Lief. p. 397 (1888).

smaller size; the length of the median line of the skull being 0,040. The skull does not narrow very markedly towards the muzzle, which is blunt. Other examples are required to show whether the inferior size of the undermentioned specimens as compared with the type of *M. stowei* is a good specific character.

Hab. South Africa.

R. 510. (*Fig.*) The skull; from the Karoo system of Tafelberg, Queenstown district. The type; figured by Owen in his 'Cat. Foss. Rept. S. Africa,' pl. xx. figs. 13, 14, and 17–20; and also by Zittel in his 'Handbuch der Palæontologie,' 1 Abth. iii. Band, 2 Lief. p. 397, fig. 387. Compared with the figure of the type skull of *M. stowei* in the 'Quart. Journ. Geol. Soc.' vol. xv. pl. xxi., and allowing for the imperfect parietal region of the latter, the two agree exactly in contour.

Presented by W. G. Atherstone, Esq., M.D.

R. 510 a. An imperfect skull, attached to a mass of matrix; from Tafelberg. The width of the interorbital bar would appear to be owing to the effects of crushing.

Presented by W. G. Atherstone, Esq., M.D.

Family DIPLOSPONDYLIDÆ [1].

Vertebræ (at least in the caudal region) of the embolomerous type [2]. A ventral armour.

This family is a provisional one, since it has been suggested that the embolomerous and rhachitomous types of vertebral structure may occur in different parts of the column of the same form [3]. The type genus *Diplospondylus*, Lydekker [4], of the Permian of Bohemia, is very imperfectly known, but the cranial bones are not pitted, and the limb-bones are penetrated by a number of nutrient foramina.

Genus **CRICOTUS**, Cope [5].

Skull (fig. 42) long and triangular, with a narrow muzzle, and the ovoid orbits situated in the hinder half; cranial bones sculptured, with a lyra. No postarticular process to the mandible. Vertebral

[1] = *Diplovertebridæ*, Fritsch. [2] See p. 140.

[3] See Fritsch, Fauna der Gaskohle, vol. ii p. 4.

[4] Geol Mag. dec. iii vol. vi. p. 325 (1889).—To replace the hybrid *Diplovertebron*, Fritsch.

[5] Proc Ac. Nat. Sci Philad. 1875, p. 405.

bodies perforated; first vertebra articulating with skull by a simple undifferentiated surface. Tail apparently long.

Hab. North America (Texas).

Cricotus heteroclitus, Cope [1].

Syn. *Cricotus discophorus,* Cope [2].

The type species. The vertebral centrum on which *C. discophorus*

Fig. 42.

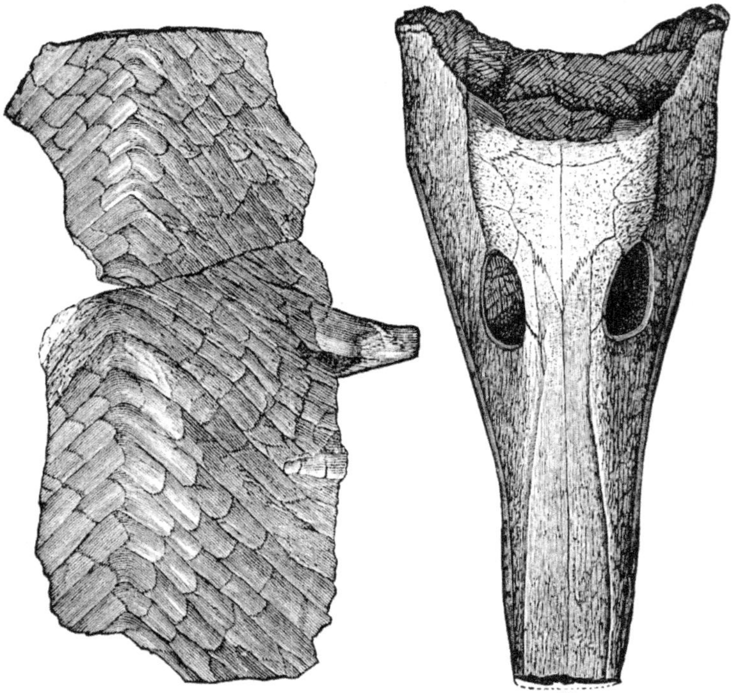

A B

Cricotus heteroclitus.—Ventral scutes (A) and frontal aspect of the cranium (B); from the reputed Permian of Texas. ½. (*After Cope.*)

was described has a length of 0,009, and a transverse diameter of 0,025. Length of skull about 0,200.

Hab. North America (Texas).

R. 577. Four vertebral centra, some or all of which are referable to

[1] Proc. Ac. Nat. Sci. Philad. 1875, p. 405.
[2] Proc. Amer. Phil. Soc. vol. xvii. p. 186 (1877).

the present species; from the reputed Permian of Texas. The larger specimen has the same dimensions as the vertebra above mentioned; and all the specimens show the characters of those figured by Cope in the 'Trans. Amer. Phil. Soc.' 1886, pl. i. (see also Zittel 'Handbuch der Palæontologie,' i. Abth. iii. Band, 2 Lief. p. 395, fig. 384).

Purchased, 1885.

Family ARCHEGOSAURIDÆ.

Vertebræ (at least in the greater part of the column) of the rhachitomous type. Skull with the occipital condyles either cartilaginous or ossified; the palatines running parallel to the maxillæ; large palatine vacuities; and an ossified sclerotic ring. Dentine of teeth generally with simple radiate foldings. A ventral armour of scutes.

It has been proposed to arrange the genera here included in this family in several families; but for the present, at least, such divisions cannot be maintained.

Genus **ARCHEGOSAURUS**, Goldfuss[1].

Skull (fig. 43) with the occipital condyles unossified, and no

Fig. 43.

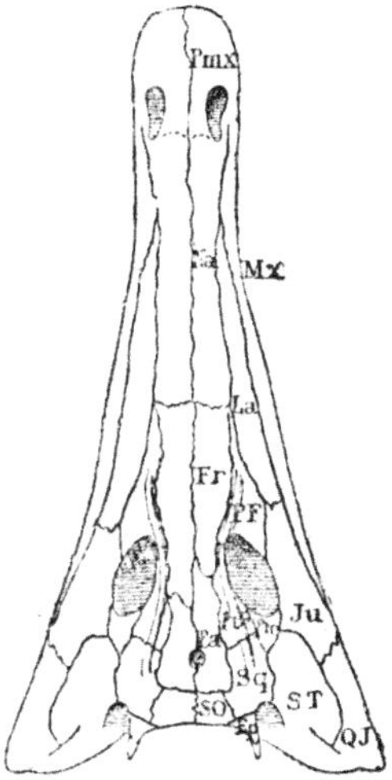

Archegosaurus decheni.—Frontal aspect of the cranium, with the sculpture omitted; from the Rothliegendes (Lower Permian) of Saarbrück. About ¼. Letters as in fig. 36. (*After Miall.*)

[1] Beitr. vorwelt. Fauna d. Steinkohlengebirges, p. 3 (1847).

teeth on the vomers or middle region of the palate; in the adult the skull long and triangular, with the orbits in the hinder third, and inclining to one another anteriorly; and the nares large, ovoid, and placed some distance behind the muzzle. Median thoracic plate long, diamond-shaped, and flattened. Neural spines of vertebræ well developed, without a horizontal expansion at the summits. Four digits in the manus.

Archegosaurus decheni, Goldfuss [1].

Syn. *Archegosaurus medius*, Goldfuss [2].
Archegosaurus minor, Goldfuss [3].

The type species. Skull of adult (fig. 43) nearly twice as long as broad, with the orbits of an elongate-oval contour and situated very far back; in large examples with a length of 0,300 (11·8 inches), but more usually about 0,200 (7·9 inches).

In the young the skull is considerably wider in proportion to its length, which does not exceed $1\frac{1}{2}$ times the width; the apparent width is frequently much increased by flattening in young specimens.

Hab. Europe (Germany).

The following specimens, or their originals, were obtained from the Coal-bearing Shales of the Lower Permian (Rothliegendes) of Lebach and Saarbrück, Rhenish Prussia; those deposits having for a long time been classed with the Carboniferous.

R. 1741. Cast of the imperfect cranium. The original is the type, and is figured by Goldfuss in his 'Beitr. vorwelt. Fauna d. Steinkohlengebirges,' pl. i. figs. 1–3, and also by Meyer in the 'Palæontographica,' vol. vi. pl. xi. fig. 5. *Purchased.*

R. 1742. Cast of a slab showing the anterior portion of the skeleton of a young individual. Original figured by Goldfuss, *op. cit.* pl. iii. fig. 1, as *A. medius* (of which it is the type), by Meyer, *op. cit.* pl. xiv. fig. 3, and in Zittel's 'Handbuch der Palæontologie,' 1 Abth. iii. Band, 2 Lief. p. 384, fig. 373. The gills remain, and the thoracic buckler is well shown. *Purchased.*

R. 1743. Cast of a slab showing the skull, thoracic buckler, and anterior vertebræ of a still smaller individual. Original

[1] Beitr. vorwelt. Fauna d. Steinkohlengebirges, p. 3 (1847).
[2] *Ibid* p. 6. [3] *Ibid.* p. 7.

figured by Goldfuss, *op. cit.* pl. iii. fig. 2, as *A. minor* (of which it is the type), and also by Meyer, *op. cit.* pl. xiv. fig. 13. *Purchased.*

R. 1744. Cast of a split nodule, showing the anterior part of the skeleton of an individual of the same size as the preceding. Original figured by Meyer, *op cit.* pl. xiv. fig. 14. *Purchased.*

R. 1745. Cast of one side of a split nodule showing the anterior portion of a skeleton agreeing in size with the type of *A. medius.* *Purchased.*

R. 1746. Cast of a slab showing the ventral armour and numerous limb-bones and ribs. Original figured by Goldfuss, *op. cit.* pl. ii. fig. 3. *Purchased.*

R. 1747. Two casts of a split nodule showing portions of the vertebral column and ventral armour. Original figured by Goldfuss, *op. cit.* pl. ii. figs. 1, 2, and also by Meyer, *op. cit.* pl. xxi. fig. 1. *Purchased.*

40162. A split slab showing the nearly entire skeleton, in an imperfect condition. The skull is badly preserved; in the middle dorsal region the intercentra are distinct; the whole of the tail remains, although its structure is obscure; the pectoral limbs are wanting. *Purchased*, 1866.

35802–3. A split slab with the imperfect skull and part of ventral armour and some of the ribs. *Presented by Sir R. Owen, K.C.B.*, 1861.

35805. A split slab with the imperfect skull and anterior portion of the vertebral column of a smaller individual. The skull lacks the muzzle and is badly preserved; the intercentra are distinct. *Presented by Sir R. Owen, K.C.B.*, 1861.

35806-7. A split slab showing an imperfect skull agreeing in size with the preceding specimen. The frontal aspect of the skull is seen. *Presented by Sir R. Owen, K.C.B.*, 1861.

35797. A split nodule showing the frontal aspect of a nearly entire skull. The skull has a length in the median line of about 0,200. The outer lamina of bone is adherent to the counterpart, so that the sculpture is not shown. *Presented by Sir R. Owen, K.C.B.*, 1861.

35800–01. A split nodule showing the frontal aspect of the hinder portion of the skull. The sculptured lamina remains in position between the orbits.
Presented by Sir R. Owen, K.C.B., 1861.

42806. A split nodule exhibiting a smaller skull from the frontal aspect. The two mandibular rami are displaced and shown on either side of the cranium. This well-preserved specimen is rather smaller than the type skull.
Van Breda Collection. Purchased, 1871.

R. 1298. A split nodule showing the skull and thoracic region of an immature individual. One side shows the under surface of the cranial bones and the external surface of the thoracic buckler and ventral scutes; while the other exhibits the frontal aspect of the skull, without the external lamina, and the impression of the ventral armour. The latter slab also exhibits the imperfect femur and tibia of either side. The length of the cranium in the middle line is 0,125, and its greatest width 0,080; the extreme length being 0,160. *By exchange*, 1888.

40043. A split nodule showing the skull and part of the ventral armour of a rather smaller individual. *Purchased*, 1866.

42754. A split nodule showing the skull, ventral armour, and the impressions of the ribs and limb-bones of an individual agreeing approximately in size with the preceding specimen. The skull appears to be relatively wider than in the latter, but this is due to the mandibular rami being flattened out alongside of the maxillæ.
Van Breda Collection. Purchased, 1871.

40004. A split nodule exhibiting the imperfect anterior portion of the skeleton of an immature individual. The sclerotic plates are well preserved, and faint traces of the vertebral column remain. *Purchased*, 1866.

35798. A split nodule with the skull and part of the ventral armour of a smaller specimen.
Presented by Sir R. Owen, K.C.B., 1861.

33069. A nodule showing the skull of a young individual. Portions of the sculptured lamina are wanting, but the specimen is otherwise well preserved; its length in the middle line is 0,080. *Purchased*, 1848.

29003. A split nodule with traces of a smaller skull. *Purchased*, 1859.

40045. A split nodule showing the imperfectly preserved skull and part of the vertebral column and ventral armour of a young individual. *Purchased*, 1866.

R. 1748. A split nodule exhibiting the skull and thoracic buckler of a young individual. Closely resembles the specimen figured by Meyer in the 'Palæontographica,' vol. vi. pl. xiv. fig. 13. *Purchased.*

42755. A split nodule with the skull and median thoracic plate of a still younger individual. Resembles the skull figured by Meyer, *op. cit.* pl. xiv. fig. 9. *Van Breda Collection. Purchased*, 1871.

R. 1749. A split nodule showing the skull and thoracic buckler of a very young individual. In this specimen the skull retains its natural contour, and is much narrower than in the preceding example. *Purchased.*

29001. A split nodule with the imperfectly preserved anterior portion of the skeleton of a rather older individual. *Purchased*, 1859.

29004. A split nodule with the imperfectly preserved skull and ventral armour of a younger specimen. The skull is much widened by flattening. *Purchased*, 1859.

42808. A split nodule showing the impression of part of the vertebral column, pelvic girdle, and dermal armour; together with a plaster cast in relief. The ilia and ischia are well shown, and accord with the larger specimens figured by Meyer in the 'Palæontographica,' vol. vi. pl. xix. *Van Breda Collection. Purchased*, 1871.

42807. A split nodule with imperfect traces of part of the vertebral column, ribs, and ventral armour. The postzygapophyses of several of the vertebræ are distinctly seen. *Van Breda Collection.*

35655. A split nodule showing portions of the vertebral column, impressions of the limb-bones, and part of the ventral armour of an immature individual. The dermal scutes are well shown. *Purchased*, 1859.

29005. Part of a split nodule exhibiting that portion of the ventral armour of an immature individual where the reversing of the direction of the lines of the scutes takes place.
Purchased, 1859.

R. 1750. A split nodule showing the imperfect median thoracic plate. *Purchased.*

R. 534. A plaster model in relief taken from a nodule showing the impression of part of the vertebral column and ribs. The portion showing the pleurocentra, intercentra (hypocentra), and ribs is figured by Fritsch in his 'Fauna der Gaskohle,' vol. ii. p. 14, fig. 123.
Presented by Dr. Anton Fritsch, 1885.

R. 1471. A plaster model in relief taken from a nodule showing the impression of the lateral aspect of three vertebræ. Original figured by Fritsch, *op. cit.* pl. lviii. fig. 13.
Purchased.

R. 1455. A plaster model in relief taken from a nodule showing the impression of the ventral aspect of four caudal vertebræ. Fritsch regards this specimen as being of an embolomerous type, the ribs being attached to the intercentra. The original is figured in pl. lviii. fig. 14 of the work cited. *Purchased.*

Genus **PLATYOPOSAURUS,** Lydekker [1].

Syn. *Platyops,* Twelvetrees [2].

Apparently closely allied in cranial characters to *Archegosaurus*, but the infoldings of the teeth sinuous instead of straight, and the orbits more rounded.

The relatively greater length of the facial portion of the skull as compared with *Archegosaurus* would not, judging from the analogy of the Crocodilia, of itself be a generic distinction.

This genus has been placed by Zittel [3] with the *Anthracosauridæ*, but Trautschold [4] figures a horseshoe-shaped intercentrum like that of *Archegosaurus*.

[1] In Nicholson and Lydekker's 'Manual of Palæontology,' 3rd ed. vol. ii. p. xi (1889).

[2] Bull. Soc. Moscou, vol. lv. pt. i. p. 117 (1880).—Preoccupied by *Platops*, Newp. 1844.

[3] Handbuch der Palæontologie, 1 Abth. iii. Band, 2 Lief. p. 399.

[4] Mém. Soc. Moscou, vol. xv. pt. i. pl. iii.

Platyoposaurus stuckenbergi (Trautschold[1]).

Syn. *Platyops stuckenbergi*, Trautschold[2].

Skull relatively longer than in *Archegosaurus decheni*, its length exceeding three times the width; total length of skull about 0,456 (18 inches). Said to be distinguished from the typical *Platyoposaurus* (*Platyops*) *rickardi* (Twelvetrees[3]) by the smaller number and elliptical section of the teeth.

Hab. Europe (Russia).

R. 694. Three fragments of sculptured bones; from the Upper Permian (Zechstein) of the Government of Perm. The sculpture resembles that of the type skull figured by Trautschold in the 'Mém. Soc. Moscou,' vol. xv. pt. i. pl. i. *Purchased*, 1886.

Genus ACTINODON, Gaudry[4].

Skull (fig. 45) much shorter and wider than in *Archegosaurus*, with the orbits placed in the middle of the length, the nares large and widely separated, the muzzle broad and rounded, the occipital condyles ossified, and a number of denticules on the vomer and

Fig. 44.

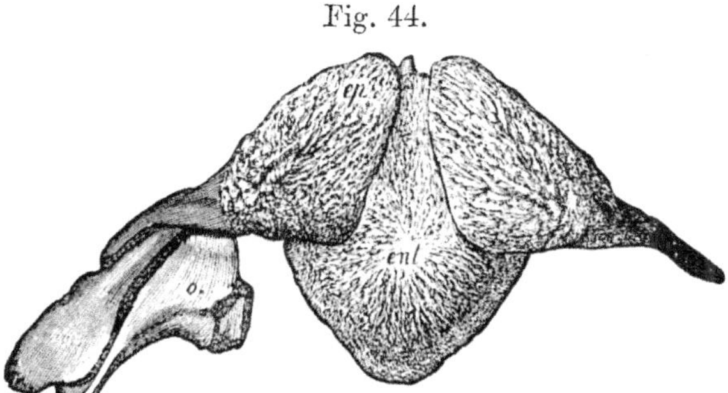

Actinodon latirostris. — Thoracic buckler; from the Rothliegendes (Lower Permian) of Autun. Reduced. *ent*, medial plate; *ep*, lateral plate; *o*, pectoral girdle. (*After Gaudry.*)

large tusks in the palatines. Middle and lateral plates of thoracic buckler (fig. 44) shorter, wider, and more curved than in *Archegosaurus*. Tarsus fully ossified. The epiotic cornua are well developed,

[1] Mém. Soc. Moscou, vol. xv. pt. i. p. 10 (1884).—*Platyops*.
[2] *Loc. cit.* [3] *Loc cit.*
[4] Nouv. Archiv. d. Muséum, vol. iii. pp. 22. 23 1867).

and separated by a deep notch from the posterior expansion of the supratemporal.

It has been suggested by Gaudry[1] that this genus is identical with *Sclerocephalus*, Goldfuss[2]; but if Zittel[3] is correct in identifying *Weissia*, Branco[4], with the latter, that genus will be distinguished by the slight development of the epiotic cornua, and the absence of a notch between the epiotic and the supratemporal. In the so-called *Weissia* the lateral borders of the cranium are convex instead of concave as in *Actinodon*.

Actinodon latirostris (Jourdan[5]).

Syn. *Archegosaurus latirostris*, Jourdan[6].
Actinodon latirostris, Gaudry[7].
Actinodon frossardi, Gaudry[8].

The type species. Skull (fig. 45) about one third longer than

Fig. 45.

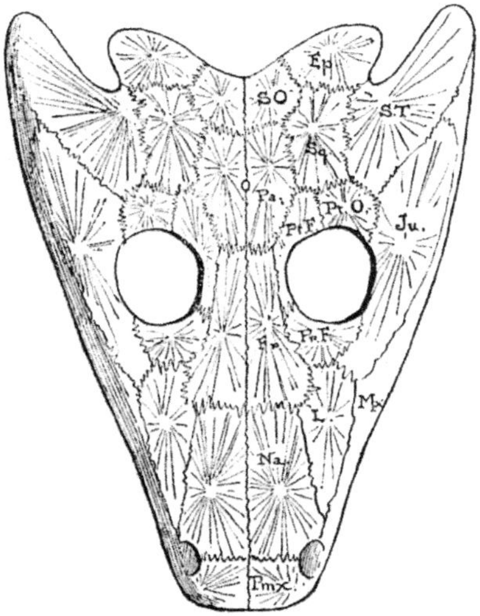

Actinodon latirostris. — Frontal aspect of the cranium, with the sculpture omitted; from the Lower Permian of Saarbrück. ⅜. *Pt.F*, postfrontal; *Pmx*, premaxilla; other letters as in fig. 31 (p. 143).

---

[1] Mém. Soc. Hist. Nat. d'Autun, vol. i. p. 29 (1888).
[2] Beitr. vorwelt. Fauna d. Steinkohlengebirges, p. 13 (1847).
[3] Handbuch der Palæontologie, 1 Abth. iii. Band, 2 Lief. pp. 391, 392.
[4] Jahrb. preuss. geol. Landesanstalt for 1886, p. 22 (1887).
[5] Verh. nat. Ver. preuss. Rheinl. vol. vi. p. 78 (1849).—*Archegosaurus*.
[6] *Loc. cit.* [7] Nouv. Archiv. d. Muséum, vol. iii. p. 31 (1867).
[8] Bull. Soc. Géol. France, sér. 2, vol. xxv. p. 577 (1868).

broad, with an extreme length of about 0,150 (6·2 inches) in adult specimens. Teeth with a small basal swelling, and foldings in the osteodentine. A ridge running from each orbit to the nares forming the boundaries of a median depression.

So far as the writer can see there appears to be no reason for departing from Gaudry's original identification of the French form with *Archegosaurus latirostris*. Fritsch[1] has, indeed, regarded the latter as inseparable from *Sclerocephalus hæuseri*, Goldfuss, but that identification falls to the ground if *Weissia* be generically identical with the latter. The elongate form of the nares in the type specimen is probably due to imperfection.

Hab. Europe (Germany and France).

R. 1299. (*Fig.*) A split nodule showing the skull of an immature individual; from the Rothliegendes (Lower Permian) of Lebach, near Saarbrück, Rhenish Prussia. In this specimen (fig. 45) the sculptured lamina of bone is adherent to the counterpart. It cannot be specifically distinguished from the larger imperfect type skull figured by Meyer in the 'Palæontographica,' vol. vi. pl. x. figs. 2, 3.
By exchange, 1888.

R. 1605. Slab of shale showing the skull and thoracic buckler, in an imperfectly preserved condition; from the Rothliegendes (Lower Permian) of Muse, near Autun (Saône-et-Loire), France. Of the skull the frontal aspect is shown, with the loss of most of the superficial layer of bone. This skull—which agrees in all respects with the one figured by Gaudry in his 'Enchaînements &c., Fossiles Primaires,' p. 265, fig. 261, as *Actinodon frossardi*,—is specifically indistinguishable from the preceding, although of larger size. In both, the orbits and nares have the same relative size and dimensions; and the preorbital ridges and median dispersion are equally clear in both. In the present specimen the boundaries of the individual bones are visible, and these agree precisely with the figure of the type of *Archegosaurus latirostris*. *Purchased*, 1889.

R. 1085. Cast of a slab of shale exhibiting the dorsal aspect of an entire skeleton, in an imperfectly preserved condition, referred by Gaudry to *Actinodon frossardi*. The original, which is preserved in the Museum at Paris, was obtained

[1] Fauna der Gaskohle, vol. i. p. 65.

from the Lower Permian of Télots, near Autun. It is figured of the natural size by Gaudry in the 'Nouv. Archiv. d. Muséum,' vol. x. pl. i., and on a reduced scale in the 'Mém. Soc. Hist. Nat. d'Autun,' vol. i. pl. ii. (1888), as *Actinodon frossardi*. The extremity of the muzzle is abnormally expanded by pressure. The orbits appear to be smaller than in the preceding specimen, but the pre-orbital ridges and median depression are very distinct. The contour of the skull seems to be more like that of the type of *A. brevis*, Gaudry.

Presented by Prof. Albert Gaudry, 1888.

Genus **COCHLEOSAURUS**, Fritsch[1].

Skull (imperfectly known) of the same general contour as in the preceding genus, but with a posterior projection from each supra-occipital, the squamosal of either side divided into an anterior and a posterior moiety, and no epiotic cornua. It is considered that the jugal entered into the formation of the outer border of the orbit.

Cochleosaurus bohemicus, Fritsch[2].

Syn. *Melosaurus bohemicus*, Fritsch[3].

The type species. Length of skull about 0,165 (6·6 inches).
Hab. Europe (Bohemia).

47493. Cast of a slab showing the dorsal aspect of the imperfect skull. The original, which is the type, was obtained from the Gaskohle (Lowest Permian) of Nyřan (Pilsen), and is figured by Fritsch in his 'Fauna der Gaskohle,' vol. ii. pl. lx. fig. 1, and diagrammatically on p. 30, fig. 137.

Presented by Dr. Anton Fritsch, 1876.

** Genus **GAUDRYA**, Fritsch[4].

Founded upon the anterior extremity of a palate, in which there are two vomers, a single row of marginal teeth, and numerous small teeth on the palatines. It does not appear how this form can be distinguished from the preceding or following genera.

[1] Fauna der Gaskohle, vol. ii. pt. i. p. 30 (1885).
[2] Sitz. k. böhm. Ges. Wiss. for 1875, p. 71 (1876).—*Melosaurus.*
[3] *Loc. cit.*
[4] Fauna der Gaskohle, vol. ii. pt. i. p. 31 (1885).

** **Gaudrya latistoma,** Fritsch [1].

The type. Apparently of the size of *Chelyosaurus vranyi.*
Hab. Europe (Bohemia).

R. 1470. Cast of a fragment of shale showing the palatal aspect of the anterior extremity of the cranium and a portion of the mandible. The original, which is the type, was obtained from the Gaskohle (Lowest Permian) of Nyřan, Bohemia, and is figured by Fritsch in his 'Fauna der Gaskohle,' vol. ii. pl. lxi. fig. 1. *Purchased*, 1888.

Genus **CHELYOSAURUS,** Fritsch [2].

Syn. *Chelyderpeton*, Fritsch [3].

Skull somewhat larger and narrower than in *Cochleosaurus*, but without posterior processes to the supraoccipitals; a single squamosal on either side, distinct epiotic cornua, nares probably very minute, anterior border of the orbit formed by the prefrontal and postorbital, and outer border by the postorbital. The occipital condyles appear to have been unossified; the vertebræ are more ossified than in *Archegosaurus*, having a basal element to the centra (hypocentrum pleurale); the intercentra are relatively very large; and the tarsus is well ossified.

Chelyosaurus vranyi, Fritsch [4].

Syn. *Chelyderpeton vranii*, Fritsch [5].

The type and only described species. Skull about one third longer than broad, its extreme length being 0,170 (6·7 inches).
Hab. Europe (Bohemia).

R. 1456. Cast of a slab showing the ventral scutes, the pelvis and part of the caudal region, and one pelvic limb. The original, which is the type, was obtained from the Rothliegendes (Lower Permian) of Oelberg, and is figured by Fritsch in the 'Sitzungsber. k. böhm. Ges. Wiss.' for 1877, p. 207, and also in his 'Fauna der Gaskohle,' vol. ii. pl. liv. *Purchased*, 1888.

[1] Fauna der Gaskohle, vol. ii. pt. i. p. 31 (1885).
[2] Sitz. k. böhm. Ges. Wiss. for 1877, p. 208 (1878).—Amended from *Chelydosaurus*.
[3] Jahresb. k. böhm. Ges. Wiss. 1877, pp. xx, xxi.
[4] Sitz. k. böhm. Ges. Wiss. for 1877, p. 208 (1878).
[5] Jahresb. k. böhm. Ges. Wiss. 1877, pp. xx, xxi.

R. 1457. Cast of a fragment of shale showing the frontal aspect of the anterior portion of the cranium. The original was obtained from the Rothliegendes of Ruppersdorf, and is figured by Fritsch in his 'Fauna der Gaskohle,' vol. ii. pl. lv. fig. 1. *Purchased*, 1888.

R. 1458. Cast of a fragment of shale showing a lateral aspect of a sacral region, referred by Fritsch to this form. The original, which is from Oelberg, is figured by Fritsch, *op. cit.* p. 24, fig. 133. Both the intercentra (hypocentra) and the basal portions of the pleuralia (hypocentra arcalia) are shown. *Purchased*, 1888.

Genus **SPHENOSAURUS**, Meyer[1].

Known only by a portion of the trunk, which is regarded by Fritsch as indicating a Labyrinthodont allied in the structure of the vertebræ to *Chelyosaurus*, although considered by Baur[2] to belong to a reptile. Cope[3] adopts the view of the Labyrinthodont affinity of the genus, and makes it the type of the family *Sphenosauridæ*, which includes *Chelyosaurus* and *Sparagmites*, and differs from the *Archegosauridæ* by the presence of the hypocentrum pleurale.

Sphenosaurus sternbergi (Fitzinger[4]).

Syn. *Palæosaurus sternbergi*, Fitzinger[5].

Apparently about one half the size of *Eryops megacephalus*.
Hab. Europe (Bohemia).

R. 1459. Cast of a slab of red sandstone exhibiting the ventral aspect of the thoracic and sacral region of the vertebral column, the pelvis, and one femur. The original, which is the type, is preserved in the Museum at Pragüe, and, according to Fritsch, was probably obtained from the Permian at the foot of the Riesengebirge, Bohemia. It is figured by Meyer in his 'Fauna der Vorwelt—Saurier aus dem Muschelkalk,' pl. lxx., where it is regarded as having been obtained from the Bunter; it is also figured by Fritsch in his 'Fauna der Gaskohle,' vol. ii. pl. lix., a diagrammatic restoration being given on p. 28, fig. 136,

[1] Fauna d. Vorwelt—Saurier aus d. Muschelkalk, p. 141 (1847).
[2] Biol. Centralblatt, vol. vi. p. 356 (1886).
[3] Amer. Nat. vol. xx. p. 77 (1886).
[4] Ann. Wien. Museums, vol. ii. p. 171 (1840).—*Palæosaurus.*
[5] *Loc. cit.*

of the same volume. If the latter restoration be correct, the intercentrum (hypocentrum arcale of Fritsch) will be greatly developed, and the central elements (pleurocentra and hypocentra pleuralia of Fritsch) proportionately reduced. *Purchased*, 1888.

Genus **SPARAGMITES**, Fritsch[1].

Syn. *Calochelys*, Fritsch[2].

Known only by a portion of the vertebral column of a small form, in which the neural spines are low and semicircular; the general structure being similar to that of *Chelyosaurus*.

Sparagmites lacertinus, Fritsch[3].

Syn. *Calochelys lacertina*, Fritsch[4].

The type and only described species.

Hab. Europe (Bohemia).

R. 1450. Electrotype from a fragment of shale, showing part of the vertebral column. The original is the type, and was obtained from the Gaskohle (Lowest Permian) of Nyřan (Pilsen); it is figured by Fritsch in his 'Fauna der Gaskohle,' vol. ii. pl. lii. fig. 1. *Purchased*, 1888.

Genus **TRIMERORHACHIS**, Cope[5].

Skull of moderate width and length, with the round orbits in the anterior half, and the nares very widely separated from one another and approximated to the orbits; occipital condyles unossified; two rows of upper and one row of lower teeth; cranial sculpture net-like. Vertebræ (fig. 46) without distinct neural spines, and the ossification of the intercentral and pleurocentral elements very superficial.

The absence of neural spines to the vertebræ distinguishes the present genus from all the others of the group, and the genus is accordingly regarded by its founder as the type of a family—*Trimerorhachidæ*.

[1] Fauna der Gaskohle, vol. ii. pt. i. p. 15 (1885).

[2] *Ibid.* vol. i. pt. i. p. 28 (1879).—Preoccupied by the Chelonian *Callichelys*, Gray (1863).

[3] *Loc. cit.*—*Calochelys*. [4] *Loc. cit.*

[5] Proc. Amer. Phil. Soc. vol. xvii. p. 524 (1878).

Trimerorhachis insignis, Cope

The type species. Length of skull about 0,160 (6·3 inches); length of intercentra 0,010 (0·4 inch); the latter dimensions indicating an animal about one third the size of *Eryops megacephalus.*

Hab. North America (Texas).

R. 576 x. The basioccipital region of the cranium, a fragment of the mandible, and four vertebræ; from the reputed Permian of Texas. These specimens agree in all respects with the corresponding elements represented in fig. 46; the neural arches retain, however, their proper relative position to the intercentra. *Purchased,* 1885.

Fig. 46.

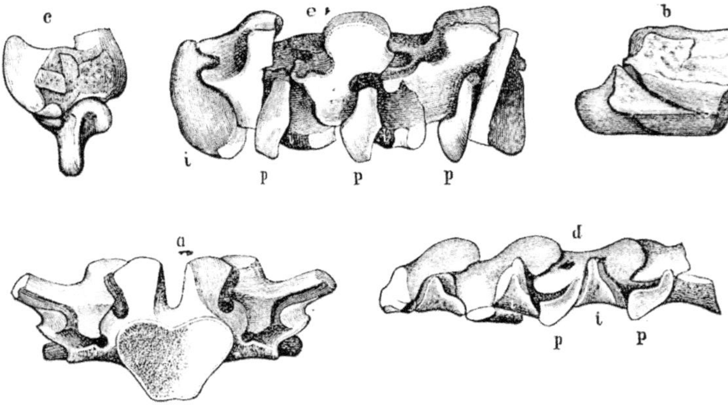

Trimerorhachis insignis.—a, occipital region of cranium; *b,* mandibular articulation; *c,* posterior view of the same; *d,* part of the vertebral column, with the neural arches crushed down; *e,* another portion of the vertebral column. From the reputed Permian of Texas. ¼. *p,* pleurocentrum; *i,* intercentrum.

R. 574. Two fragments of jaws probably belonging to this species; from Texas. *Purchased,* 1885.

R. 574 x. Three imperfect bones, apparently associated with the preceding; from Texas. The expanded bone is probably part of the pectoral girdle, while the other two specimens appear to belong to the limbs. *Purchased,* 1885.

[1] Proc. Amer. Phil. Soc. vol. xvii. p. 524 (1878).

R. 574 y. The distal half of a femur; from Texas. This specimen accords in contour with the much larger femur of *Eryops* figured by Cope in the 'Proc. Amer. Phil. Soc.' vol. xix. pl. iv. figs. 15, 16, 19, being nearly of the same absolute size as the figure. *Purchased*, 1885.

Family Position Uncertain.

Rhytidosteus differs from the typical *Archegosauridæ* in the absence of large palatal vacuities, and thereby approximates to the *Anthracosauridæ* and *Nyraniidæ*, and also agrees with the latter in the presence of denticules on the palate. The presence of denticules on the mandible appears to connect *Rhytidosteus* with *Eryops*, and if the latter had a closed palate of similar type the family name *Eryopidæ*, proposed by Cope, should be adopted for this group.

Genus **ERYOPS**, Cope [1].

Syn. *Rhachitomus*, Cope [2].
Epicordylus, Cope [3].
Parioxys, Cope [4].

Skull [5] vaulted, comparatively short and wide, with large postero-lateral expansions, the round and rather small orbits in the posterior half, the nares ovoid, widely separated, and placed some distance behind the extremity of the broad and blunt muzzle, and the occipital condyles ossified. Mandible with alveolar parapet, but without postarticular process. Teeth subconical and pointed; anterior ones enlarged; hinder ones small; minute denticules on the mandibular ramus internally to the alveoli; probably similar denticules on the palate; upper teeth in a single row. Cranial sculpture coarsely pitted and ridged. The synonymy is taken from Cope, 'Trans. Amer. Phil. Soc.' vol. xvi. p. 286.

Eryops megacephalus, Cope [6].

Syn. *Rhachitomus valens*, Cope [7].

The type species; of very large size. Skull with the length in the median line equal to the width, the muzzle very broad and blunt, and the orbits forming the most prominent point in the profile; extreme length about 0,575 (22·5 inches), length in median line about 0,465 (18·3 inches).

Hab. North America (Texas).

[1] Proc. Amer. Phil. Soc. vol. xvii. p. 188 (1877). [2] *Ibid.* p. 526 (1878).
[3] *Ibid.* p. 515 (1878). [4] *Ibid.* p. 521 (1878).
[5] See Amer. Nat. vol. xviii pp 34, 35, figs. 5, 6.
[6] Proc. Amer. Phil. Soc. vol. xvii. p. 188 (1877). [7] *Ibid.* p. 526 (1878).

The following specimens were obtained from the reputed Permian of Texas. Purchased, 1885.

R. 565. Fragment from the anterior extremity of the cranium, showing the broken bases of three large teeth. This and the following specimens of teeth are referred to this species on account of their large size.

R. 570. Two fragments of the cranium, each with the base of a large tooth.

R. 570 a. Fragment of a mandibular ramus. This specimen shows the broken bases of three very large and some smaller teeth, internally to which the summit of the inner surface of the bone carries a number of minute knob-like denticules. It seems to have belonged to the same individual as No. R. 565.

Fig. 47.

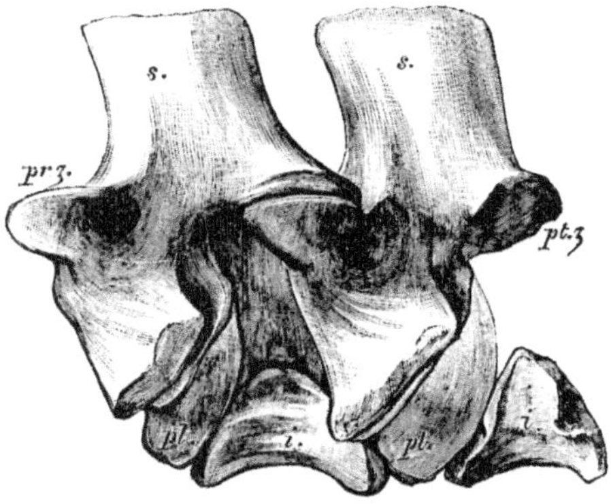

Eryops megacephalus.—Left lateral aspect of two trunk-vertebræ; from the reputed Permian of Texas. ⅜. *s*, neural spine; *pr.z*, prezygapophysis; *pt.z*, postzygapophysis; *pl*, pleurocentrum; *i*, intercentrum.

R. 570 b. Part of a mandibular ramus, which appears to be from the posterior region of the same jaw as that to which the preceding specimen belonged. The inner part is wanting.

R. 570 c. Fragment of a mandibular ramus with nine imperfect teeth. This specimen, in which the inner lamina is wanting, may have belonged to the opposite side of the jaw to which the two preceding specimens pertained.

R. 576. (*Fig.*) Several associated vertebræ, two of which are represented in woodcut, fig. 47. These specimens accord with the portion of the vertebral column figured by Cope in the 'Proc. Amer. Phil. Soc.' vol. xix. pl. iii. The intercentra distinctly show in several cases a capitular rib-facet on the posterior border of their lateral surface. The close resemblance of these elements to the intercentra of *Embolophorus* (No. R. 569, p. 109) appears to be conclusive that Cope is right in regarding them as the homologues of the latter, and that they are not (as has been contended) part of the centrum of the vertebra.

Eryops africanus, Lydekker [1].

Known by part of the mandible, which has an approximate length of about 0,400 (15·8 inches), and apparently indicates a species of somewhat smaller size than *E. megacephalus*, with the symphysial teeth relatively less enlarged and the intercentrum less fully ossified. The generic reference is provisional.

Hab. South Africa.

R. 466. (*Fig.*) The imperfect right ramus of the mandible; from the Karoo system of South Africa. The glenoidal region and the greater part of the dentary bone are entire, but the region between the two is imperfect. This specimen is the type, and the dentary region is figured by the writer in the 'Quart. Journ. Geol. Soc.' vol. xlvi. pl. xii. fig. 2. The absence of a postarticular process is clearly shown. To the inner side of the alveolar tract the surface of the bone carries a band of minute denticules precisely similar to those found in the mandible (No. R. 570 *a, b*) referred to *Eryops megacephalus* (p. 192); and there are no characters by which the present specimen can be generically distinguished from the latter.

Presented by Sir R. Owen, K.C.B., 1884.

R. 470. An intercentrum, apparently associated with the preceding; from South Africa. Noticed by the writer in the 'Palæontologia Indica,' ser. 4, vol. i. pt. iv. p. 10, where it is referred to *Rhytidosteus*; and figured in the 'Quart. Journ. Geol. Soc.' vol. xlvi. pl. xii. figs. 3, 4. This specimen is rather smaller than the intercentra (No. R. 576) of *Eryops megacephalus*, and therefore agrees in relative size with the mandible; it is also less ossified

[1] Quart. Journ. Geol Soc. vol. xlvi. pt. 2 (1890).

internally, and is thus more crescent-like. It exhibits facets for the capitula of the ribs on the posterior border of the upper extremity of the outer surface. Rhachitomous vertebræ from the Karoo system of the Cape are preserved in the Museum at Munich (as the writer is informed by Dr. E. Fraas), which may belong either to the present form or to *Rhytidosteus*.

Presented by Sir R. Owen, K.C.B., 1884.

Genus **RHYTIDOSTEUS**, Owen [1].

Skull depressed and relatively short; orbits situated in the posterior half, and separated from one another by the whole width of the cranium; nares placed far behind the muzzle, directed laterally, and likewise separated by the whole width of the cranium; occipital region unknown; palate extensively ossified, without vacuities near the medial line (so far as shown). Mandible without alveolar parapet, but with a postarticular process. A row of comparatively small premaxillary and maxillary teeth, and on either side of the hinder maxillary teeth an irregular series of denticules, while smaller denticules spread out over the palate; one large tusk in advance of, and another behind, the posterior nares; a series of denticules on either side of the marginal row of mandibular teeth. Teeth with complex foldings and large pulp-cavity. Sculpture coarsely radiate, with an indistinct lyra between orbits and nares.

This genus is referred to the present group on account of the presence of denticules on the jaws, by which it appears to be allied to *Eryops*. The full ossification of the palate differs, however, so remarkably from that of *Archegosaurus*, that the genus should probably be referred to another family. Whether, however, it should form the type of a distinct family (in which *Eryops* should probably be included) or should be classed with the *Nyraniidæ*, cannot be determined until the relations of the bones of the palate are fully known. The position of the orbits in the present genus is quite peculiar.

Rhytidosteus capensis, Owen [2].

The type and only described species. Length of skull probably about 0,250 (9·8 inches); muzzle very broad and blunt.

Hab. South Africa.

R. 455. (*Fig.*) The flattened skull, wanting the posterior portion of the cranium, and with the mandible imperfect; from the Karoo system of Beersheba, Orange Free State. The type specimen; figured by Owen in the 'Quart. Journ. Geol.

[1] Quart. Journ. Geol. Soc. vol. xl. p. 333 (1884).

[2] *Loc. cit.*

Soc.' vol. xl. pls. xvi., xvii., all the figures being reversed. In the figure of the palate (pl. xvii. fig. 1) the aperture marked *vc* is a fracture. The median bone which Owen terms the palatine is more probably the anterior extremity of the parasphenoid: while it is probable that at least a considerable portion of the bones termed vomers represent the palatines. If this be so the structure of the palate will approximate very decidedly to that of *Nyrania*. The absence of a postarticular process to the mandible is well shown; in the figure too short a space is allowed for the missing portion of the ramus.

By exchange with the Blomfontein Museum, 1884.

R. 503. Four fragments of bones of the cranium apparently referable to this species, and probably associated with the preceding specimen: from Beersheba. One of these specimens includes a bone which appears to be the greater portion of the quadrate, attached to which are fragments of two other bones.

By exchange with the Blomfontein Museum, 1884.

Incertæ sedis.

Genus **PHOLIDOGASTER**, Huxley[1].

Very imperfectly known, the frontal aspect of the skull being concealed in the type. There is complete ventral armour; the ventral surface of the vertebræ closely approximates to that of *Sphenosaurus*, and suggests that the present genus indicates an allied rhachitomous type.

Pholidogaster pisciformis, Huxley[2].

The type and only described species. Length of skull about 0,177 (7 inches); total length of skeleton about 1,116 (44 inches).

Hab. Europe (Scotland).

30534. (*Fig.*) Slab of shale showing the ventral aspect of the skeleton; from the Lower Carboniferous of Gilmerton, near Edinburgh. The type specimen; figured by Huxley in the 'Quart. Journ. Geol. Soc.' vol. xviii. pl. xi. figs. 3, 4. The mandible, lateral thoracic plates, ventral armour, and the ventral aspect of the later thoracic vertebræ are well shown. The contour of the vertebræ so strongly resembles that of *Sphenosaurus* as to suggest that they are really intercentra.

Purchased, 1856.

[1] Quart. Journ. Geol. Soc. vol. xviii. p. 294 (1862).

[2] *Loc. cit.*

Suborder *MICROSAURIA*.

Salamandriform Labyrinthodonts, in which the vertebral centra are elongated, and constricted in the middle, with traces of the notochord internally; and the ribs generally long, curved, and double-headed. Teeth with large pulp-cavities, and without plications of the dentine. Occipital region of the skull and the pubis ossified. Scutes may be present on the dorsal as well as on the ventral aspect; and the carpus and tarsus may be ossified or cartilaginous.

Family UROCORDYLIDÆ.

Stout and long-tailed forms, with the epiotic cornua frequently much produced, and pitted cranial bones; neural spines and chevrons of caudal vertebræ much dilated at their extremities, and pectinated; no caudal ribs. No dorsal scutes.

Genus **UROCORDYLUS,** Huxley [1].

The type genus. Skull triangular, truncated behind, with rounded muzzle and aborted epiotic cornua. Teeth slender, pointed, and smooth, with a slight backward curvature; pulp-cavity large; no trace of labyrinthic structure. Neural spines of caudal vertebræ tall, slender, and expanded in a fan-like manner; chevrons usually shorter and wider. Tail deep and strong, with some 80 vertebræ. Ribs strong, three times the length of the vertebræ, with distinct double heads. Middle thoracic plate thin and shield-like; ventral scutes oat-like.

Urocordylus wandesfordi, Huxley [2].

Syn. *Urocordylus reticulatus,* Hancock and Atthey [3].

The type species. Length of skull about 0,033 (1·3 inches), length of body 0,595 (19·5 inches).

Hab. Europe (Ireland and England).

41851 e. Slab of shale showing the impression of part of the skeleton; from the Coal-Measures (Upper Carboniferous) of Jarrow Colliery, Kilkenny, Ireland. Accords with the type skeleton figured by Huxley in the 'Trans. Roy. Irish Academy,'

[1] Trans. Roy. Irish Academy, vol. xxiv. p. 359 (1867).

[2] *Loc. cit.*

[3] Ann. Mag. Nat. Hist. ser. 4, vol. iv. p. 182 (1869); see also Nat. Hist. Trans. Northumb. and Durham, vol. iii. p. 310 (1870).

vol. xxiv. pl. xx. The anterior caudal region is well shown; and the ventral scutes are also exhibited.
Purchased, 1870.

41851 f. Slab showing the impression of the anterior portion of a skeleton apparently referable to this form; from Jarrow.
Purchased, 1870.

Urocordylus scalaris, Fritsch [1].

Distinguished from the type species by the greater expansion of the summits of the neural spines and chevrons of the caudal vertebræ.

Hab. Europe (Bohemia).

R. 355. Electrotype from a fragment of shale, showing portions of the skeleton. The original, which is one of the types, was obtained from the Gaskohle (Lowest Permian) of Nyřan (Pilsen), and is figured by Fritsch in his 'Fauna der Gaskohle,' vol. i. p. 129, fig. 76. *Purchased*, 1882.

R. 356. Electrotype from a fragment of shale, showing portions of the skeleton. Original from Nyřan. *Purchased*, 1882.

Genus **CERATERPETUM**, Huxley [2].

Syn. *Scincosaurus*, Fritsch [3].

Skull parabolic and of great width, with very long epiotic cornua. Maxillary teeth short and smooth. Neural spines of caudal vertebræ low and wide, overlapping one another, and in close contact; chevrons of same similarly expanded. Tail with some 40 vertebræ. Ribs four times the length of the vertebræ. Middle thoracic plate stout and triangular; ventral scutes oblong.

Ceraterpetum galvani, Huxley [4].

The type species. Length of skull in middle line about 0,026 (1 inch). Total length of skeleton about 0,253 (10 inches).

Hab. Europe (Ireland).

41851 a. Slab of shale showing the impression of part of the vertebral column; from the Coal-Measures (Upper Carboniferous) of Jarrow Colliery, Kilkenny.
Purchased, 1870.

[1] Sitz. k. bohm. Ges. Wiss. for 1875, p. 73 (1876).
[2] Trans. Roy. Irish Academy, vol. xxiv. p. 354 (1867).—*Keraterpeton.*
[3] Sitz. k. bohm. Ges Wiss. for 1875, p. 72 (1876).
[4] *Loc. cit.*

41851 b. Slab showing the imperfect skeleton of a small individual; from Jarrow. This specimen is much smaller than either of the two types figured in the 'Trans. Roy. Irish Academy,' vol. xxiv. pl. xix. *Purchased*, 1870.

41851 c. Slab with the imperfect skeleton of a somewhat larger individual; from Jarrow. *Purchased*, 1870.

41851 d. Slab showing the impression of a pelvic limb not improbably belonging to this form. *Purchased*, 1870.

Ceraterpetum crassum, Fritsch [1].

Syn. *Scincosaurus crassus*, Fritsch [2].

The orbits smaller, and the skull smaller in proportion to the body than in the type species.

Hab. Europe (Bohemia).

The originals of the following specimens were obtained from the Gaskohle (Lowest Permian) of Nýřan (Pilsen). The specimens were purchased in 1882.

R. 351. Electrotype from a fragment of shale, showing portions of the skeleton, among which is the median thoracic plate. Original figured by Fritsch in his 'Fauna der Gaskohle,' vol. i. pl. xxvii. fig. 4, and the thoracic plate in pl. xxix. fig. 7.

R. 352. Electrotype showing the dislocated skeleton. A pes is figured by Fritsch, *op. cit.* pl. xxviii. fig. 10.

R. 353. Electrotype showing the imperfect skeleton. The sacral region and pelvic limbs figured by Fritsch, *op. cit.* pl. xxx. fig. 1.

R. 354. Electrotype exhibiting the thoracic and caudal regions. Nine of the ventral scutes are figured by Fritsch, *op. cit.* pl. xxx. fig. 5.

[1] Sitz. k bohm. Ges. Wiss. for 1875, p 72 (1876).—*Scincosaurus.*

[2] *Loc. cit.*

FAMILY UNCERTAIN.

Genus **LEPTERPETUM,** Huxley [1].

Tail long. Skull long and triangular, with the orbits in the middle of the length; mandibular symphysis elongated. Thoracic buckler unknown; ventral scutes long and rhomboidal. Ribs thin and curved. Caudal vertebræ with long chevrons, about 25 in number. Carpus and tarsus unossified.

Lepterpetum, sp.

The following specimen indicates an individual of larger size than the type specimen of the typical *L. dobbsi*, Huxley [2] (which measures 6 inches in length), but there is not sufficient evidence to show whether it is specifically distinct.

Hab. Europe (Ireland).

41851. Slab of shale showing an imperfect skeleton apparently referable to a member of this genus; from the Coal-Measures (Upper Carboniferous) of Jarrow Colliery, Kilkenny. The length of the vertebral centra in the dorsal region is 0,060. *Purchased*, 1870.

Lepterpetum (?), sp.

The undermentioned specimen has been provisionally referred by Fritsch to this genus.

Hab. Europe (Bohemia).

R. 374 a. Electrotype from a fragment of shale, showing portions of the skull. The original was obtained from the Gaskohle (Lowest Permian) of Nyřan (Pilsen), and is described and figured by Fritsch in his 'Fauna der Gaskohle,' vol. i. p. 172, pl. xli. fig. 10. *Purchased*, 1882.

Family LIMNERPETIDÆ.

Body longer and tail shorter than in the preceding family, the skull being broad and frog-like, with smooth bones; the neural spines and chevrons of the caudal vertebræ simple, and caudal ribs present. Teeth small, with their summits either smooth or grooved, and the pulp-cavity large. Ventral scutes sculptured; no dorsal armour.

[1] Trans. Roy. Irish Academy, vol. xxiv. p. 362 (1867).—*Lepterpeton*.

[2] *Loc. cit.*

Genus LIMNERPETUM, Fritsch[1].

Syn. *Microdon*, Fritsch[2].

The type and only described genus. All the species are small.

Limnerpetum elegans, Fritsch[3].

Ventral scutes small, ovoid, with concentric and oblique sculpture; less wide than in *L. laticeps*.

Hab. Europe (Bohemia).

R. 357. Electrotype from a fragment of shale, showing the greater part of the skeleton. The original is the type, and was obtained from the Gaskohle (Lowest Permian) of Nyřan (Pilsen). The specimen forms the subject of pl. xxxiv. of vol. i. of Fritsch's 'Fauna der Gaskohle.'

Purchased, 1882.

Limnerpetum laticeps, Fritsch[4].

Syn. *Microdon laticeps*, Fritsch[5].

Ventral scutes small, much broader than long, with a thickened posterior border and concentric sculpture. Total length about 0,100 (3·9 inches).

Hab. Europe (Bohemia).

R. 358. Electrotype from a fragment of shale, exhibiting the nearly entire skeleton. The original, which is the type, was obtained from the Gaskohle (Lowest Permian) of Třemošna (Pilsen), and is figured by Fritsch in his 'Fauna der Gaskohle,' vol. i. pl. xxxvi fig. 1. *Purchased*, 1882.

Limnerpetum obtusatum, Fritsch[6].

Ventral scutes of medium size, ovoid, with radiating sculpture; those of thoracic region with thickened posterior border, which disappears in the pelvic region. Body relatively larger, and tail shorter than in *L. laticeps*; total length about 0,160 (6·3 inches).

Hab. Europe (Bohemia).

R. 359. Electrotype from a fragment of shale, showing the nearly entire skeleton. The original is the type, and was obtained from the Gaskohle (Lowest Permian) of Nyřan (Pilsen). It is figured by Fritsch in his 'Fauna der Gaskohle,' vol. i. pl. xxxv. figs. 1–7; the ventral scutes being figured on pp. 155–156, figs. 95–98. *Purchased*, 1882.

[1] Fauna der Gaskohle, vol. i. pt. iii. p 147 (1881) —*Limnerpeton*.
[2] Sitz. k. bohm. Ges. Wiss for 1875, p. 73 (1876) —Preoccupied.
[3] Fauna der Gaskohle, vol. i. pt. iii. p. 152 (1881).
[4] *Ibid.* pt. i. p. 28 (1879).—*Microdon*.
[5] *Loc. cit.*
[6] *Ibid.* pt. iii. p. 154 (1881).

Family HYLONOMIDÆ.

Body slender and lizard-like; skull narrow, with smooth or faintly sculptured bones; neural spines of vertebræ well developed, and long and slender ribs. Teeth smooth, or with grooved summits. The whole body covered with sculptured scutes (fig. 48). Internal gills may be developed. For *Smilerpetum* and *Hylerpetum* see Addenda.

Genus **HYLONOMUS,** Owen[1] (*ex* Dawson).

Syn. *Stelliosaurus*, Fritsch[2].
Hyloplesion, Fritsch[3].

The type genus. Skull broad, narrowing in front, with rounded muzzle, and the orbits in the anterior half; cranial bones smooth, with a few irregular pits. *Hyloplesion* was identified with *Holonomus* by Credner[4].

For the type species see Addenda.

Hylonomus longicostatus (Fritsch[5]).

Syn. *Stelliosaurus longicostatus*, Fritsch[6].
Hyloplesion longicostatum, Fritsch[7].

Total length about 0,085 (3·6 inches).
Hab. Europe (Bohemia).

R. 372. Electrotype from a fragment of shale, showing the postcephalic portion of the skeleton. The original was obtained from the Gaskohle (Lowest Permian) of Třemošna (Pilsen), and is figured by Fritsch in his 'Fauna der Gaskohle,' vol. i. pl. xxxvi. fig. 2. *Purchased*, 1882.

R. 373. Two electrotypes from fragments of shale, exhibiting imperfect skeletons; the originals were obtained from the Gaskohle (Lowest Permian) of Nyřan (Pilsen), and are figured by Fritsch, *op. cit.* pl. xxvii. fig. 5, and pl. xlv. fig. 6. *Purchased*, 1882.

[1] Quart. Journ. Geol. Soc. vol. xviii. p. 238 (1862).
[2] Sitz. k. böhm. Ges. Wiss. for 1875, p. 72 (1876).—Withdrawn.
[3] Fauna der Gaskohle, vol. i. pt. iv. p. 160 (1883).
[4] Zeitschr. deutsch geol. Ges. vol. xxxvii p. 734 (1885)
[5] Sitz. k. böhm. Ges. Wiss. for 1875, p. 72 (1876).—*Stelliosaurus*
[6] *Loc. cit.*
[7] Fauna der Gaskohle, vol. i. pt. iv. p. 160 (1883).

Genus **SEELEYA**, Fritsch[1].

Head large and tail short (fig. 48). Skull long and triangular, with the premaxillary teeth larger than those of the maxilla, and small teeth on the bones of the palate. Ribs long, stout, and curved.

Fig. 48.

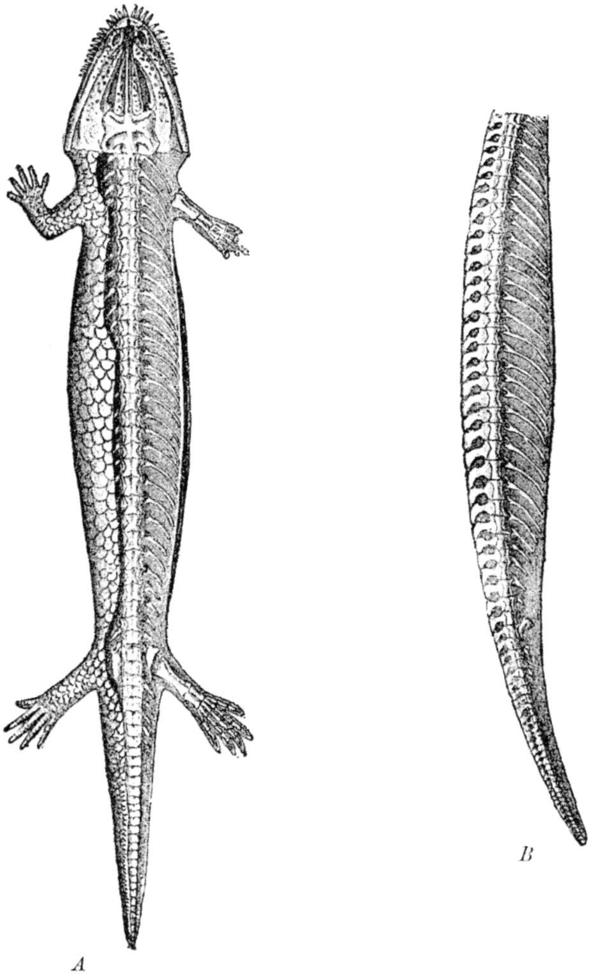

Seeleya pusilla.—A. Restoration of the ventral aspect, with the scutes removed from the left side; *B.* Lateral aspect of the thoracic and caudal region; from the Gaskohle (Lowest Permian) of Bohemia. $\frac{5}{1}$. (*After Fritsch.*)

[1] Fauna der Gaskohle, vol. i. pt. iv. p. 165 (1883).

Seeleya pusilla, Fritsch[1].

The type and only described species; total length 0,023 (0·9 inch).

Hab. Europe (Bohemia).

R. 343. Electrotype from a fragment of shale, showing the entire skeleton. The original is the type, and was obtained from the Gaskohle (Lowest Permian) of Nyřan (Pilsen); it is figured by Fritsch in his 'Fauna der Gaskohle,' pl. xl. fig. 1. *Purchased*, 1882.

Genus **RICNODON,** Fritsch[2].

Body short, with a relatively large parabolic skull, of which the bones are marked by minute pits. The pectoral limbs nearly as large as the pelvic. Denticules on the parasphenoid and pterygoids. Scutes large, with thickened posterior border.

Ricnodon trachylepis, Fritsch[3].

Imperfectly known; distinguished from *R. copei* and *R. dispersus*, Fritsch, by the characters of the scutes.

Hab. Europe (Bohemia).

R. 374. Electrotype from a fragment of shale, showing the ventral scutes. The original is one of the types, and was obtained from the Gaskohle (Lowest Permian) of Nyřan (Pilsen); it is figured by Fritsch in his 'Fauna der Gaskohle,' vol. i. pl. xliv. fig. 13. *Purchased*, 1882.

Genus **ORTHOPLEUROSAURUS,** Lydekker (*n. nom.*).

Syn. *Orthocosta*, Fritsch[4].

Neural spines of vertebræ longer than in *Seeleya*, and the ribs short and straight.

Orthopleurosaurus microscopicus (Fritsch[5]).

Syn. *Orthocosta microscopica*, Fritsch[6].

The type and only described species. Very minute.

Hab. Europe (Bohemia).

[1] Fauna der Gaskohle, vol. i. pt. iv. p. 165 (1883).
[2] *Ibid.* p 167.
[3] *Ibid.* p 170.
[4] *Ibid.* pt. i p. 28 (1879).—Hybrid.
[5] *Loc. cit.*—*Orthocosta.*
[6] *Loc. cit.*

R. 349. Electrotype from a fragment of shale, showing the imperfect hinder half of the skeleton. The original, which is the type, was obtained from the Gaskohle (Lowest Permian) of Nyřan (Pilsen), and is figured by Fritsch in his 'Fauna der Gaskohle,' vol. i. pl. xliv. figs. 1, 2. *Purchased*, 1882.

Family MICROBRACHIDÆ.

Small, slender forms, with very short pectoral limbs, strongly sculptured cranial bones, and scutes covering the whole body [1]. Ribs stout and curved, and of nearly equal length throughout the series. Middle thoracic plate very broad, with a long posterior process.

Genus **MICROBRACHIS,** Fritsch [2].

The type genus.

Microbrachis pelicani, Fritsch [3].

The type species. Total length about 0,125. Twenty-eight upper teeth; cranial sculpture very elaborate; teeth on stem of parasphenoid.

Hab. Europe (Bohemia).

R. 369. Three electrotypes from fragments of shale, one showing the postcephalic portion of the skeleton, the second the skull, and the third the imperfect skeleton. The originals were obtained from the Gaskohle (Lowest Permian) of Nyřan (Pilsen), and are figured by Fritsch in his 'Fauna der Gaskohle,' vol. i. pls. xlv. and xlviii. *Purchased*, 1882.

R. 370. Two electrotypes from a fragment of shale, showing the ventral and dorsal aspects of part of the vertebral column. The original was obtained from Nyřan, and is figured by Fritsch, *op. cit.* pl. xlviii. fig. 8. *Purchased*, 1882.

[1] In the definition of the family in the 'Fauna der Gaskohle,' vol. i. p. 173, it is stated that there are only ventral scutes; but in the fig. on p. 175 the whole body is scutellated.

[2] Sitz. k. bohm. Ges. Wiss. for 1875, p. 72 (1876).

[3] *Loc. cit.*

Microbrachis mollis, Fritsch[1].

Twelve upper teeth; cranial sculpture simpler than in type species.

Hab. Europe (Bohemia).

R. 371. Electrotype from a fragment of shale, showing the greater portion of the skeleton, the caudal region wanting. The original was obtained from the Gaskohle (Lower Permian) of Nyřan (Pilsen); the skull is figured by Fritsch in his 'Fauna der Gaskohle,' vol. i. pl. xlviii. fig. 1.

Purchased, 1882.

Suborder *AISTOPODA*.

Body long and snake-like, without limbs, and apparently without pectoral or pelvic girdles. Vertebræ with elongated centra and aborted neural spines. Ribs slender, and barbed like those of fishes. Teeth smooth, without plications of the dentine. External gills probably persistent.

Family DOLICHOSOMATIDÆ.

All the members of the suborder are included in this family, which is equivalent to both the *Phlegethonidæ* and *Molgophidæ* of Cope.

Genus **DOLICHOSOMA,** Huxley[2].

Skull small, and tapering from occiput to muzzle, with smooth bones, extensive vacuities in the preorbital region, and the nasals, parietals, and frontals anchylosed together; mandible as long as cranium, with very slender rami. Vertebræ about 150 in number, with very strongly developed transverse processes. Ribs at first angularly bent, then straight, with two processes. No evidence of dermal armour.

Dolichosoma longissimum, Fritsch[3].

Vertebræ exceeding 150 in number; the ribs twice the length of the vertebræ; total length about 0,707 (28 inches).

Hab. Europe (Bohemia).

[1] Sitz. k. bohm. Ges. Wiss. for 1875, p 179 (1876).
[2] Trans. Roy. Irish Academy, vol. xxiv. p. 366 (1867).
[3] Sitz. k. bohm. Ges. Wiss. for 1875, p. 72 (1876).

R. 360. Electrotype from a fragment of shale, showing a considerable portion of the skeleton, without the skull. The original was obtained from the Gaskohle (Lowest Permian) of Nyřan (Pilsen), and is figured by Fritsch in his 'Fauna der Gaskohle,' vol. i. pl. xvii. fig. 1. *Purchased*, 1882.

R. 361. Electrotype from a fragment of shale, exhibiting the skull and cervical vertebræ. Original from Nyřan; figured by Fritsch, *op. cit.* pl. xviii. fig. 1. *Purchased*, 1882.

Dolichosoma (?) angustatum, Fritsch [1].

Founded on an imperfect skull with anchylosed frontals, which is provisionally referred to this genus, but may belong to *Ophiderpetum*.

Hab. Europe (Bohemia).

R. 362. Electrotype from a fragment of shale, showing the skull. The original is the type, and was obtained from the Gaskohle (Lowest Permian) of Nyřan (Pilsen); it is figured by Fritsch in his 'Fauna der Gaskohle,' vol. i. pl. xxi. fig. 1. *Purchased*, 1882.

Genus OPHIDERPETUM, Huxley [2].

Skull imperfectly known, but probably shorter and wider in front than in *Dolichosoma*. Vertebræ about 100 in number. Ribs with a dorsal and a ventral process near the proximal extremity. Ventral armour in the form of oat-like scutes; dorsal surface of the body covered with shagreen-like tubercles; toothed plates in the neighbourhood of the cloaca.

Ophiderpetum brownriggi, Huxley [3].

Syn. *Ophiderpeton nanum*, Hancock & Atthey [4].

The type species. Length of skull about 0,042 (1·6 inches); total length of large examples exceeding 0,534 (21 inches).

O. nanum was founded on a small specimen from Northumberland which cannot be specifically distinguished from the present form.

Hab. Europe (Ireland and England).

41851 l. Fragment showing the impression of the anterior part of the skeleton of a small individual; from the Coal-Measures

[1] Fauna der Gaskohle, vol. i. pt. ii. p. 117 (1880).

[2] Trans. Roy. Irish Academy, vol. xxiv. p. 364 (1867).—*Ophiderpeton*.

[3] *Loc. cit.*

[4] Ann. Mag. Nat. Hist. ser. 4, vol. i. p. 276. See also Nat. Hist. Trans. Northumb. & Durham, vol. iii. p. 79 (1869).

(Upper Carboniferous) of Jarrow Colliery, Kilkenny, Ireland. The head is badly preserved, but appears to have been of a parabolic shape. Except for its smaller dimensions this specimen cannot be distinguished from the type skeleton figured by Huxley in the 'Trans. Roy. Irish Academy,' vol. xxiv. pl. xxii. fig. 1. The impressions of the ventral scutes are distinctly visible. *Purchased*, 1870.

Ophiderpetum granulosum, Fritsch[1].

Length of ventral scutes half that of the vertebral centra; transverse processes of vertebræ occupying one third the length of the centra. Smaller than *O. brownriggi*.

Hab. Europe (Bohemia).

R. 363. Electrotype from a fragment of shale, exhibiting the middle region of the skeleton, with the ventral scutes and dorsal granules. The original was obtained from the Gaskohle (Lowest Permian) of Nyřan (Pilsen); and is figured by Fritsch in his 'Fauna der Gaskohle,' vol. i. pl. xvii. fig. 2. *Purchased*, 1882.

Ophiderpetum pectinatum, Fritsch[2].

Ventral scutes rugose, and three times the length of the vertebral centra. Smaller than *O. brownriggi*.

Hab. Europe (Bohemia).

R. 364. Electrotype from a fragment of shale, showing the ventral scutes. The original, which is one of the types, was obtained from the Gaskohle (Lowest Permian) of Nyřan (Pilsen); and is figured by Fritsch in his 'Fauna der Gaskohle,' vol. i. pl. xx. fig. 1. *Purchased*, 1882.

R. 365. Electrotype from a fragment of shale, exhibiting the plates in the cloacal region. Original from Nyřan; figured by Fritsch, *op. cit.* pl. xx. fig. 5. *Purchased*, 1882.

Ophiderpetum vicinum, Fritsch[3].

Ventral scutes as long as the vertebral centra; transverse processes of vertebræ occupying more than half the length of the centra. Nearly as large as the type species.

Hab. Europe (Bohemia).

[1] Fauna der Gaskohle, vol. i. pt. i p. 27 (1879).
[2] *Loc. cit.*
[3] *Loc. cit.*

R. 367. Electrotype from a fragment of shale, showing the ventral scutes. The original, which is one of the types, was obtained from the Rothliegendes (Lower Permian) of Kounova; and is figured by Fritsch in his 'Fauna der Gaskohle,' pl. xix. fig. 2. *Purchased*, 1882.

Ophiderpetum corvini, Fritsch[1].

Cloacal plates very large, much curved and strongly toothed.

Hab. Europe (Bohemia).

R. 366. Electrotype from a fragment of shale, showing one of the cloacal plates. The original, which is one of the types, was obtained from the Permian of Kounova; and is figured by Fritsch in his 'Fauna der Gaskohle,' vol. i. pl. xx. fig. 12. *Purchased*, 1882.

Ophiderpetum zieglerianum, Fritsch[2].

Characterized by the great length of the ventral scutes, which are quite smooth.

Hab. Europe (Bohemia).

R. 368. Electrotype from a fragment of shale, showing a mass of ventral scutes. The original, which is the type, was obtained from the Gaskohle (Lowest Permian) of Zieglerschacht, near Nyřan (Pilsen); and is figured by Fritsch in his 'Fauna der Gaskohle,' vol. i. pl. xx. fig. 14. Each scute consists of three pieces. *Purchased*, 1882.

Suborder *BRANCHIOSAURIA*.

Short-tailed Salamandriform Labyrinthodonts, in which the vertebral centra are barrel-shaped, with a notochordal canal which expands in the middle of each centrum; ribs short and straight, with a simple expanded head. Teeth with large pulp-cavities, and no plication of the dentine. Occipital region of the skull and the pubis cartilaginous. There are no dorsal scutes;' and the carpus and tarsus are unossified.

Family APATEONIDÆ.

In this family the skull is generally longer than in the *Protritonidæ*; the teeth have flutings near the summit; and all the vertebræ carry ribs. Both the median and lateral thoracic plates are stalked. The form of the parasphenoid is of the same general type as in the *Protritonidæ*, but the stalk is relatively shorter.

[1] Fauna der Gaskohle, vol. i. pt. ii. p. 124 (1880). [2] *Loc. cit.*

Genus **MELANERPETUM**, Fritsch[1].

The characters in which this genus differs from the typical but imperfectly known *Apateon*, Meyer[2], of the German Carboniferous, are not indicated by its founder. The median thoracic plate has a long stem.

Melanerpetum pulcherrimum, Fritsch[3].

Of larger size than the typical *M. pusillum*, with a shorter and wider skull. It is suggested[4] that the one may be the adult of the other.

Hab. Europe (Bohemia).

R. 350. Electrotype from a fragment of shale, showing the entire skeleton. The original is the type, and was obtained from the Rothliegendes (Lower Permian) of Ruppersdorf, near Braunau; and is figured by Fritsch in his 'Fauna der Gaskohle,' vol. i. pl. xiv. fig. 1. *Purchased*, 1882.

Fig. 49.

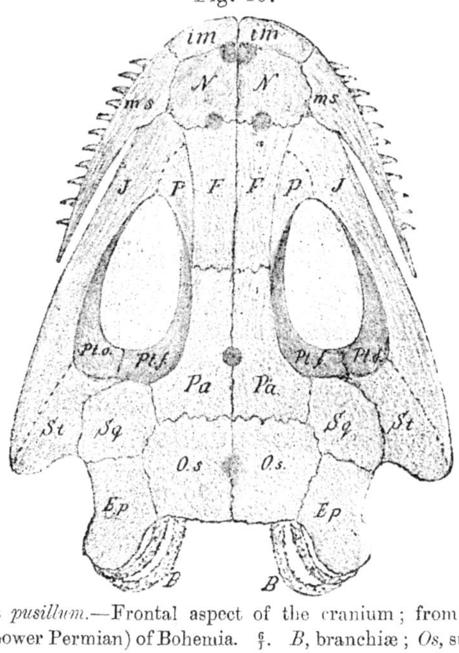

Melanerpetum pusillum.—Frontal aspect of the cranium; from the Rothliegendes (Lower Permian) of Bohemia. $\frac{6}{1}$. *B*, branchiæ; *Os*, supraoccipital; other letters as in fig. 29, p. 140. (*After Fritsch.*)

---

[1] Fauna der Gaskohle, vol. i. pt. i p. 27 (1879).—*Melanerpeton.*
[2] Neues Jahrb. 1844, p. 49.
[3] *Loc. cit.*
[4] Fritsch, *loc. cit.*

Family PROTRITONIDÆ[1].

Skull (fig. 29) broad and parabolic, with the small palatines[2] lying in the middle line immediately behind the vomers, and separated from the maxillæ; parasphenoid with a long stalk and a shield-like posterior extremity; teeth smooth, with a large pulp-cavity. Ventral scutes small, pointed, sculptured, and arranged in rows.

Genus **PROTRITON,** Gaudry[3].

Syn. *Salamandrella*, Gaudry[4].
Branchiosaurus, Fritsch[5].
(?) *Pleuronoura*, Gaudry[6].

The type genus. Skull wide, with very large orbits, and some 30 bony plates in the sclerotic; cranial bones very slightly pitted; no teeth on the parasphenoid, pterygoids, or palatines, but according to Fritsch a few minute ones on the vomers. Median thoracic plate rhomboidal, without a stem, rounded anteriorly, and the middle of the outer surface granulated. Four digits in manus[7].

Internal gills were developed in the young, in which there was no ventral armour.

The identity of *Branchiosaurus* with *Protriton* is pointed out by Geinitz and Deichmüller[8]; Fritsch[9] also showed that *Protriton* was founded on a very early larva, and *Pleuronoura* on a somewhat later larva, apparently inseparable from the form to which he applies the name *Branchiosaurus*. The same view is also adopted by Credner[10], who retains, however, the name *Branchiosaurus*[11].

Protriton petrolei, Gaudry[12].

Syn. *Salamandrella petrolei*, Gaudry[13].
? *Pleuronoura pellati*, Gaudry[14].

---

[1] =*Branchiosauridæ*, Fritsch.

[2] See the figure in Zittel's 'Handbuch der Palæontologie,' 1 Abth. iii. Band, 2 Lief. p. 370.

[3] Bull. Soc. Géol. France, sér. 3, vol. iii. p. 300 (1875).

[4] Comptes Rendus, vol. lxxx. p. 442 (1875).—Preoccupied.

[5] Sitz k. bohm. Ges. Wiss. for 1875, p. 72 (1876).

[6] Bull. Soc. Géol. France, sér. 3, vol. vii. p. 72 (1879).

[7] *Teste* Credner. In the restoration given by Fritsch, the manus is represented with five digits.

[8] See Zeitschr. deutsch. geol. Ges. vol. xxxvi. p. 685 (1884).

[9] Fauna der Gaskohle, vol. i. p. 94.

[10] Zeitschr. deutsch. geol. Ges. vol. xxxviii. pp. 577–578 (1886).

[11] See Geol. Mag. decad. iii. vol. iv. p. 276 (1887).

[12] Comptes Rendus, vol. lxxx. p. 442 (1875).—*Salamandrella.*

[13] *Loc. cit.*

[14] Bull. Soc. Géol. France, sér. 3, vol. iii. p 300 (1875).

Branchiosaurus gracilis, Credner [1].
Branchiosaurus amblystomus, Credner [2].

The type species. Length of adult about 0,202 (8 inches).

Fig. 50.

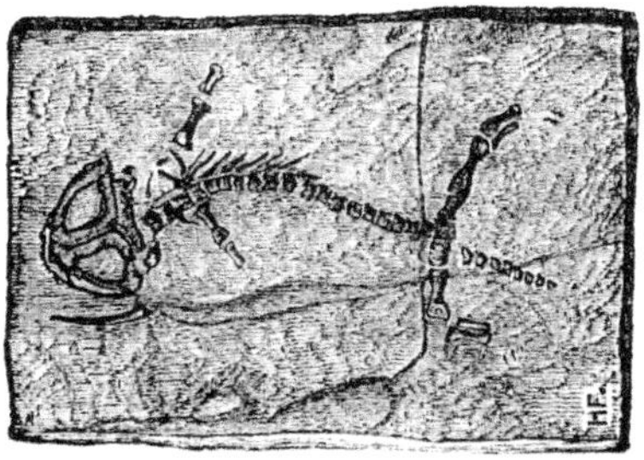

Protriton petrolei.—Skeleton of larva; from the Rothliegendes (Lower Permian) of Autun. ¼. (*After Gaudry.*)

The specific identity of the larval *B. gracilis* with the adult *B. amblystomus*, of the Rothliegendes of Saxony, is admitted by their describer [3]; while the impossibility of separating the former from the larval *P. petrolei* of the corresponding beds of France was pointed out by Geinitz and Deichmüller [4].

Hab. Europe (France, Saxony, and Thuringia).

R. 489. Fragment of shale showing the skeleton of a larva; from the Rothliegendes (Lower Permian) of Millery, near Autun (Saône-et-Loire), France. This specimen agrees in all respects with the one represented in fig. 50.
Presented by Sir R. Owen, K.C.B., 1884.

R. 1300. Slab of shale with several larval skeletons; from Millery.
Purchased, 1888.

Protriton salamandroides (Fritsch [5]).

Syn. *Branchiosaurus salamandroides* (Fritsch [6]).

The type of *Branchiosaurus*. Distinguished from the preceding species by the stouter vertebræ of the larva, in which the proportion of the width to the length is 1 to 8 instead of 1 to 13. The transverse processes of the vertebræ are also stouter and wider.

Hab. Europe (Bohemia).

[1] Zeitschr. deutsch. geol. Ges. vol. xxxiii. p. 306 (1881). [2] *Ibid.* p. 575.
[3] *Ibid.* vol. xxxviii. pp. 577, 578 (1886).
[4] *Vide ibid.* vol. xxxvi. pp. 685–686 (1884).
[5] Sitz. k. bohm. Ges. Wiss. for 1875, p. 72 (1876).—*Branchiosaurus.*
[6] *Loc. cit.*

R. 348. Electrotype from a fragment of shale, showing the skeleton of a larva. The original was obtained from the Gaskohle (Lowest Permian) of Nyřan (Pilsen); and is figured by Fritsch in his 'Fauna der Gaskohle,' vol. i. pl. i. fig. 1. Except for its better state of preservation, this specimen can scarcely be distinguished from the larvæ of *P. petrolei*. *Purchased*, 1882.

R. 1604. Fragment of shale with the skeleton of a larger individual; from the Gaskohle of Tremosna (Pilsen). *Purchased*, 1889.

Family Position Uncertain.

The following genera are included by Fritsch in the preceding family, but only the skull is known. The parasphenoid has the expanded base characteristic of *Branchiosaurus*.

Genus **SPARODUS**, Fritsch[1].

Syn. *Batrachocephalus*, Fritsch[2].

Skull with strong marginal teeth, and a number of conical teeth on the anterior bones of the palate. According to Fritsch's restoration the palatines are interposed between the vomers and maxillæ.

Sparodus validus, Fritsch[3].

Syn. *Batrachocephalus validus*, Fritsch[4].

The type species. Length of skull about 0,033; 17 mandibular teeth, of which the first three are double the size of the others; 27 teeth in each vomer, and 11 in each maxilla; outer surface of mandible smooth.

Hab. Europe (Bohemia).

R. 345. Electrotype from a fragment of shale, showing the palatal aspect of the skull. The original, which is one of the types, was obtained from the Gaskohle (Lowest Permian) of Nyřan (Pilsen); and is figured by Fritsch in his 'Fauna der Gaskohle', vol. i. pl. viii. fig. 1. *Purchased*, 1882.

R. 344. Electrotypes from two fragments of shale, exhibiting the mandible and part of the palate. Originals from Nyřan; figured by Fritsch, *op. cit.* pl. viii. figs. 4, 5. *Purchased*, 1882.

[1] Fauna der Gaskohle, vol. i. pt. i. p. 84 (1879).
[2] Sitz. k. böhm. Ges. Wiss. for 1875, p. 76 (1876).—Preoccupied by Bleeker, 1846. [3] *Loc. cit.*—*Batrachocephalus*. [4] *Loc. cit.*

Sparodus crassidens, Fritsch [1].

Syn. *Batrachocephalus crassidens*, Fritsch [2].

Twenty-five maxillary teeth, which are smaller than in the type species, and of which the anterior ones are not enlarged; vomerine teeth apparently less numerous; outer surface of mandible pitted.

Hab. Europe (Bohemia).

R. 347. Electrotype from a fragment of shale, showing the palatal aspect of the skull and the middle thoracic plate. The original, which is one of the types, was obtained from the Rothliegendes (Lower Permian) of Kounova; and is figured by Fritsch in his 'Fauna der Gaskohle,' vol. i. pl. x. fig. 1. *Purchased*, 1882.

R. 346. Electrotypes from two fragments of shale, showing parts of the upper and lower jaws. Originals from Kounova; figured by Fritsch, *op. cit.* pl. ix. figs. 4, 9. *Purchased*, 1882.

Genus **DAWSONIA,** Fritsch [3].

The vomers much smaller than in *Sparodus*, and having only a few teeth at their antero-external angles; small denticules on the stem of the parasphenoid and the anterior portions of the pterygoids, and long teeth on the outer border of the latter; cranial bones strongly sculptured.

Dawsonia multidens, Lydekker (*n. nom.*).

Syn. *Dawsonia polydens*, Fritsch [4].

The type and only described species. Length of skull about 0,038.

Hab. Europe (Bohemia).

The originals of the following specimens, which include the types, were obtained from the Rothliegendes (Lower Permian) of Kounova. The specimens were purchased in 1882.

R. 340. Electrotypes from two fragments of shale, showing the parasphenoid and fragments of the skull. Originals figured by Fritsch in his 'Fauna der Gaskohle,' vol. i. pl. xi. fig. 6, and pl. xii. fig. 7.

[1] Sitz. k. bohm. Ges. Wiss. for 1875, p. 76 (1876).—*Batrachocephalus.*

[2] *Loc. cit.*

[3] Fauna der Gaskohle, vol. i. pt. i. p. 89 (1879). [4] *Ibid.* p. 90.—Hybrid.

R. 342. Electrotypes from two fragments of shale, exhibiting portions of the skull. Originals figured, *op. cit.* pl. xi. figs. 3, 5.

R. 341. Electrotype from a fragment of shale, showing the middle thoracic plate. Original figured, *op. cit.* pl. xi. fig. 13.

SUBORDINAL POSITION UNCERTAIN.

It is impossible to assign the following forms to any definite serial position, even if the whole of them are rightly referred to the Labyrinthodontia.

Anthracerpetum crassosteum, Owen [1].

Known by fragments of teeth, and cranial and other bones.
Hab. Europe (Wales).

R. 674. (*Fig.*) Four fragments of rock with imperfect bones; from the Coal-Measures of Llantrissent, Glamorganshire. The types; figured by Owen in the 'Geol. Mag.' decad. i. vol. ii. pl. i. figs. 1–4. *Presented by J. E. Lee, Esq.*, 1885.

Lepidotosaurus duffi, Hancock & Howse [2].

Known by the undermentioned specimen, which indicates a reptile of the approximate size of *Sphenospondylus*; it is stated in the 'Rep. Brit. Assoc.' for 1874, p. 187, that the evidence for referring this form to the Labyrinthodontia is insufficient.
Hab. Europe (England).

R. 723. One half of a split slab, showing part of the thoracic region in a badly preserved condition; from the Marl-slate (Middle Permian) of Midridge, Durham. This specimen is the counterpart of the type slab figured in the 'Quart. Journ. Geol. Soc.' vol. xxvi. pl. xxxviii.
Presented by W. C. Stobart, Esq., 1886.

GENUS *non det.*

Known only by the undermentioned specimen, which indicates a form about one-third the size of *Eryops megacephalus*.
Hab. Tasmania.

R. 500. Cast of a femur; from strata of unknown age in Tasmania. The contour of the distal half approximates to that of the femur of *Eryops* figured by Cope in the 'Proc. Amer. Phil. Soc.' vol. xix. pl. iv. The length of the specimen is 0,065. *Presented by Sir R. Owen, K.C.B.*, 1884.

[1] Geol. Mag. decad. i. vol. ii. p. 6 (1865).—*Anthrakerpeton.*
[2] Quart. Journ. Geol. Soc. vol. xxvi. p. 557 (1870).

ICHNITES.

A. Specimens which are probably in most cases of Labyrinthodont origin.

CHIROSAURUS, Kaup[1].

Syn. *Chirotherium*, Kaup[2].

The impressions of both feet are pentadactylate, with distinct nails; those of the manus being rather more than half the size of those of the pes. The pollex is short, and projects nearly at right angles to the axis of the third digit; the third digit of the pes is the largest.

The Labyrinthodont nature of these footprints was suggested by Owen[3]; Miall[4] considered that some of them might by Dinosaurian; Winkler[5] regards them as formed by *Trematosaurus brauni*, but the larger size of *Capitosaurus nasutus* is in favour of those of the type species of *Chirosaurus* having been made by the latter animal. The name *Chirosaurus* is earlier than either *Capitosaurus* or *Trematosaurus*.

The so-called *Chirotherium minus* appears to indicate another genus (*vide infrà*).

Chirosaurus barthi, Kaup[6].

Syn. *Chirotherium barthi*, Kaup[7].
Chirotherium majus, Sickler[8].

The type species, occurring typically in the Bunter of Hessberg. The footprints of the pes usually have a length of about eight inches in large individuals.

The name *Chirotherium geinitzi*, Hornstein[9], has been applied to an animal indicated by footprints of the same general form, but differing decidedly in contour; the type specimen being from the Bunter of Karlshafen.

Hab. Europe (Germany)

[1] Neues Jahrb. 1835, p. 328. As it was considered that these footprints were probably Mammalian, the name *Chirotherium* was proposed, but in the event of their proving to be Saurian the alternative name *Chirosaurus* was also given. [2] *Loc. cit.*

[3] See Palæontology, 2nd ed. p. 190 (1861).

[4] Rep. Brit. Assoc. for 1873, p. 244.

[5] Archiv. Mus. Teyler, sér. 2, vol. ii. p. 187 (1886).

[6] Neues Jahrb. 1835, p 328. [7] *Loc cit.*

[8] Die vorzuglichsten Farhten-Abdrucke, etc., von Hildburghausen (1836).—*Not seen.*

[9] See Archiv. Mus. Teyler, sér. 2, vol. ii. p. 174.

R. 728. (*Fig.*) Slab showing casts of footprints; from the Bunter Sandstone (Lower Trias) of Hessberg, near Hildburghausen, on the flanks of the Thuringerwald. Noticed and one of the casts of the hind feet figured by Buckland in his 'Geology and Mineralogy' (Bridgewater Treatise), 1st ed. p. 265, pl. 26″. *Purchased*, 1835.

R. 728 a. Plaster cast of the same slab, showing the impressions. *Made in the Museum.*

Chirosaurus stortonensis (Morton[1]).

Syn. *Chirotherium stortonense*, Morton[2].

The impressions of the pes of this form differ from those of *C. barthi* in being stouter and shorter, with the digits less widely separated and the first digit placed nearer to the heel. Their usual length is about 8·7 inches; and they approximate in form to the much larger impressions from the Keuper of Torperley (Cheshire) described by Egerton[3] as *Chirosaurus* (*Chirotherium*) *herculis*. The latter impressions measure upwards of 15 inches in length, and were probably made by an animal of the size of *Mastodonsaurus giganteus*.

Hab. Europe (England).

The following specimens were obtained from the Lower Keuper (Upper Trias) of Cheshire.

44826. Slab showing impressions of the feet; from Storeton (Stourton) Hill, near Birkenhead. *Presented by B. Bright, Esq.*, 1873.

R. 729. Slab with casts of impressions of the feet; from Storeton. *Presented by J. Tomkinson, Esq.*

R. 730. Two slabs with casts of footprints; from Storeton. *No history.*

R. 398. Ten slabs of sandstone showing footprints, or casts of the same; from Storeton. *Presented by C. Westendarp, Esq.*, 1884.

R. 414. Impression of a hind foot; from Storeton. *Purchased*, 1884.

21831. Slab showing casts of footprints; from Lymm, near Warrington. *Purchased*, 1848.

[1] Proc. Liverpool Geol. Soc. 1863, p. 17.—*Chirotherium.* [2] *Loc. cit.*
[3] Lond. and Edinb. Phil. Mag. 1838, p. 492.

R. 731. Two slabs with the impression of a foot and a cast of the same; from Lymm. *No history.*

R. 295. Six slabs of sandstone showing the impressions of feet; from Lymm. *Egerton Collection. Purchased*, 1882.

39162. Slab with cast of an impression of one of the hind feet; from Cheshire. *Purchased*, 1865.

37721. Natural casts of impressions of feet; from Cheshire. *Saul Collection. Purchased*, 1863.

19621. Slab showing numerous impressions of the feet in a single track; from Lymm. *Presented by Lord F. Egerton*, 1845.

R. 732. A small slab showing the casts of one medium-sized and one smaller footprint; from Lymm. *No history.*

R. 733. A slab with casts of the impressions of the feet of a small individual; probably from Cheshire. *No history.*

GENUS *non det.*

(*Chirotherium minus*, Sickler[1].)

In this form the pollex and hallux appear to have been very minute, and were probably situated in the middle of the carpus or tarsus, so that the footprints appear to be tetradactylate; the other digits are longer and narrower than in *Chirosaurus barthi*, with much longer nails. The space between the individual footprints is relatively greater than in the latter. The prints of the pes have a length of 3·5 and a width of 2·5 inches.

Hab. Europe (Germany).

R. 419. Slab of rock showing impressions of footprints; from the Bunter (Lower Trias) of Vogelsberg, near Johannisburg, East Prussia. Similar impressions are figured by Winkler in the 'Archiv. Mus. Teyler,' sér. 2, vol. ii. pl. vi. *Purchased*, 1884.

GENUS *non det.*

The following pentadactyle footprints differ considerably from those of *Chirosaurus*; they have no lateral appendage, and the digits are wide apart and slender, with well-marked nails. The prints have a length of 2·3, and a width of 1·7 inches.

Hab. Europe (Germany).

[1] Die vorzuglichsten Farhten-Abdrucke, etc., von Hildburghausen (1836).—*Not seen.*

R. 734. Plaster-cast of a slab with footprints. The original was obtained from the Bunter Sandstone (Lower Trias) of Hessberg, near Hildburghausen, on the flanks of the Thuringerwald; it is described and figured by Sickler in his 'Die vorzüglichsten Farhten-Abdrücke, etc., von Hildburghausen,' p. 8, pl. vii.; the cast being noticed and figured by Buckland in his 'Geology and Mineralogy' (Bridgewater Treatise), 1st ed. p. 265, pl. 26'''. Similar footprints are figured by Winkler in the 'Archiv. Mus. Teyler,' sér. 2, vol. ii. pl. viii. *No history.*

CHELICHNUS, Jardine[1].

Chelichnus duncani (Owen[2]).

Syn. *Testudo duncani*, Owen[3].

Founded upon pentadactylate footprints, showing distinct nails, and having a diameter of from 1 to 2 inches, which were regarded by their describer as of Chelonian origin.

Hab. Europe (Scotland).

43576. Two slabs with footprints; from the Bunter Sandstone (Lower Trias) of Corncockle Muir, near Applegarth, Dumfriesshire. *Purchased*, 1872.

Genus *non det.*

(*Chirotherium reiteri*, Moore[4].)

The small footprints of the undermentioned type show no signs of affinity with the genus to which they have been referred.

Hab. North America.

40328. A small slab showing impressions of the foot of a comparatively small animal, apparently corresponding with those described under the above name; from the Coal-Measures of Phenixville, Pennsylvania. *Purchased*, 1867.

Genus RHYNCHOSAURUS (*suprà*, i. p. 296).

It has been suggested that the small tracks of the undermentioned type may be those of *Rhynchosaurus*.

[1] Ann. Mag. Nat. Hist. ser. 2, vol. vi. p. 209 (1850).
[2] Rep. Brit. Assoc. for 1841, p. 160 (1842).—*Testudo.* [3] *Loc. cit.*
[4] Amer. Journ. ser. 3, vol. v. p. 292 (1873).

21834. A small slab containing numerous footprints; from the Keuper sandstone (Upper Trias) of Staunton, two and a half miles from Burton-on-Trent, Staffordshire.
Purchased, 1848.

33156. A small slab with similar tracks; from Staunton.
Purchased, 1858.

38805. Small slab with similar impressions; from the Keuper of Coven, near Brewood, South Staffordshire.
Presented by Rev. H. Houseman, 1862.

38803. Fragment of rock with similar tracks; from Coven.
Presented by Rev. H. Houseman, 1862.

40154. A very large slab, with numerous impressions of the feet of the same form as the preceding; from Coven.
Presented by Rev. H. Houseman, 1862.

SAURICHNITES, Geinitz [1].

Founded upon the evidence of small footprints with short digits which are regarded as those of Labyrinthodonts. While some of these footprints are pentadactylate (*S. lacertoides*), others (*S. salamandroides*) have but four digits in the manus.

Saurichnites rittlerianus, Hochstetter [2].

Saurichnites perlatus, Fritsch (MS.).

Hab. Europe (Bohemia).

R. 1474. Cast of a slab showing two perfect and three imperfect footprints of these two forms. Original from the Lower Permian of Semil, Bohemia. *Purchased*, 1888.

MACROPTERNA, Hitchcock [3].

Three or four digits in the manus, and four in the pes; the manus much smaller than the pes.

Macropterna divaricans, Hitchcock [4].

Of small size. Length of middle toe 0·7 inch.
Hab. North America.

[1] Neues Jahrb. 1861, p. 65. [2] Verh. geol Reichsanstalt, 1868, p. 432.
[3] Mem. Amer. Acad. ser. 2, vol. iii. p. 233 (1848). [4] *Loc. cit.*

28105. Three slabs showing footprints; from the reputed Trias of Turner's Falls, Connecticut river, Massachusetts.
Presented by Prof. C. U. Sheppard, 1852.

B. *Specimens which are probably for the most part of Reptilian origin, although some may possibly be Avian.*

BRONTOZOUM, Hitchcock[1].

Founded on large tridactylate footprints, in which all the digits are directed forwards, the inner digit being the shortest, and the terminal claw long and tubular. There are no impressions of the manus.

Brontozoum giganteum, Hitchcock[2].

Syn. *Ornithichnites giganteus*, Hitchcock[3].
Ornithoidichnites giganteus, Hitchcock[4].

The type species; of very large size, the length of the middle digit being 12·5 inches, and that of the whole foot 14 to 18 inches.

Hab. North America.

20042. Two slabs with impressions of feet; from the reputed Trias of Greenfield, Massachusetts. *Purchased*, 1846.

Brontozoum sillimani, Hitchcock[5].

Syn. *Ornithoidichnites sillimani*, Hitchcock[6].
Brontozoum sillimanium, Hitchcock[7].

Founded on smaller tracks than those of the type, in which the middle toe has a length of 6 inches, the total length of the foot being 8 inches.

Hab. North America.

36593. Slab showing the impression of a foot; from the reputed Trias of the Connecticut river, Massachusetts.
Presented by Prof. Hitchcock, 1862.

Brontozoum validum, Hitchcock[8].

Hab. North America.

[1] Mem. Amer. Acad. ser. 2, vol. iii. p. 169 (1848).
[2] Amer. Journ. ser. 1, vol. xxix. p. 316 (1836).—*Ornithichnites.*
[3] *Loc. cit.*
[4] Final Rep. on Geology of Massachusetts, vol. ii. p. 484 (1841).
[5] Amer. Journ. ser. 1, vol. xlvii. p. 317 (1844).—*Ornithoidichnites.*
[6] *Loc. cit.* [7] Mem. Amer. Acad. ser. 2, vol. iii. p. 171 (1848).
[8] ? Ichnology of New England (1858).—*Not seen.*

20045 a. Slab showing tracks of this and other forms; from the reputed Trias of Greenfield, Massachusetts.
Purchased, 1846.

ANISOPUS, Hitchcock[1].

Manus and pes with four digits each; the pes being nearly twice as large as the manus.

Anisopus gracilis, Hitchcock[2].

The type species. Length of the foot 0·9 inch.
Hab. North America.

36594. Slab of sandstone with footprints; from the reputed Trias of the Connecticut river, Massachusetts.
Presented by Prof. Hitchcock, 1862.

Tracks of this form are also shown on the slabs 18322 and 20045 (*infrà*).

ANOMŒPUS, Hitchcock[3].

Manus apparently with five digits, pes with three, all of which are directed forwards. Heel long.

Anomœpus intermedius, Hitchcock[4].

Hab. North America.

28106. Slab with a footprint; from the reputed Trias of Turner's Falls, Connecticut river, Massachusetts.
Presented by Prof. C. H. Sheppard, 1852.

Anomœpus curvatus, Hitchcock[5].

Anomœpus gracillimus, Hitchcock[6].

Syn. *Brontozoum gracillimum*, Hitchcock[7].

Hab. North America.

18322. Two large slabs of shale with impressions of the feet of these and the preceding forms, together with those of *Anisopus gracilis*, *Grallator cuneatus*, and *Tridentipes elegantior*; from Turner's Falls. *Purchased*, 1844.

[1] Mem. Amer. Acad. ser. 2, vol. iii. p. 226 (1848).—Preoccupied.
[2] *Loc. cit.* [3] *Loc. cit.*
[4] Supplement to Ichnology of New England, p. 2 (1865).
[5] *Loc. cit.* [6] *Loc. cit.*
[7] Mem. Amer. Acad. ser. 2, vol. iii. p. 175 (1848).

20045. A small slab with several tracks of *A. intermedius* and *A. gracillimus*, and also of *Anisopus gracilis*; from the reputed Trias of Greenfield, Massachusetts.

Purchased, 1846.

Tracks of *A. intermedius* and *A. gracillimus* are also shown on the slab 20045 a (*suprà*, p. 22).

GRALLATOR, Hitchcock [1].

Small forms founded upon impressions of the tridactylate hind feet.

Grallator cuneatus, Hitchcock [2].

Hab. North America.

Tracks of this form are shown on the slabs 20045 a and 18322 (p. 22).

TRIDENTIPES, Hitchcock [3].

Another genus founded upon tridactylate prints of the hind feet.

Tridentipes elegantior, Hitchcock [4].

Hab. North America.

Tracks of this form are shown on the slab 18322 (p. 221).

[1] Ichnology of New England, p. 72 (1858).—*Not seen.*

[2] *Loc. cit.*

[3] *Ibid.* p. 88.—*Not seen.*

[4] *Loc. cit.*

ADDENDA TO LABYRINTHODONTIA.

(*The following specimens were overlooked until the preceding portion of the text was in type; the whole of them were determined by Sir J. W. Dawson.*)

Family DENDRERPETIDÆ (p. 170).

Genus **DENDRERPETUM** (p. 170).

Dendrerpetum acadianum, Lyell & Dawson [1].

The type species.
Hab. Nova Scotia.

434–438. Fragments of rock showing numerous imperfect remains; from the Coal-Measures of Nova Scotia.
By exchange, 1884.

Dendrerpetum oweni, Dawson [2].

Readily distinguished from *D. acadianum* by its smaller size, more delicate cranial structure, longer and more curved teeth, and narrower and more pointed ventral scutes.
Hab. Nova Scotia.

R. 439. Fragment of rock showing part of a jaw, cranial bones, and ventral scutes; from the Coal-Measures of Nova Scotia.
By exchange, 1884.

Family HYLONOMIDÆ (p. 201).

Genus **HYLONOMUS** (p. 201).

Hylonomus lyelli, Dawson [3].

The type species. Length from 0,127 to 0,152 (5 to 6 inches). Teeth numerous.
Hab. Nova Scotia.

443–445. Fragments of rock showing numerous fragmentary remains; from the Coal-Measures of South Joggins, Nova Scotia. No. R. 445 shows part of a maxilla and palatine.
By exchange, 1884

[1] Quart. Journ. Geol. Soc. vol. ix. p. 58 (1853).
[2] Air-Breathers of the Coal-Period, p. 32 (1863).
[3] Quart. Journ. Soc. vol. xvi. p. 274 (1859).

Hylonomus wymani, Dawson[1].

Smaller and more slender than the type species, with the teeth less numerous (about 22 in mandible), shorter and more obtuse.

Hab. Nova Scotia.

R. 446. Fragments of rock with imperfect remains; from the Coal-Measures of South Joggins, Nova Scotia. *By exchange*, 1884.

Hylonomus latidens, Dawson[2].

Said to be of stouter build than the type species; with relatively shorter limbs. Generic position doubtful.

Hab. Nova Scotia.

R. 447. Fragment of rock with part of a jaw and other imperfect remains; from the Coal-Measures of South Joggins, Nova Scotia. *By exchange*, 1884.

Genus **SMILERPETUM**, Dawson[3].

Very imperfectly known. Said to be distinguished from *Hylonomus* by the wedge-shaped teeth, the presence of numerous teeth on the vomers and palatines, and the form of the presumed caudal vertebæ.

Smilerpetum aciedentatum, Dawson[4].

Syn. *Hylonomus aciedentatus*, Dawson[5].

The type and only described species.

Hab. Nova Scotia.

R. 433. Three fragments of rocks showing part of a jaw and some other imperfect remains; from the Coal-Measures of Nova Scotia. *By exchange*, 1884.

Genus **HYLERPETUM**, Owen[6].

Imperfectly known. Cranial bones finely striated. Marginal teeth conical, smooth at the base, but finely grooved near the summit; numerous small palatal teeth, with some larger ones. Vertebræ with short and cylindrical centra, and well-developed spines and zygapophyses.

[1] Quart. Journ. Geol. Soc. vol. xvi. p. 276 (1859).

[2] Phil. Trans. for 1882, p. 637 (1883).

[3] *Ibid.* p. 638.—*Smilerpeton.*

[4] Quart. Journ. Geol. Soc. vol. xvi. p. 275 (1859).—*Hylonomus.*

[5] *Loc. cit.*

[6] Quart. Journ. Geol. Soc. vol. xviii. p. 281 (1862).—*Hylerpeton.*

Hylerpetum dawsoni, Owen [1].

The type species. Length of mandible 0,044.

Hab. Nova Scotia.

R. 441–442. Four fragments of rock, showing a tooth and numerous dermal scutes; from the Coal-Measures of Nova Scotia. The scutes resemble the specimens figured by Dawson in the 'Phil. Trans.' for 1882, pl. xli. *By exchange*, 1884.

Hylerpetum longidentatum, Dawson [2].

Said to be distinguished from the type species by the longer and more slender teeth, and the greater breadth of the median thoracic plate.

Hab. Nova Scotia.

R. 440. Four fragments of rock, showing part of a jaw, an imperfect median thoracic plate, and numerous dermal scutes; from the Coal-Measures of South Joggins, Nova Scotia. *By exchange*, 1884.

Family Uncertain.

Genus **FRITSCHIA,** Dawson [3].

Comprises a small lizard-like form, characterized by its slender rod-like dermal scutes, which resemble those of *Ophiderpetum.* The genus may perhaps be allied to *Urocordylus.*

Fritschia curtidentata, Dawson [4].

The type species.

Hab. Nova Scotia.

R. 449. Fragment of rock, showing part of a jaw and other imperfect remains; from the Coal-Measures of South Joggins, Nova Scotia. *By exchange*, 1884.

[1] Quart. Journ. Geol. Soc. vol. xviii. p. 241 (1868)
[2] Phil. Trans. for 1882, p. 640 (1883).
[3] *Ibid.* p. 641. [4] *Loc. cit.*

SUPPLEMENT.

Order ORNITHOSAURIA (*suprà*, i. p. 2).

Family RHAMPHORHYNCHIDÆ (*suprà*, i. p. 27).

Genus **SCAPHOGNATHUS** (*suprà*, i. p. 27).

Scaphognathus purdoni, Newton[1].

Of large size, the length of the skull being approximately 0,190 (7·5 inches). Teeth apparently more numerous than in type species; preorbital vacuity more ovoid, and the whole cranium less depressed, with a distinct median channel between the prefrontals and frontals.

Hab. Europe (England).

R. 1613. Cast of the cranium, showing the brain *in situ*, but wanting the greater portion of the rostrum. The original was obtained from the Upper Lias of Whitby, Yorkshire, and is figured by Newton in the 'Phil. Trans.' for 1888, pls. lxxvii., lxxviii. Some remarks on the relations of the bones are made by Baur in the 'Geol. Mag.' decad. iii. vol. vi. p. 171. *Made in the Museum*, 1888.

Genus **RHAMPHORHYNCHUS** (*suprà*, i. p. 29).

Rhamphorhynchus gemmingi (*suprà*, i. p. 31).

R. 1692. Cast of a slab of lithographic limestone, showing the bones and the impression of the patagium of the left pectoral limb. The original was obtained from the Lower Kimeridgian of Winterberg, near Eichstadt, Bavaria; and is preserved in the Museum at Munich. It is described and figured by Zittel in the 'Palæontographica,' vol. xxix. p. 51, pl. x. *By exchange*, 1889.

Rhamphorhynchus jessoni, Lydekker (n. sp.).

A provisional species, apparently allied to *R. gemmingi*. The cervical vertebræ show a distinct facet on the centrum for the

[1] Proc. Roy. Soc. vol. xliii. p. 436 (1888).

articulation of a cervical rib. The pelvis and sacrum exhibit the characteristic features of the genus; the ischium being anchylosed to the ilium, and the latter attached to four vertebræ of the sacrum by anchylosed ribs.

Hab. Europe (England).

R. 1755. Portions of the skeleton; from the Oxford Clay, St. Ives, Huntingdonshire. The types. These comprise several vertebræ, from the cervical, dorsal, and sacral regions; the imperfect ilia and ischia; one femur, and the distal half of another; and some other fragmentary bones.

Presented by T. Jesson, Esq., 1890.

Order CROCODILIA (*suprà*, i. p. 42).

The *Metriorhynchinæ* (*infrà*) show that the presence of dermal scutes and the absence of sclerotic plates are not invariable characteristics of the order.

Family CROCODILIDÆ (*suprà*, i. p. 44).

Generically Undetermined Remains.

From the Pliocene of the Siwalik Hills, India (suprà, i. p. 57). *These specimens belong to the Cautley Collection.* *Presented*, 1840.

R. 954. A slightly imperfect right coracoid, remarkable for its extreme curvature.

R. 952. The imperfect proximal extremity of the right humerus of a very large individual.

R. 952 a. The proximal half of a smaller left humerus.

R. 947. The distal half of a right humerus. The radial condyle is very strongly developed.

R. 947 a. The distal half of a similar right humerus.

R. 948. The distal extremity of a right humerus. The radial condyle is less prominent than in the preceding specimens.

R. 949. The distal extremity of a rather smaller right humerus.

R. 945. The distal portion of the left humerus of a very large individual, in a somewhat waterworn condition.

R. 946. The distal portion of a rather smaller left humerus, much waterworn.

R. 950. The distal portion of a still smaller left humorus.

R. 1589. The distal extremity of a very small left humerus.

R. 1588. The proximal extremity of a right femur.

R. 1589 a. The distal portion of a left femur.

R. 1589 b. The distal portion of a left femur agreeing in size with the preceding, but wanting the outer condyle.

R. 1590. The distal portion of a left tibia.

R. 1590 a. The distal portion of a smaller left tibia.

R. 578. A first sacral vertebra, wanting the ribs and all the zygapophyses excepting the left prezygapophysis.

Genus **CROCODILUS** (*suprà*, i. p. 52).

Crocodilus spenceri (*suprà*, i. p. 60).

R. 1753. (*Fig.*) The nearly entire skull of an adult individual; from the London Clay of Sheppey. The type of *C. toliapicus*; figured by Owen in his 'Reptilia of the London Clay' (Mon. Pal. Soc.), vol. i. pl. ii. *A.* *No history.*

Genus **TOMISTOMA** (*suprà*, i. p. 62).

Tomistoma (?) macrorhynchus (*suprà*, i. p. 64).

Koken[1], after regarding this species as probably referable to *Tomistoma*, concludes that it really belongs to *Thoracosaurus*; although differing from the type species of the latter in the absence of the preorbital vacuity, and in the junction of the nasals with the premaxillæ. The species is recorded by this writer from the Maastricht Cretaceous.

Genus **GARIALIS** (*suprà*, i. p. 65).

Garialis (?) dixoni (*suprà*, i. p. 69).

33238 a. The imperfect left humerus of an immature Crocodilian which may belong to this species; from the Middle Eocene of Bracklesham, Sussex.
Hastings Collection. Purchased, 1855.

[1] Zeitschr. deutsch. geol. Ges. vol. xl. p. 754 (1889).

Family GONIOPHOLIDIDÆ (*suprà*, i. p. 76).

Genus **GONIOPHOLIS** (*suprà*, i. p. 79).

Goniopholis (*cf.* **crassidens,** *suprà*, i. p. 79).

The cervical vertebræ of this genus are readily distinguished from those of *Metriorhynchus* by their longer centra, in which the hæmal carina is either slight or absent.

26007. A cervical vertebra wanting the greater part of the neural arch; from the Wealden of Sussex.
Mantell Collection. Purchased, 1853.

R. 140. The centra of two cervical vertebræ; from the Wealden of the Isle of Wight. These and the preceding specimens precisely resemble the cervical of No. R. 605 (*suprà*, i. p. 81), although they are of smaller size.
Fox Collection. Purchased, 1882.

R. 1614. A nearly entire lumbar vertebra of a very large individual; from the Wadhurst Clay near Hastings. On the right side the neural arch is entire with the exception of the greater part of the transverse process.
Dawson Collection. Purchased, 1888.

R. 1614 a. A bone which, judging from its mineralogical condition, appears to have been associated with the preceding, and which may be an "intermediate" rib; from the same locality. *Dawson Collection.*

R. 1729. An imperfect scute probably belonging to this genus, and perhaps associated with the preceding; from the same locality. From the great height of the ridge and the irregular contour and small size of the horizontal plate, it is probable that this specimen is one of the irregular scutes of the nuchal region.
Dawson Collection. Purchased, 1889.

36529. The imperfect left coracoid of a smaller Crocodilian probably belonging to this genus; from the Wealden of Cuckfield, Sussex. *Mantell Collection. Purchased*, 1853.

Goniopholis minor, Koken[1].

Founded upon a cervical vertebra from the Wealden of Germany, characterized by its small dimensions. It has yet to be proved

[1] Pal. Abhandl. vol. iii. pt. v. p. 332 (1887).

that this form is distinct from *G. tenuidens* of the Purbeck (*suprà*, i. p. 83).

Hab. Europe (Germany and England).

R. 214. A series of associated specimens, comprising the nearly entire second dorsal vertebra, right coracoid, right humerus, distal portion of right femur, and some fragmentary bones; from the Wealden of the Isle of Wight. The dorsal vertebra agrees precisely in size with the type cervical figured by Koken in the 'Pal. Abhandl.' vol. iii. pl. xxxv. fig. 5; the length of the centrum being 0,025 (0·98 inch), and its transverse diameter 0,029 (1·15 inch). The agreement in the length of the dorsal with that of the type cervical would seem to indicate that the larger centrum figured on page 333, fig. 12 of the memoir cited as a last cervical belongs to a different form. The length of the coracoid is 0,052 (2·05 inches), and that of the humerus 0,072 (2·8 inches). *Fox Collection. Purchased*, 1882.

Family TELEOSAURIDÆ (*suprà*, i. p. 91).

Subfamily Metriorhynchinæ (*suprà*, i. p. 91).

The following additional characters may be given[1]:—Sclerotic plates in the orbit; no dermal scutes; mandible, at least in most cases, without lateral vacuity. For the distinctive characters of the vertebral column, limbs, and pectoral and pelvic girdles, see Hulke, Proc. Zool. Soc. 1888, pp. 417–442[2].

Genus **GEOSAURUS** (*suprà*, i. p. 271).

Syn. *Dacosaurus* (*suprà*, i. p. 92).
Cricosaurus, in parte (*suprà*, i. p. 95).

Carinæ of the teeth serrated, and the enamel smooth.

Geosaurus maximus (*suprà*, i. p. 92).

= *Dacosaurus maximus* (*l. c.*).

41802. An imperfect cervical vertebra probably referable to this species; from the Kimeridge Clay of Ely. This specimen

[1] See Quart. Journ. Geol. Soc. vol. xlv. p. 57. It is there stated, on the evidence of Deslongchamps's figures, that the mandible of *Metriorhynchus* has a vacuity, but Mr. Leeds's specimens show this to be incorrect.

[2] The imperfect bone represented in fig. 1, p. 428, of that memoir as a scapula is really a coracoid, with the bar of the fontanelle broken away.

agrees in size with the vertebræ of No. 40103 *b* (*suprà*, i. p. 93), but appears to have occupied a more anterior position in the neck. Only a portion of the right side of the neural arch remains. The specimen closely resembles in structure the cervicals of *Metriorhynchus*.

Purchased, 1869.

Genus **SUCHODUS**, Lydekker[1].

Skull with the general characters of *Metriorhynchus*, but shorter and wider, with fewer teeth, and no distinct expansion of the extremity of the mandible, and without a long interval between the 4th and 5th tooth. Teeth with compressed and carinated crowns, having nearly smooth enamel, and without serrations on the carinæ.

This genus appears to connect *Metriorhynchus* by dental characters with *Geosaurus*, while in the form of the mandible and the reduced number of teeth it approximates to *Machimosaurus*. *Metriorhynchus brachyrhynchus*, Deslongchamps[2], of the Oxfordian of Normandy, approximates to this genus in its relatively short rostrum.

Suchodus durobrivensis, Lydekker[3].

The type and only described species. Of the approximate size of *Metriorhynchus brachyrhynchus*, but with a broader rostrum, relatively shorter and wider nasals, which do not reach the premaxillæ, and a narrower orbital notch, in which the fronto-prefrontal suture is almost a continuation of the posterior border of the orbit; 12 or 13 lower teeth; cranial sculpture slightly developed.

The type of this species is part of a mandible without teeth in the collection of Mr. A. N. Leeds.

Hab. Europe (England).

R. 1711. Cast of a tooth. The original belongs to an imperfect skull from the Oxford Clay near Peterborough, preserved in the collection of A. N. Leeds, Esq., of Eyebury, near that town. *Made in the Museum*, 1889.

Genus **METRIORHYNCHUS** (*suprà*, i. p. 95).

Syn. *Cricosaurus, in parte* (*loc. cit.*).

The enamel of the teeth has well-marked vertical rugæ, and

[1] Quart. Journ. Geol. Soc. vol. xlvi. pt. ii. p. 288 (1890)

[2] See 'Notes Paléontologiques,' p. 333, pl. xxiii. (1869).

[3] *Loc. cit.*

there are no serrations on the carinæ. The mandible has a distinct terminal expansion, with a long interval between the 4th and 5th tooth.

[In fig. 14, Pt. I., *la* = prefrontal.]

Metriorhynchus superciliosum[1] (*suprà*, i. p. 96).

Syn. ? *Metriorhynchus blainvillei*, Deslongchamps[2].
= *Teleosaurus blainvillei*, Deslongchamps[3].
Steneosaurus dasycephalus, Seeley[4].

In the type cranium[5] the prefrontals have only a very short sutural junction with the frontals and a straight inner border; whereas in the type of *M. blainvillei*[6], of the Kelloway rock, the prefrontals have a long sutural junction with the frontals and a convex inner border. The large series of skulls in the collection of Mr. A. N. Leeds shows, however, such a great variation in these respects that they cannot be regarded as of specific value.

Hab. Europe (France and England).

The following specimens were obtained from the Oxford Clay near Peterborough, and, unless the contrary is stated, were presented by A. N. Leeds, Esq., 1889.

R. 1665. Cast of the frontal region of the cranium. The original, which is a portion of the type of *Steneosaurus dasycephalus*, is preserved in the Woodwardian Museum, Cambridge; it indicates a comparatively large individual. The contour of the prefrontals is somewhat intermediate between that obtaining in the type of the present species and that of the so-called *M. blainvillei*.

Made in the Museum, 1889.

R. 1666. The nearly entire skull of a smaller individual. The contour of the prefrontals is almost exactly the same as in the type cranium.

R. 1530. The imperfect skull of an individual agreeing in size with the preceding. The posterior half of the right prefrontal is entire, and has the short sutural junction with the

1 Amended from *superciliosus*.
2 Notes Paléontologiques, p. 295 (1869).
3 Bull. Soc. Linn. Norm. sér. 2, vol. i. p. 146 (1867).
4 Index to Aves etc. in Cambridge Museum, p. 140 (1869).—No description.
5 Notes Paléontologiques, pl. xx. fig. 2.
6 *Ibid.* pl. xx. fig. 1.

frontal and nearly straight inner border characteristic of the type.

R. 1529. The imperfect skeleton of a smaller individual apparently referable to this species. The prefrontals have a convex inner border. In this specimen the sutures are very indistinct and thereby suggest maturity; but since the same feature occurs in all the smaller skulls of this type in the collection of Mr. Leeds, it would seem that this is not the case.

Specifically Undetermined Specimens.

24805. The centrum of a cervical vertebra; from the Kimeridge Clay of Devizes, Wiltshire. Closely resembles the cervicals of *M. superciliosum*, No. R. 1529.
Cunnington Collection. Purchased, 1849.

42094. Two imperfect cervical vertebræ; from the Neocomian bone-bed of Potton, Bedfordshire. These specimens, which were probably derived from the Kimeridge Clay, have longer centra, with a more prominent hæmal carina, than the preceding specimen. *Purchased*, 1870.

42096. An imperfect cervical vertebra of similar type; from Potton.
Purchased, 1870.

46805. The centrum of a dorsal vertebra; from Devizes.
Cunnington Collection. Purchased, 1875.

46805 a. The crushed centrum of a dorsal vertebra; from Devizes.
Cunnington Collection.

46805 b. The centrum of a lumbar vertebra; from Devizes.
Cunnington Collection.

46805 c. The centrum and base of the arch of a caudal vertebra; from Devizes. *Cunnington Collection.*

44179. Three imperfect centra of dorsal vertebræ; from the Kimeridge Clay of Weymouth, Dorsetshire. These specimens are the only recorded instance of the occurrence of *Metriorhynchus* at this locality. *Purchased*, 1873.

47990. The three teeth entered in Pt. I. p. 94 under the head of *Dacosaurus* are referable to *Metriorhynchus*.

Incertæ Sedis.

Genus **CROCODILÆMUS,** Jourdan (MS.[1]).

Syn. *Chælosaurus*, Jourdan [2]. *Alligatorium*, Jourdan [3].

Characterized by the extreme complexity of the dermal skeleton, in which the scutes are pitted, and by the nearly equal size of the teeth.

Crocodilæmus jourdani, Lydekker (n. sp.).

The type species; typically of small size.

Hab. Europe (France).

The cast of the type specimen (No. 40344) is entered in Part I. p. 98, as *Metriorhynchus* (?), sp.

Subfamily Teleosaurinæ (*suprà*, i. p. 102).

Genus **STENEOSAURUS** (*suprà*, i. p. 108).

Steneosaurus brevidens (*suprà*, i. p. 115).

R. 1477. A number of associated bones; from the Great Oolite near Northampton. The chief specimens are portions of the skull, the glenoidal portion of the right coracoid, two imperfect cervical and several dorsal vertebræ, the imperfect femora, portions of ribs, and part of a dermal scute. In the skull the greater portion of the cranial and a part of the mandibular rostrum are preserved. In size the cranium agrees with the one figured on p. 186 of Phillips's 'Geology of Oxford.' The two cervical vertebræ agree in size with No. 4821 (*suprà*, i. p. 115), and belong to the anterior part of the neck. The anterior cervical has a blunt hæmal carina, and closely resembles the one figured by Phillips, *op. cit.* p. 200, diagram lviii. The dorsals are smaller than No. R. 79 (*suprà*, i. p. 115), which may indicate a distinct and larger species. *Purchased*, 1889.

R. 1477 a. An imperfect anterior cervical vertebra agreeing in size with the cervicals of the preceding specimen; from the Great Oolite near Northampton. The centrum is entire, but the greater portion of the neural spine is wanting. The hæmal surface is deeply concave, and excavated above the level of the transverse processes. *Purchased*, 1889.

[1] Quoted by Deslongchamps, 'Notes Paléontologiques,' p. 119 (1867).

[2] Extr. Procès-Verb. Soc. Agric. Lyon, vol. vi. p. vii (1862).

[3] *Ibid.* p. xii.

The following specimens may indicate an immature individual of this species.

R. 1478. A series of associated bones; from the Great Oolite of Northampton. These comprise several imperfect cervical and dorsal vertebræ, fragments of jaws, ribs, and other bones, together with some teeth which may or may not be associated. The cervical vertebræ, although smaller, agree in form with those of No. R. 1477; and are quite different from the smaller and more elongated type figured by Phillips, *loc. cit.* fig. 1, and referred to *Teleosaurus subulidens*. *Purchased*, 1889.

Order PARASUCHIA (*suprà*, i. p. 123).

According to Koken [1] there is no middle Eustachian canal in this group, the channel described as such being incorrectly named.

Recent observations show that the skull of *Phytosaurus* is constructed so essentially on the Rhynchocephalian plan as to entail the necessity of raising the Parasuchia to the rank of a distinct order. The Crocodilia can, therefore, be much more exactly defined than was previously possible.

In the skull of *Phytosaurus* (Pt. I. pp. 124–5) the arrangement of the temporal arcades is strictly the same as in *Sphenodon*; and a similar resemblance occurs in the structure of the palate—notably in the junction of the vomer with the pterygoids, and the relations of the palatines, pterygoids, and quadrates; while in both forms there is a vacuity on the occipital aspect between the quadrate and quadrato-jugal. The great development of the premaxillary rostrum of *Phytosaurus* is similar to the condition obtaining in the Rhynchocephalian genus *Champsosaurus* and the Ichthyopterygia. The abdominal ribs are of the chevron-like type found in *Sphenodon*, and quite different from those of the Crocodilia.

The Parasuchia, therefore, resemble the Rhynchocephalia and Ichthyopterygia in the structure of the skull, abdominal ribs, and probably of the clavicular arch; while they agree with the Crocodilia in the characters of the vertebræ and the dermal armour; their thecodont dentition being likewise an Archosaurian character. It depends on the relative degree of importance attached to cranial as opposed to vertebral and dermal characters whether the order should be retained in the Archosaurian or transferred to the Streptostylic Branch. The Aetosauria should probably be included in the Parasuchia.

[1] Zeitschr. deutsch. geol. Ges. vol. xl. p. 764 (1889).

Order DINOSAURIA (*suprà*, i. p. 131).

Suborder *SAUROPODA* (*suprà*, i. p. 131).

Family CETIOSAURIDÆ (*suprà*, i. p. 133).

Genus **CARDIODON**, Owen [1].

Founded upon teeth, which are of comparatively small size, with short, broad, and highly incurved crowns, of which the enamel is marked by distinct vertical rugæ.

It has been suggested that these teeth belong to *Cetiosaurus oxoniensis* (*suprà*, ii. p. x), one of them having been figured on page 253 of Phillips's 'Geology of Oxford' under that name—in which case the present name should supersede *Cetiosaurus*. They would, however, agree better in relative size with the vertebra described as *Bothriospondylus robustus* (*suprà*, i. p. 171), which appears to be of a longer type and has a shallower fossa than the dorsals of *C. oxoniensis*.

Cardiodon rugulosus, Owen [2].

Syn. (?) *Bothriospondylus robustus* (*suprà*, i. p. 171).

The type species.

Hab. Europe (England).

R. 1527. The imperfect crown of a tooth; from the Great Oolite (Lower Jurassic) of Cirencester. When entire this specimen must have agreed closely with the type tooth figured in Owen's 'Odontography,' pl. 75 A. figs. 7, *a*, *b*.

Presented by Prof. G. S. Boulger, 1889.

Genus **MOROSAURUS**, Marsh [3].

Apparently nearly allied to *Cetiosaurus*, but with a relatively shorter humerus. In *M. lentus*, Marsh[4], the arches of the dorsal vertebræ are extremely long.

[1] Odontography, pt. iii. p. 291 (1845).

[2] *Ibid.* pl. 75 A. fig. 7 (1845). This plate was issued with the third part of the work.

[3] Amer Journ. ser. 3, vol. xv. p. 242 (1878).

[4] *Ibid.* vol. xxxvii. p. 333 (1889).

Morosaurus brevis (Owen).

Syn. *Cetiosaurus brevis*, Owen (*suprà*, i. p. 139).
Pelorosaurus becklesi, Mantell [1].
Bothriospondylus elongatus, Owen (*suprà*, i. p. 142).
Morosaurus becklesi, Marsh [2].

As already noticed by the writer [3] there is no evidence to show that the limb-bones on which *P. becklesi* was founded, together with the dorsal vertebra described as *Bothriospondylus elongatus*, are not referable to the present form, which was founded upon the evidence of caudal vertebræ. The whole are therefore provisionally regarded as belonging to a single species. If rightly associated the humerus will be shorter in proportion to the vertebræ than in the typical American species. The ilium No. 2576 (*suprà*, i. p. 143) presents all the characters of that of *Morosaurus* and probably also belongs to the same form.

Genus **PLEUROCŒLUS**, Marsh [4].

Represented by small forms, which are characterized by the crowns of the teeth being less spoon-shaped and more like compressed cones than those of *Morosaurus*; and the elongation of the centra of the cervical and dorsal vertebræ (fig. 52), the latter having a very long and deep lateral cavity gradually shelving upwards towards the neural arch, and with the opisthocœlous character very strongly marked.

Pleurocœlus valdensis, Lydekker [5].

Typified by teeth (fig. 51), which from their small size and contour are provisionally referred to this genus. The undermentioned vertebræ approximate to those of the typical species.

Possibly this type of tooth may indicate more than one species.

To this species belong the teeth entered in Part I. p. 185 under the heading of *Hylæosaurus*.

Hab. Europe (England).

[1] Proc. R. Inst. vol. i. p. 153 (1852).—No description.
[2] Amer. Journ. ser. 3, vol. xxxvii. p. 325 (1889).
[3] Nicholson & Lydekker, 'Manual of Palæontology,' 3rd ed. vol. ii. p. 1179 (1889).
[4] Amer. Journ. ser. 3, vol. xxxv. p. 90 (1888).
[5] Geol. Mag. dec. iii. vol. vi. p. 325 (1889).

3562. The upper portion of a tooth belonging either to this or an allied form; from the Wealden of Cuckfield. This specimen is remarkable for the expansion of the lateral angles

Fig. 51.

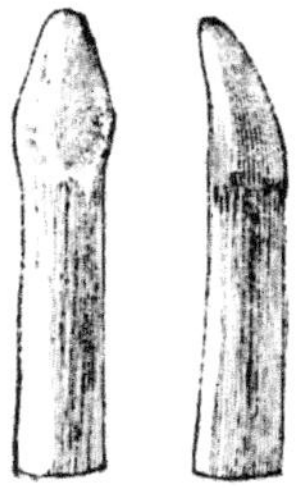

Pleurocœlus valdensis.—Outer and profile views of a tooth; from the Wealden of Sussex. $\frac{1}{1}$.

of the crown, which thus assumes a somewhat trefoil-like shape. *Mantell Collection. Purchased*, 1838.

3534. The flattened upper portion of a more normally shaped tooth; from Cuckfield. *Mantell Collection.*

R. 1616. The imperfect centrum of a dorsal vertebra probably (*Fig.*) belonging to a somewhat immature individual of this

Fig. 52.

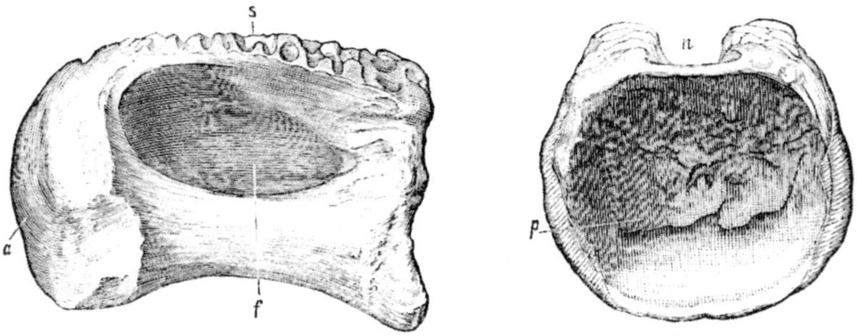

A. B.

Pleurocœlus nanus.—Right lateral and posterior aspects of the centrum of a dorsal vertebra; from the Potomac beds of North America. $\frac{1}{2}$. *s*, surface for articulation of neural arch; *f*, lateral pit; *n*, neural canal; *a*, anterior ball; *p*, posterior cup. (*After Marsh.*)

form; from the Wealden of Sussex. This specimen, of which the anterior half is wanting, is described and figured by the writer in the 'Quart. Journ. Geol. Soc.' vol. xlvi.

pt. 2, pl. ix. figs. 1, 1 *a*. In its elongated form, deep posterior cup, and large upwardly-shelving lateral cavity, it resembles the dorsal of the typical *P. nanus* (fig. 52). The length when entire was approximately 0,076 (3 inches) against 0,097 (3·8 inches) in the latter. The smooth outer surface of the bone is in marked contrast to the roughness found in the extremely young vertebræ described as *Bothriospondylus suffosus* (see p. 243).

Mantell Collection.

R. 1730. An imperfect adult vertebra probably belonging to the same form as the preceding, or one closely allied thereto; from the Wealden of Brook, Isle of Wight. Described and figured by the writer, *op. cit.* pl. ix. figs. 2, 2 *a*. The lower half of the centrum is wanting, the terminal surfaces are imperfect, and the outer lamina of bone is more or less completely gone from the upper part of the lateral surfaces of the centrum, by which the contour of the upper half of the lateral cavity is wanting. The fracture of the centrum extends through the central pit of the lateral cavity. When entire the length of the centrum was some 4 inches. *Presented by the Dorsetshire Museum*, 1889.

The following specimen may belong either to the present or the next family.

R. 1621. The distal extremity of a right femur; from the Wealden, probably of Brook, Isle of Wight. This specimen, which has been somewhat flattened by crushing, presents the general characters of the femur of *Atlantosaurus*, but is of considerably smaller size. In its colour and mineral condition it is similar to No. 36559 (*suprà*, i. p. 139).

Baber Collection. Purchased, 1889.

Family ATLANTOSAURIDÆ (*suprà*, i. p. 143).

This family name is provisional. It has been stated that the undermentioned forms are not referable to this family, but this was probably due to a misconception of the nature of the pelvis.

Genus **PELOROSAURUS** (*suprà*, i. p. 145).

Syn. *Neosodon*, De La Moussaye [1].
(?) *Gigantosaurus* (*suprà*, i. p. 146).

[1] Bull. Soc. Géol. France, sér. 3, vol. xIII. p. 51 (1885).

Includes very large forms which appear to be closely allied to the American *Brontosaurus.*

The accompanying table gives the dimensions of the bones of the species of this genus compared with those of allied forms.

	Atlantosaurus.	*Brontosaurus.*	*Pelorosaurus conybeari.*	*Pelorosaurus humerocristatus* and *Leedsi.*	*Hoplosaurus.*	*Morosaurus* (type).	*Cetiosaurus.*	*Morosaurus brevis.*
Length of scapula	...	60	...	...	...	45	54	...
" humerus.........	...	50 ?	54	57	...	36	51·5	24
" ischium	...	40	...	35·5	27	30·5	39 ?	...
" femur	96	70 ?	...	...	...	46	64	...
Width of dorsal centrum.	...	13	...	...	8·5	...	...	6
" lumbar " ...	...	14	...	11·5 [1]	...	...	...	7
" caudal " ...	...	12	...	...	...	...	? 10	7

Pelorosaurus conybeari (*suprà*, i. p. 145).

R. 1610. (*Fig.*) The crown of a tooth, perhaps referable to this species; from the Wealden of Kent. This specimen (fig. 53), which

Fig. 53.

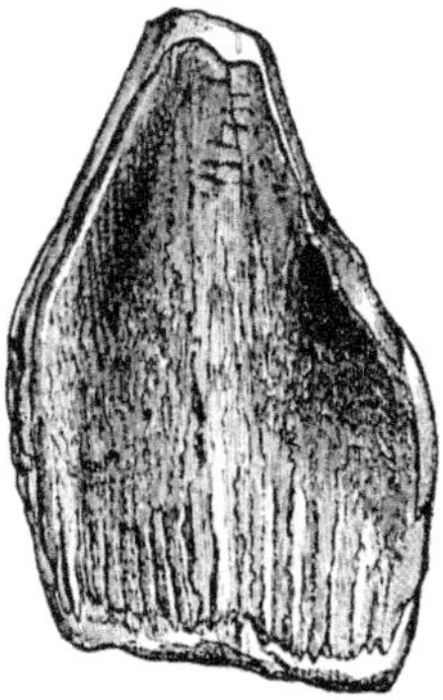

(?) *Pelorosaurus conybeari.*—Inner aspect of the crown of a tooth; from the Wealden of Kent. 1/1 (From the 'Quart. Journ. Geol. Soc.')

[1] Caudal vertebræ of a larger individual are nearly equal to those of *Brontosaurus* in point of size.

has been somewhat water-worn, is figured by the writer in the 'Quart. Journ. Geol. Soc.' vol. xlv. p. 243, fig. 7, where it is provisionally described as *Ornithopsis.* It differs in contour from the tooth of *Hoplosaurus armatus*; but if referable to the present form must be one of the smaller teeth, since it is inferior in dimensions to those mentioned under the head of *P. humerocristatus.*

Presented by H. Willett, Esq., 1888.

Pelorosaurus humerocristatus (Hulke).

Syn. *Ornithopsis humerocristatus* (*suprà*, i. p. 151).

The teeth from the Portlandian of Boulogne described by Sauvage as *Iguanodon precursor* [1] and subsequently as *Caulodon precursor* [2], and by De la Moussaye as *Neosodon* [3], probably indicate a form closely allied to or identical with the present [4].

R. 1484. Cast of the proximal extremity of a right tibia which may be referable to this form. The original was obtained from the Kimeridge Clay at Ely, and is preserved in the Woodwardian Museum at Cambridge.

Made in the Museum, 1888.

R. 1485. Cast of the centrum of a late caudal vertebra, probably specifically identical with the preceding. History the same as that of the latter. *Made in the Museum*, 1888.

Pelorosaurus manseli (Hulke).

Syn. *Ornithopsis manseli* (*suprà*, i. p. 152).

R. 1486. Cast of an imperfect sacral vertebra, which from its comparatively small size may be referable to this species. The original, which was obtained from the Kimeridge Clay of Stretham (Cambridgeshire), is preserved in the Woodwardian Museum at Cambridge. It is mentioned by Seeley in his 'Index to Remains of Aves &c. in Cambridge Mus.' pp. 94, 95, as one of the types of *Gigantosaurus*, and provisionally referred to the cervical region.

Made in the Museum, 1888.

[1] Bull. Soc Géol. France, sér. 3, vol. iv. p. 438 (1876).
[2] *Ibid.* vol. xvi. p. 626 (1888).
[3] *Ibid.* vol xiii. p. 51 (1885).
[4] See Quart. Journ. Geol. Soc. vol. xlv. pp. 244, 245 (1889).

Pelorosaurus leedsi (Hulke[1]).

Syn. *Ornithopsis leedsi*, Hulke[2].

Of gigantic size, and only separable from *P. humerocristatus* on account of its lower geological horizon. The pelvis so closely resembles that of *Brontosaurus* (*suprà*, i. p. 144, fig. 23) as to show that the two forms were closely allied.

The types were originally described as being from the Kimeridge instead of the Oxford Clay, while the ventral aspect of the pubis and ischium was mistaken for the dorsal[3].

R. 1716. Cast of the imperfect centrum of a lumbar vertebra. The original was obtained from the Oxford Clay near Peterborough, in association with the type pelvis, and is preserved in the collection of A. N. Leeds, Esq., of Eyebury, near that town. The portion of the body of the centrum above the cavity is wanting, but the contour of the posterior terminal face is fairly well preserved. The width of the posterior face is about 0,292 (11·5 inches); the length of the associated ischium being 0,897 (35·5 inches). The corresponding dimensions of *Brontosaurus excelsus* are 0,354 (14 inches) and 1,012 (40 inches). The neural arch belonging to the present specimen has the same expanded spine as in *Brontosaurus*. Caudal vertebræ in the collection of Mr. Leeds from the same deposits, which, although not associated, doubtless belong to this species, are indistinguishable from those of *Brontosaurus*. One of them has a total height of 0,658 (26 inches), a width across the centrum of 0,292 (11·5 inches), and a height above the latter of 0,261 (10·3 inches); the corresponding dimensions in *Brontosaurus* being 0,726 (28 inches), 0,305 (12 inches), and 0,267 (10·5 inches). The presence of well-defined prezygapophyses and a single chevron-facet as well as the expanded neural spine at once distinguishes these vertebræ from the caudals of *Cetiosaurus*. (See table of measurements on p. 240 for comparative dimensions of the bones of this species.)

Made in the Museum, 1889.

** **Bothriospondylus suffossus** (*suprà*, i. p. 170).

Incorrectly referred in Part I. to the Theropoda (see Introduction to Pt. II. p. ix).

[1] Quart. Journ. Geol. Soc. vol. xliii. p. 695 (1887).—*Ornithopsis*
[2] *Loc. cit.* [3] *Ibid.* vol. xlv. p. 391.

The vertebræ on which this species (the type of *Bothriospondylus*) was founded indicate an immature and perhaps fœtal Dinosaur, which may be the young of *Pelorosaurus humerocristatus* or of *P. manseli*; or may perhaps be a smaller form allied to *Pleurocœlus*.

Genus **HOPLOSAURUS**, Gervais [1].

Syn. *Ornithopsis*, Seeley (*suprà*, i. p. 146).

The ischium is wider in proportion to the pubis than in *Pelorosaurus leedsi*, a difference which may probably be regarded as indicating the right of the present form to stand as a distinct genus

Hoplosaurus armatus, Gervais [2].

Syn. *Ornithopsis hulkei*, Seeley (*suprà*, i. p. 146).
Ornithopsis eucamerotus, Hulke (*suprà*, i. p. 146).
Pelorosaurus armatus, Lydekker [3].

The type species: founded on the tooth represented in Pt. I. p. 147, fig. 24. The dimensions of some of the bones are given in the table on p. 240.

Suborder *THEROPODA* (*suprà*, i. p. 154).

Family CŒLURIDÆ (*sup rà*,

In the definition of this family it should be stated that the middle and posterior cervical vertebræ may be amphicœlous.

Genus **CALAMOSPONDYLUS**, Lydekker [4].

Known only by cervical vertebræ, which are relatively shorter than in the type genus, and were probably all opisthocœlous.

Calamospondylus foxi, Lydekker [5].

The type and only described species.
Hab. Europe (England).

[1] Zool. et Pal. Françaises, 1st ed. p. 263 (1852).—*Oplosaurus*.
[2] *Loc. cit.*
[3] Geol. Mag. dec. iii. vol. vi. p. 325 (1889).
[4] *Ibid.* p. 121 (1889).
[5] *Loc. cit.*

R. 901. (*Fig.*) Two associated cervical vertebræ, one of which has lost the greater portion of the neural arch; from the Wealden of the Isle of Wight. The types; the entire specimen is figured by the writer in the 'Geol. Mag.' decad. 3, vol. vi. p. 120, reproduced in fig. 54.

Fox Collection. Purchased, 1882.

Fig. 54.

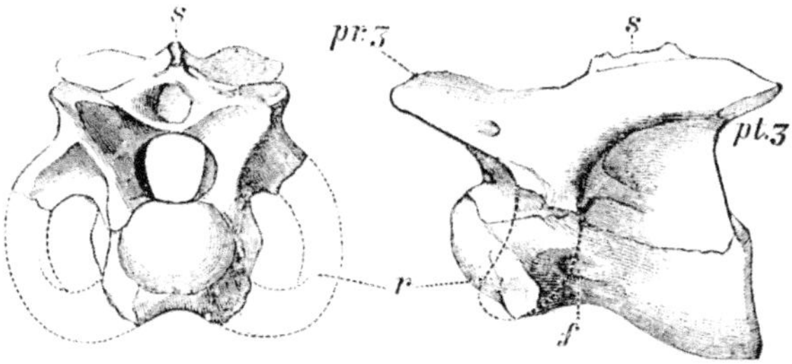

Calamospondylus foxi.—Anterior and left lateral aspects of a cervical vertebra; from the Wealden of the Isle of Wight. ⅔.

Family MEGALOSAURIDÆ (*suprà*, i. p. 157).

Genus **MEGALOSAURUS** (*suprà*, i. p. 157).

Megalosaurus dunkeri (*suprà*, i. p. 163[1]).

This species was founded upon a tooth of comparatively large size presenting the characters noticed in Part I. Referring the undermentioned specimen to this species on account of its relatively large size and its geological horizon, it will be further characterized by its elongated metatarsals, in which the second was much longer than the fourth, both of these bones having their antero-internal borders rounded, while the distal surface of the fourth is comparatively narrow.

The type specimen was obtained from the lower division of the Hastings beds, equivalent either to the Wadhurst Clay or the underlying Ashdown beds.

R. 1525. (*Fig.*) The second left metatarsal; from Hollington quarry, near Hastings; belonging to the same individual as the fourth metatarsal, No. R. 604 *d* (*suprà*, i. p. 167). The two bones

[1] The authority for this species should be Dames, Sitzungsb. Ges. nat. Berlin, 1884, p. 187.

are figured together by the writer in the 'Quart. Journ. Geol. Soc.' vol. xlvi. p. 46, fig. 4. The present specimen was obtained in 1889 at a distance of about 180 yards to the eastward of the spot where No. R. 604 *d* was found in 1884. The vertebra, No. R. 604 *a* (*suprà*, i. p. 166), the tibia R. 604 *c* (*ibid.* p. 167), and No. R. 604 *d* were all found together. *Dawson Collection. Purchased*, 1889.

R. 1525 a. An imperfect later caudal vertebra apparently associated with the preceding; from Hollington quarry. The neural arch is crushed down upon the centrum, and the transverse processes are imperfect. *Dawson Collection.*

Megalosaurus oweni, Lydekker[1].

Smaller than the preceding species, with the second metatarsal only slightly longer than the fourth; both those bones having sharp antero-internal borders, and the distal surface of the fourth being broad and squared.

This species is founded on the metatarsus No. 2559 (*suprà*, i. p. 167), which really belongs to the right side, the metatarsal described by Owen as the 2nd being the 4th, and *vice versâ*. The metatarsal which was compared in Part I. with the 4th of *M. dunkeri* is really the 2nd of the right side. No. 2661 (*suprà*, i. p. 168) will be the 4th of the left side; while Nos. 2680 and 2574 (*loc. cit.*) belong to the right side.

The type specimen is from the Upper part of the Tunbridge beds, which overlies the Wadhurst Clay, and it is probable that most or all the specimens from these beds at Cuckfield and the Weald Clay of the Isle of Wight, entered in Part I. under the head of *M. dunkeri*, belong to this species.

Hab. Europe (England).

Genus *non det.*

It is suggested that the undermentioned specimens may possibly belong to *Megalosaurus*.

R. 1412. The axis vertebra and intercentrum of a Dinosaur; from (*Fig.*) the Wealden of the Isle of Wight. Described and figured

[1] Geol. Mag. dec. iii. vol. vi. p. 325 (1889).

Fig. 55.

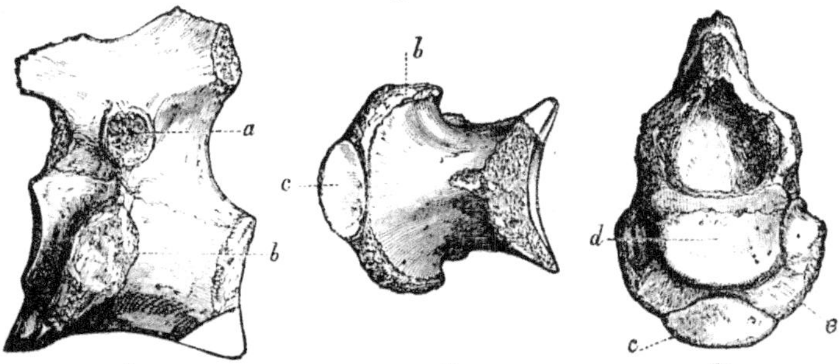

Left lateral (A), anterior (B), and hæmal (C) aspects of the axis vertebra of a Dinosaur; from the Wealden of the Isle of Wight. ½. *a*, upper (diapophysis) costal articulation; *b*, lower do. (parapophysis); *c*, axial intercentrum (hypapophysis); *d*, articulation for centrum of atlas (odontoid process); *e*, articulation for inferior ring of atlas. (From the 'Quart. Journ. Geol. Soc.').

by the writer in the 'Quart. Journ. Geol. Soc.' vol. xlv. p. 44, fig. 2; the figure being reproduced in the accompanying woodcut. *Fox Collection. Purchased*, 1882.

Family ANCHISAURIDÆ (*suprà*, i. p. 174).

In addition to their more or less oblique and continuous serration the teeth are characterized by their short and somewhat thickened crowns, of which the posterior border is either slightly convex or nearly straight.

In the *Megalosauridæ* the crowns are more compressed and taller, with a distinctly concave posterior border; *Epicampodon* (*suprà*, i. p. 174, fig. 29) approximating to this type.

Genus **THECODONTOSAURUS** (*suprà*, i. p. 174).

The teeth with the posterior border distinctly convex.

It is possible that *Anchisaurus* may prove to be inseparable from this genus, and in that, if not in any, case the family name *Thecodontosauridæ* should be adopted.

Thecodontosaurus platyodon (Riley and Stutchbury [1]).

Syn. *Palæosaurus platyodon*, Riley and Stutchbury [2].

---

[1] Proc. Geol. Soc. vol. ii. p. 398 (1836, vol. dated 1838).—*Palæosaurus*.

[2] *Loc. cit.*

Larger than *T. antiquus* (*suprà*, i. p. 175), with broader teeth in which the serrations appear to be less oblique.

Referred to this genus by Huxley in the 'Quart. Journ. Geol. Soc.' vol. xxvi. pp. 43–44, where it is stated that no difference can be detected in the direction of the serrations between the teeth of this from and *T. antiquus*; but the undermentioned specimens certainly have them less oblique than in the figure of the type of the latter, in which respect they resemble the original figure of the type of the present species.

Hab. Europe (England).

R. 1695. The crowns of two small teeth; from the Upper Trias (? Lower Keuper) of Somersetshire. The larger, although of considerable dimensions, precisely resembles the figure of the type tooth given in the 'Trans. Geol. Soc.' ser. 2, vol. v. pl. xxix. fig. 5[1]. Although the lower portion of the anterior border is imperfect, the serrations can still be detected continuing to the base of the crown.

Presented by Prof. Jäger, 1889.

Specimens which may be referable either to T. antiquus *or* T. platyodon.

R. 1539. Mass of rock showing the dorsal aspect of the imperfect right ilium, and a fragment of another bone; from the Upper Triassic (Lower Keuper) bone-bed near Bristol. The ilium accords with the specimen figured by Huxley in the 'Quart. Journ. Geol. Soc.' vol. xxvi. pl. iii. fig. 7, from the inner side. *By exchange*, 1888.

R. 1544. The imperfect distal half of a femur; from the Bristol bone-bed. *By exchange*, 1888.

R. 1551. Mass of rock showing the ventral surface of a phalangeal of the pes; from the Bristol bone-bed.

By exchange, 1888.

R. 1535. Fragment of rock showing the ventral aspect of the centrum of a caudal vertebra; from the Bristol bone-bed.

By exchange, 1888.

[1] In the reproduction of this figure given in Part I. p. 174, the serrations are made too oblique.

R. 1534. Fragment of rock showing the ventral aspect of a later caudal vertebra; from the Bristol bone-bed.

By exchange, 1888.

Other specimens from the same deposit which cannot be satisfactorily determined were obtained at the same time as the preceding.

Genus **MASSOSPONDYLUS**[1], Owen (*emend.* Lydekker[2]).

Imperfectly known. Centra of dorsal vertebræ much constricted and inferiorly excavated, with oval terminal faces. The teeth (figs. 56, 57) probably belonging to this genus have the posterior border of the crown either sinuous or nearly straight.

The genus was founded by Owen on the evidence of caudal vertebræ, but since the description is insufficient and there is no figure, it has been suggested that it may be permissible to regard the original of the undermentioned vertebra as the type.

Massospondylus carinatus, Owen[3] (*emend.* Lydekker[1]).

The type species. Apparently about two thirds the size of *Megalosaurus dunkeri*. Teeth unknown.

Fig. 56.

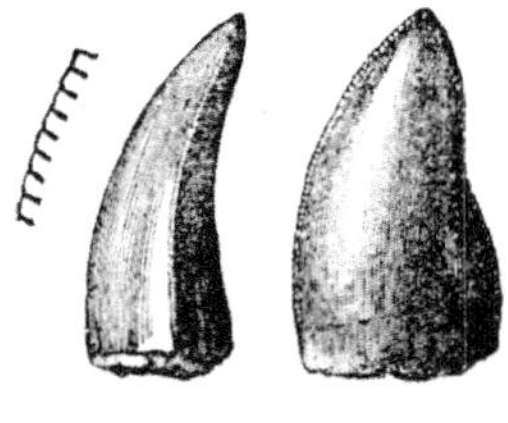

Massospondylus hislopi.—Posterior and external aspects of the crown of a tooth; from the Maleri stage of the Indian Gondwanas. 1/1. (From the 'Palæontologia Indica.')

Fig. 57.

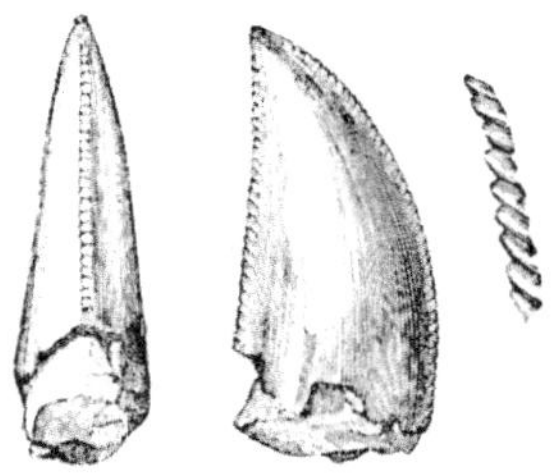

Massospondylus (?) *rawesi.*—Posterior and external aspects of the crown of a tooth; from the Lameta beds near Nágpur. 1/1.

The teeth referred to the Indian *M. hislopi*, Lydekker[5] (fig. 56), have comparatively thick crowns; that form being probably of approximately the same age as the present one.

Hab. Africa.

[1] Cat. Foss. Rept. Mus. R. Coll. Surgeons, p. 97 (1854).

[2] Rec. Geol. Surv. Ind. vol. xxi. p. 146 (1888).

[3] *Loc. cit.* *Loc. cit.* [5] *Op. cit.* vol. xxiii. pt. i. (1890).

R. 1312. Cast of the centrum and base of the arch of a dorsal vertebra. The original (fig. 58), which is preserved in the Museum of the Royal College of Surgeons (No. 336), is taken as the type, and was obtained from the Stormberg[1] beds of the Karoo system, at Harrismith, near the Drakensberg range, in Basutoland. It is described and figured by the writer in the 'Rec. Geol. Surv. Ind.' vol. xxi. p. 147, fig. 3. *Made in the Museum*, 1888.

Fig. 58.

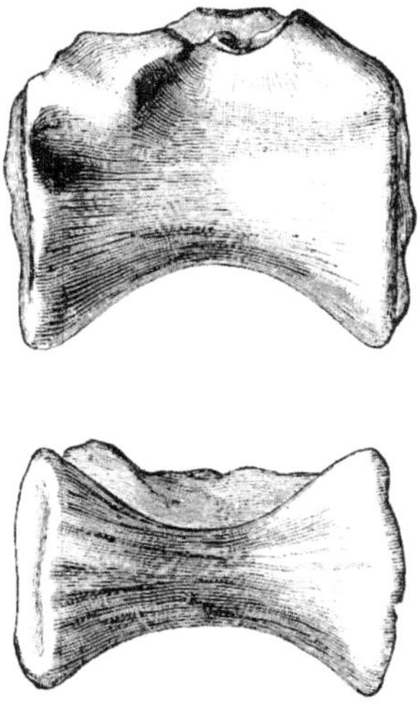

Massospondylus carinatus.—Lateral and hæmal aspects of the centrum of a dorsal vertebra; from the Stormberg beds of the Karoo system of Basutoland. $\frac{1}{2}$. (From the 'Rec. Geol. Surv. Ind.')

R. 1312 a. Cast of a phalangeal of the pes probably referable to this form. The history of the original is the same as that of the preceding specimen; and it is figured by the writer *loc. cit.* *Made in the Museum*, 1888.

Massospondylus (?) rawesi, Lydekker[2].

Known by a tooth which has a more compressed and straighter crown than in the teeth referred to *M. hislopi.* Generic reference provisional.

Hab. India.

[1] See Quart. Journ. Geol. Soc. vol. xxiii. p. 144.
[2] Rec. Geol. Surv. Ind. vol. xxiii. p. 22 (1890).

R. 1736. Cast of the crown of a tooth. The original, which is the type, is preserved in the Museum of the Geological Society, and was obtained at Tákli, near Nagpur, Central Provinces, having probably been derived from the Infratrappean Lameta beds, which are approximately of Middle Cretaceous age. It is noticed by Hislop in the 'Journ. Bombay As. Soc.' vol. vi. p. 196 (1861), and also in the 'Quart. Journ. Geol. Soc.' vol. xx. p. 281 (1864); and is described and figured by the writer in the 'Rec. Geol. Surv. Ind.' vol. xxiii. p. 21, fig. 1, from which woodcut fig. 57 is reproduced. *Made in the Museum*, 1889.

Family Uncertain.

Genus **ARCTOSAURUS,** Adams[1].

Known only by an imperfect cervical vertebra (fig. 59). The

Fig. 59.

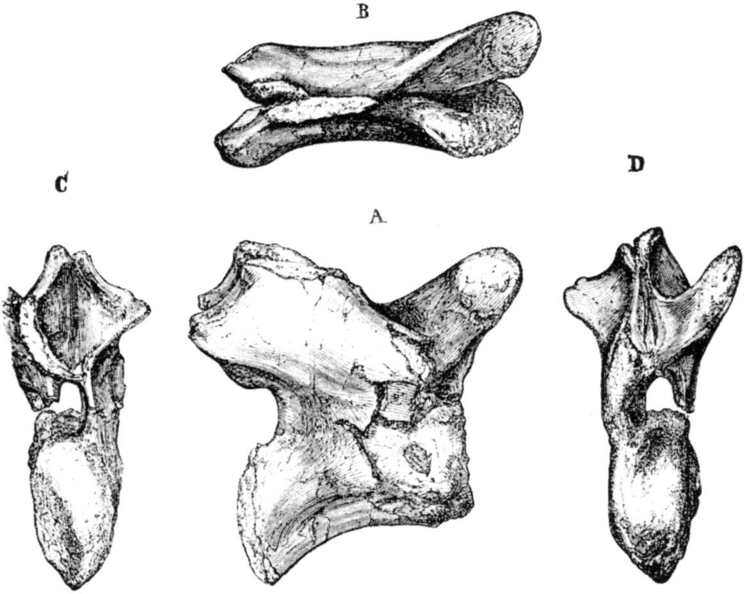

Arctosaurus osborni.—Right lateral (A), neural (B), posterior (C), and anterior (D) aspects of an imperfect cervical vertebra; from Bathurst Island. $\frac{1}{1}$. (From the 'Proc. R. Irish Academy.')

[1] Proc. R. Irish Academy, ser. 2, vol. ii. p. 177 (1875).

centrum of that specimen is amphicœlous, with a sharp hæmal carina, considerable lateral compression, and a highly curved ventral profile. The neck of the animal to which it belonged was evidently much arched; and from the general characters of the vertebra, and especially the deep median incisions between the zygapophyses, it is probable that it belonged to a Dinosaur allied to the *Anchisauridæ*. This is confirmed by the longitudinal fissure in the centrum, which is probably due to the crushing in of a central cavity.

Arctosaurus osborni, Adams [1].

The type and only described species. Apparently somewhat smaller than *Calamospondylus foxi*.

Hab. Arctic America.

R. 1611. Cast of a crushed and imperfect cervical vertebra. The original (fig. 59), which is the type, and is preserved in the Museum of Science and Art, Dublin, was obtained from beds of unknown age at Rendezvous Mountain, at the north end of Bathurst Island. It is described and figured by Adams in the Proc. R. Irish Academy, ser. 2, vol. ii. p. 177 *et seq.*; and also by the present writer in the 'Geol. Mag.' dec. iii. vol. vi. pp. 352, 353. The neural spine, the right prezygapophysis, and the costal facets of both sides are wanting; and a great part of the outer surface of the left side of the centrum is likewise absent. *Presented by the Director of the Science and Art Museum, Dublin,* 1889.

Suborder *ORNITHOPODA* (*suprà*, i. p. 175).

Family STEGOSAURIDÆ.

=Omosauridæ (*suprà*, i. p. 176).

Genus **STEGOSAURUS,** Marsh [2].

Syn. *Omosaurus*, Owen (*suprà*, i. p. 177). Preoccupied by Leidy, 1856.

The absence of an inner trochanter in the American forms is not regarded by Marsh ('Amer. Journ.' ser. 3, vol. xxxvii. p. 327) as a generic distinction.

[1] Proc. R. Irish Academy, ser. 2, vol. ii. p. 177 (1875).

[2] Amer. Journ. ser. 3, vol. xiv. p. 513 (1877).

Stegosaurus armatus (Owen).

Syn. *Omosaurus armatus*, Owen (*suprà*, i. p. 177).

Stegosaurus hastiger (Owen).

Syn. *Omosaurus hastiger*, Owen (*suprà*, i. p. 179).

Genus **EUSCELESAURUS,** Huxley [1].

Imperfectly known and provisionally referred to the *Stegosauridæ*. Limb-bones solid [2]; femur with a large inner trochanter; tibia with small proximal expansion, and apparently suturally connected at its extremities with the fibula; astragalus and calcaneum respectively united closely with the tibia and fibula.

The caudal vertebræ, phalangeals, and a pelvic bone which appears to be the preacetabular portion of a pubis, figured by Fischer in the 'Arch. du Muséum,' sér. 1, vol. vi. pl. xi. (1870), were obtained with the undermentioned specimens, and doubtless belong either to the present or the next genus. The vertebræ appear to be of an Iguanodont type.

Euscelesaurus browni, Huxley [3].

The type species. Approximating in size to *Iguanodon mantelli*, the femur measuring about one yard in length.

Hab. South Africa.

The following specimens include the types, and were obtained at Aliwal North [4], *in the Stormberg range, from the Stormberg beds of the Karoo system. They were presented to the Museum in* 1888 *by Professor T. H. Huxley.*

R. 1625. The imperfect right femur. The type specimen; described by Huxley in the 'Quart. Journ. Geol. Soc.' vol. xxiii. pp. 1–4. Both articular extremities are wanting, and the greater portion of the inner trochanter is broken away. In some respects the contour of this bone is nearer to the femur of *Iguanodon* than to that of *Stegosaurus*.

R. 1625 a. The proximal extremity of the left tibia and fibula. Noticed by Huxley, *op. cit.* p. 4. The part regarded in that notice as the cnemial crest appears to be the head of

[1] Quart. Journ. Geol. Soc. vol. xxiii. p. 4 (1867).—*Euskelesaurus.*
[2] *Vide infrà.* [3] *Loc. cit.*
[4] See Quart. Journ. Geol. Soc. vol. xxiii. p. 144.

the fibula. This specimen agrees in relative size with the type, and probably belonged to the same individual; it approximates to the corresponding bones of *Stegosaurus*, there being apparently a bony union between the tibia and fibula.

R. 1625 b. The distal extremity of the left tibia and fibula, with the astragalus attached to the former. Noticed by Huxley, *op. cit.* p. 4, as belonging to the right side. This specimen is probably the distal portion of the preceding. The tibia is much less expanded than in *Iguanodon*, and the three bones appear to be united, as in *Stegosaurus*.

R. 1625 c. Fragmentary undetermined bones. Two of these specimens have been cut and polished in order to exhibit their internal structure, which appears to be completely solid. It is stated in the original description that these specimens show a large medullary cavity, but this statement seems to be due to a misapprehension of their structure.

Genus **ORINOSAURUS**, Lydekker[1].

Syn. *Orosaurus*, Huxley[2].

Distinguished from the preceding by the much greater expansion of the head of the tibia, which has a large cnemial crest like that of *Iguanodon*. Apparently no bony union between the tibia and fibula. The genus may form a link connecting the *Stegosauridæ* and *Iguanodontidæ*.

Orinosaurus capensis, Lydekker[3].

Syn. *Orosaurus*, sp., Huxley[4].

The type species. Considerably larger than *Euscelesaurus browni*. *Hab.* South Africa.

R. 1626. The proximal extremity of the left tibia; from the Stormberg beds of the Karoo system at Aliwal North, in the Stormberg range. The type specimen. Described by Huxley in the 'Quart. Journ. Geol. Soc.' vol. xxiii. pp. 4, 5, as the distal extremity of the femur; but redetermined

[1] Geol. Mag. dec. iii. vol. vi. p. 353 (1889).
[2] Quart. Journ. Geol. Soc. vol. xxiii. p. 5 (1867).—Preoccupied by *Oreosaurus*, Peters, Abh. Ak. Berlin, 1862, p. 201.
[3] *Loc. cit.* [4] *Loc. cit.*

by the writer in the 'Geol. Mag.' dec. iii. vol. vi. p. 353. This bone accords so closely with the proximal extremity of the tibia of *Iguanodon*, that there is no doubt but that it is the corresponding bone; and were it not apparently solid throughout it might be referred to the same family.

Presented by Professor T. H. Huxley, 1888.

R. 1626 a. The middle portion of a right femur agreeing in relative size with the preceding; from the same locality. Noticed by Huxley, *op. cit.* p. 1, as a left femur of *Euscelesaurus*. It is possible that this bone may belong to the present form. *Presented by Professor T. H. Huxley*, 1888.

Family SCELIDOSAURIDÆ (*suprà*, i. p. 180).

Genus **SYNGONOSAURUS**, Seeley[1].

Imperfectly known, and provisionally referred to this family. Dorsal vertebræ with compressed centra, to which the arches are firmly united, tall neural canal, and no rib-facets on the arches. Terminal faces of anterior dorsals triangular, and a sharp hæmal ridge to centrum, which disappears later in the series.

Syngonosaurus macrocercus, Seeley[2].

The type and only described species. The following specimens agree with the type dorsals, but it is not easy to see how they differ from those described in the memoir cited (p. 613) as *Eucercosaurus*.

Hab. Europe (England).

R. 460. (*Fig.*) An imperfect dorsal vertebra; from the Cambridge Greensand. Described and figured by the writer in the 'Quart. Journ. Geol. Soc.' vol. xlv. p. 43, fig. 1, and provisionally referred to this species. The figure is reproduced in woodcut fig. 60. The tall and narrow neural canal is a feature in addition to those mentioned in the description, in which this vertebra agrees with the *Scelidosauridæ*.

Presented by Sir R. Owen, K.C.B., 1884.

R. 460 a. A more imperfect dorsal vertebra, which has been transversely fractured and the broken surfaces polished: from

[1] Quart. Journ. Geol. Soc. vol. xxxv. p. 621 (1879).

[2] *Loc. cit.*

the Cambridge Greensand. Probably associated with the preceding; noticed by the writer, *l. c.* p. 42.
Presented by Sir R. Owen, K.C.B., 1884.

R. 460 b. The centrum and base of the arch of a nearly similar dorsal vertebra, which has been longitudinally bisected in

Fig. 60.

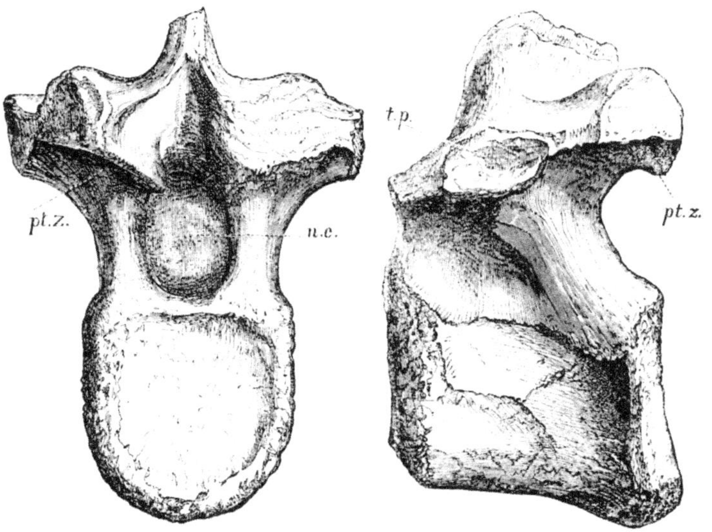

? *Syngonosaurus macrocercus.*—Posterior and left lateral aspect of an imperfect dorsal vertebra; from the Cambridge Greensand. ¼. *t.p*, transverse process; *pt.z*, postzygapophysis; *n.c*, neural canal. (From the 'Quart. Journ. Geol. Soc.')

a vertical plane; from the Cambridge Greensand. Probably associated with the preceding; noticed by the writer, *l. c.* *Presented by Sir R. Owen, K.C.B.*, 1884.

R. 460 c. An imperfect dorsal vertebra probably belonging to the same individual as the preceding; from the Cambridge Greensand. *Presented by Sir R. Owen, K.C.B.*, 1884.

46376. The centrum of a sacral vertebra which belongs to the present or an allied form; from the Cambridge Greensand.
Cunnington Collection. Purchased, 1875.

GENUS *non det.*

The following specimen probably indicates a Dinosaur allied to the American Cretaceous forms described as *Ceratops*, and the European Neocomian *Struthiosaurus* (*Cratæomus*).

R. 1656. (*Fig.*) An imperfect bone which is provisionally regarded as one of the horn-cores of the skull; from the Wealden of Brook, Isle of Wight. This specimen, which is described and figured by the writer in the 'Quart. Journ. Geol. Soc.' vol. xlvi. pt. 2, is extensively hollowed at the base, and presents a striking resemblance to the horn-cores figured by Marsh in the 'Amer. Journ.' ser. 3, vol. xxxvi. pl. xi., as the types of *Ceratops*, and in a less marked degree to the larger bone figured by Seeley in the 'Quart. Journ. Geol. Soc.' vol. xxxvii. pl. xxvii. fig. 4, as *Cratæomus*, and which is stated by Marsh, in the 'Geol. Mag.' dec. iii. vol. vi. p. 207, to belong to the skull.

Presented by the Dorsetshire Museum, 1889.

Family IGUANODONTIDÆ (*supra*, i. p. 191).

Genus **CRYPTODRACO**, Lydekker [1].

Syn. *Cryptosaurus*, Seeley [2].

Typically known by the femur, which is characterized by its straight and stout shaft, on which the inner trochanter forms a ridge, and the absence of an intercondylar groove on the anterior aspect. On the proximal aspect of this bone the head is larger than the great trochanter.

Cryptodraco eumerus (Seeley [3]).

Syn. *Cryptosaurus eumerus*, Seeley [4].

The type species. Of comparatively small size, the length of the femur being 0,312 (12·25 inches).

Hab. Europe (England).

R. 1609. Cast of the right femur. The original, which is preserved in the Woodwardian Museum, Cambridge, and is the type, was obtained from the Oxford Clay, locality unknown. It is figured by Seeley in the 'Quart. Journ. Geol. Soc.' vol. xxxi. pl. vi., and is noticed by the writer on p. 45 of

[1] Quart. Journ. Geol. Soc. vol. xlv. p. 46 (1889).
[2] *Ibid* vol. xxxi. p. 149 (1875).—Preoccupied by Geoffroy (1831).
[3] *Loc. cit.*—*Cryptosaurus*.
[4] *Loc. cit.*

vol. xlv. of the same serial. The inner trochanter and the outer condyle are broken away.

Made in the Museum, 1888.

R. 1617. An imperfect dorsal vertebra which would agree in relative size with the type, and may be referable to the present form; from the Oxford Clay of Weymouth. This specimen, which has been bisected in a vertical longitudinal plane, shows the centrum and the base of the neural canal. The length of the former is 0,061 (2·4 inches), and its transverse diameter 0,051 (2 inches). It has somewhat the same degree of lateral compression as in the dorsals of *Iguanodon dawsoni*, and the neural canal is of the low type characteristic of the *Iguanodontidæ* as distinguished from the *Scelidosauridæ*.

Presented by R. Lydekker, Esq., 1889.

Genus **CAMPTOSAURUS**, Marsh[1].

Syn. *Camptonotus*, Marsh[2].
Including *Cumnoria*, Seeley (*suprà*, i. p. 195).

Teeth simpler than in the typical group of *Iguanodon*. Cervical vertebræ opisthocœlous; sacrals flattened inferiorly and not anchylosed. Manus with five normal digits. Ilium (*suprà*, i. p. 192, fig. 35) typically deep, with short and pointed pre- and postacetabular processes, the latter having a distinct ventral plate; pubis relatively stout, and as long as ischium. Femur slightly longer than tibia, with curved shaft and pendent inner trochanter; typically four functional digits in pes.

The following forms are provisionally referred to this genus, their small size distinguishing them from all known members of the Proiguanodont group of *Iguanodon*.

Camptosaurus valdensis, Lydekker[3].

Of the approximate size of *C. leedsi*, but doubtless, on account of its higher geological horizon, specifically distinct.

Hab. Europe (England).

Type, the femur No. R. 167, *suprà*, i. p. 195.

The mandibular ramus, No. 180, *suprà*, i. p. 227, is provisionally referred to this species.

[1] Amer. Journ. ser. 3, vol. xxix. p. 169 (1885).
[2] *Ibid.* vol. xviii. p. 501 (1879).—Preoccupied by Uhl for a genus of Orthopterous Insects in 1864.
[3] Quart. Journ. Geol. Soc. vol xlv. p. 48 (1889).

Camptosaurus prestwichi (Hulke).

Syn. *Iguanodon prestwichi*, Hulke (*suprà*, i. p. 196).
Cumnoria prestwichi, Seeley (*suprà*, i. p. 196).

Provisionally referred to this genus. Distinguished from the type species (*suprà*, i. p. 192) by the presence of a long preacetabular process to the ilium.

Camptosaurus leedsi, Lydekker [1].

Of small size, the length of the femur being 0,280 (11·2 inches); inner trochanter of latter in proximal half of shaft.

Hab. Europe (England).

R. 1608. Cast of the left femur. The original (fig. 61), which is

Fig. 61.

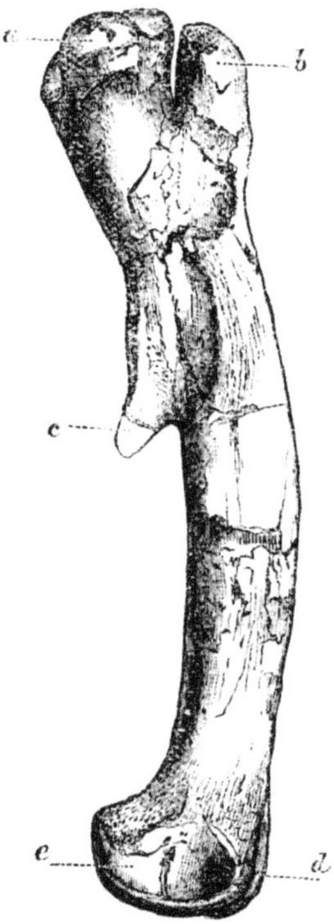

Camptosaurus leedsi.—Inner aspect of the left femur; from the Oxford Clay, near Peterborough. $\frac{1}{3}$. *a*, head; *b*, lesser trochanter; *c*, inner do.; *d*, intercondylar groove; *e*, entocondyle. (From the 'Quart. Journ. Geol. Soc.')

---

[1] Quart. Journ. Geol. Soc. vol. xlv. p. 48 (1889).

the type and only known specimen, was obtained from the Oxford Clay (Middle Jurassic) near Peterborough, Northamptonshire, and is preserved in the collection of A. N. Leeds, Esq., of Eyebury, near that town. It is described and figured by the writer in the 'Quart. Journ. Geol. Soc.' vol. xlv. pp. 45–48, fig. 3.

Made in the Museum, 1888.

Genus **IGUANODON** (*suprà*, i. p. 195).

Proiguanodont Group (suprà, i. p. 195).

I. fittoni shows that the flattening of the hæmal surfaces of the sacral vertebræ is not characteristic of all the members of this group. The ilium is always relatively deep, without reflection of the upper border. *Iguanodon prestwichi* is transferred from this group to *Camptosaurus*, see p. 258.

Iguanodon dawsoni (*suprà*, i. p. 196).

Ilium deep, without reflection of superior border; the preacetabular process being comparatively shallow, with a broad horizontal roof-like inward extension at its origin from the preacetabular notch, and an outward inclination of its lower border near the extremity; postacetabular portion long, deep, and rounded terminally, with slight inflection of inferior moiety; pubic process directed forward. Femur (*infrà*) with nearly straight shaft, and head set approximately at right angles to the latter; form of inner trochanter unknown.

These characters are given as supplementary to those mentioned in Pt. I. The type ilium is refigured in the 'Quart. Journ. Geol. Soc.' vol. xlvi. p. 37, fig. 1 B.

The specimens, Nos. R 604, 604 *a*, 811, 811 *a* (*suprà*, i. pp. 198–200), are referred below to *I. hollingtoniensis*, so that the sacrum and ischium of the present form are unknown.

R. 1627. An associated series of bones; from the Wadhurst Clay of Brede, Sussex. These specimens comprise the left scapula (*a*) with the articular surfaces imperfect; the shaft of the left humerus (*b*); the distal two thirds of the left ulna (*c*); the middle portion of the right ilium (*d*); the imperfect right femur (*e*), with the ventral surface embedded in matrix; the shafts of the two tibiæ, and the distal extremity of the left tibia (*f*): the third left metatarsal (*g*): and three imperfect caudal vertebræ. The ilium, which exhibits the acetabulum, the portion immediately above the same, and the preacetabular notch, agrees precisely with that of the type, and affords the grounds of the specific reference. The femur has lost the distal extremity, and the portion of the shaft on which the inner trochanter is situated is very imperfect. The length from the interval between the greater trochanter and the upper extremity of the intercondylar fissure is 0,834 (29 inches), and the total approximate length about 0,937 (37 inches). The head is placed nearly at right angles to the shaft, and there is but a slight convexity of the anterior border of the latter, in both of which respects this bone resembles the femur of *I. mantelli.*

These specimens are noticed by the writer in the 'Quart. Journ. Geol. Soc.' vol. xlvi. p. 38.

Dawson Collection. Purchased, 1889.

R. 1734. Cast of an almost entire late cervical vertebra, not improbably belonging to this species. The original was found in a nodule of hardened Wealden Clay on the beach at Hastings, and is in the possession of P. Rufford, Esq., of Hastings. It is described and figured by the writer in the 'Quart. Journ. Geol. Soc.' vol. xlvi. p. 44, fig. 3. With the exception of some damage to the rim of the posterior cup of the centrum, the specimen is perfect. The total height is 0,330 (13 inches), the height from the base of the neural canal to the summit of the spine 0,177 (7 inches), and the transverse diameter of the posterior cup of the centrum 0,164 (6·5 inches).

Made in the Museum, 1889.

Iguanodon fittoni, Lydekker[1].

Apparently of somewhat smaller dimensions than *I. dawsoni.* Ilium very deep, with its superior border convex and not reflected and the external surface very concave; the preacetabular process very deep, and probably comparatively short, without horizontal inward extension at its origin, and no outward inclination of the inferior border; postacetabular portion deep and pointed terminally, with the inferior moiety much inflected so as to form a wide shelf placed nearly at right angles to the lateral surface; preacetabular

[1] Geol. Mag. dec. iii vol. vi. p. 354 (1889).

notch very shallow, and pubic process deflected; interval between pre- and postacetabular notches short. Sacral vertebræ with rounded hæmal surfaces and anchylosed together. Femur with 'crested' inner trochanter placed lower down than in *I. mantelli.*

With the exception of its long preacetabular process, the contour of the ilium approximates to that of the type species of *Camptosaurus* (*suprà*, i. p. 192).

Hab. Europe (England).

R. 1635. (*Fig.*) The nearly entire left ilium; from the Wadhurst Clay of Shornden, near Hastings, Sussex. The type specimen; noticed by the writer in the Geol. Mag. dec. iii. vol. vi. p. 354, and figured in the 'Quart. Journ. Geol. Soc.' vol. xlvi. p. 37, fig. 1 C. The preacetabular process has been broken, and the greater portion of the pubic process is wanting. The portion remaining shows, however, that the latter was deflected, as in *Camptosaurus.* The ilium is proportionately deeper and more concave externally than in *I. dawsoni*, while the ischial attachment is less defined. This specimen was obtained at a higher horizon than the types of *I. dawsoni*, which also came from the same quarry. *Dawson Collection. Purchased*, 1889.

R. 1635 a. The proximal extremity of the left ischium, apparently belonging to the same individual as the ilium. This specimen was found in the same stratum at a distance of 25 yards westward of the spot where the ilium lay [1]. It is of the same form as the ischium of *I. hollingtoniensis.* *Dawson Collection.*

R. 1635 b. The imperfect sacrum, apparently belonging to the same individual as the preceding. The vertebræ, although larger, are of the same form as in the sacrum referred to *I. mantelli* (No. 37685, *suprà*, i. p. 220). This specimen was found 25 yards westward of the ischium, or 50 yards from the ilium. *Dawson Collection.*

R. 1635 c. The centrum of a caudal vertebra, found with the ischium. *Dawson Collection.*

R. 1635 d. Three imperfect and much worn upper teeth found in the neighbourhood of the type ilium. *Dawson Collection.*

[1] Associated specimens are often found far apart in these deposits; thus the two metatarsals of *Megalosaurus dunkeri* (p. 245) lay about 180 yards from one another.

Iguanodon hollingtoniensis, Lydekker[1].

Considerably smaller than *I. dawsoni*, the limb-bones being of the approximate size of those of *I. mantelli*, but the vertebræ somewhat larger. Ilium (imperfect) deep, with its superior border not reflected; the preacetabular process less deep than in *I. fittoni*, without inner extension; contour of postacetabular portion unknown; preacetabular notch deep, and pubic process directed forwards; interval between pre- and postacetabular notches long. Ischium as in *Camptosaurus*. Sacral vertebræ with flattened hæmal surfaces, and not anchylosed together. Femur (length 0,880 = 32 inches) with curved shaft, 'pendent' inner trochanter, placed in lower half of the bone, and the head set obliquely to the shaft. Anterior and middle dorsal vertebræ with compressed centra; rib-facet in middle dorsals rising to summit of neural platform. The scapula of the skeletons, Nos. R. 33, R. 1636, is relatively larger than in other species; that bone in the latter skeleton being remarkable for its length and slenderness.

In the structure of the sacrum, ischium, and femur this species agrees with *Camptosaurus*, from which it is, however, at once distinguished by the modification of the phalangeal of the pollex into a spine, as in typical species of *Iguanodon*.

Hab. Europe (England).

It is probable that the imperfect skeleton No. R. 33 (*suprà*, i. p. 226), from Hollington quarry, belongs to this species. The bone catalogued as a fragment of an ilium proves to be the imperfect right coracoid.

R. 1148 (*suprà*, i. p. 217). These specimens with the following (*Fig.*) associated ones are the types. The right femur, which shows the entire inner trochanter, is figured by the writer in the 'Quart. Journ. Geol. Soc.' vol. xlvi. p. 41, fig. 2.

R. 1629. An associated series of bones belonging to the same individual as the preceding; from the Wadhurst Clay of Hollington quarry, near Hastings, Sussex. These comprise the imperfect scapulæ (*a*); the crushed left radius and ulna (*b*); the phalangeal spine of the pollex (*c*); the left femur (*d*); the imperfect left tibia (*e*); the proximal portion of the left fibula (*f*); the second left metatarsal and the associated proximal phalangeal (*g*); and several fragments of ribs and other bones. The femur, which has

[1] Geol. Mag. dec. iii. vol. vi. p. 355 (1889).

been crushed and broken into several fragments, and has lost the extremity of the inner trochanter, agrees approximately in size with that of *I. mantelli* (*e. g.* No. 2650, *suprà*, i. p. 223); its total length being 0,880 (32 inches). The convexity of the shaft and the obliquity of the head are well shown. The length from the upper extremity of the intercondylar fissure to the interval between the greater trochanter and the head is 0,644 (25·5 inches).

Dawson Collection. Purchased, 1889.

R. 1632. Ten associated imperfect sacral and caudal vertebræ; from Hollington quarry. These specimens were obtained a short distance from the preceding, and almost certainly belong to the same individual; they are noticed by the writer in the 'Quart. Journ. Geol. Soc.' *op. cit.* p. 40. The sacrals, which have flattened hæmal surfaces and were not anchylosed together, precisely resemble those of No. R. 811 (*infrà*). *Dawson Collection. Purchased*, 1889.

R. 1632 a. Three imperfect and much crushed cervical vertebræ and the distal portion of a metatarsal, associated with the preceding. The cervicals resemble those of R. 33.

Dawson Collection.

R. 1632 b. The terminal phalangeal of the second digit of the left pes, associated with the preceding. *Dawson Collection.*

R. 604–811. These specimens (*suprà*, i. pp. 198, 199), which came (*Fig.*) from the Hollington quarry, are referred to this species on account of their resemblance to the preceding, and also on the ground that the undermentioned ilium is different from that of *I. dawsoni*. The left ischium is figured in Part I. p. 199, in association with the ilium of the latter species. The dorsal vertebræ agree with those of No. R. 1148.

R. 811 b. The imperfect and crushed left ilium, associated with the preceding specimens; from Hollington quarry. Noticed by the writer in the 'Quart. Journ. Geol. Soc.' vol. xlvi. p. 42[1]. The extremities of the preacetabular and pubic processes are wanting, and the postacetabular portion is imperfect. The preacetabular process differs from that of *I. dawsoni* in the absence of an inner extension. Compared with the ilium of *I. fittoni*, this specimen differs by the shallower preacetabular process, the longer interval

[1] Incorrectly said to be the specimen figured on p. 37, fig. 1 E.

between the pre- and postacetabular notches, the deeper preacetabular notch, and apparently by the more forward direction of the pubic process. It apparently closely resembles the ilium of No. R. 1636; its diameter being probably somewhat increased by pressure.

Dawson Collection. Purchased, 1884.

R. 1636. (*Fig.*) An imperfect skeleton, probably referable to this species; from the Wadhurst Clay of Shornden, near Hastings. The bones preserved comprise a number of imperfect vertebræ and ribs; the imperfect right scapula and coracoid; a humerus, radius, and ulna; part of the right ilium; the left femur, with the shaft very imperfect; the head of a tibia; part of a fibula; an imperfect astragalus, with other bones of the feet; and a sternal (?) bone. The ilium is figured (reversed) by the writer in the 'Quart. Journ. Geol. Soc.' vol. xlvi. p. 37, fig. 1 E[1].

Dawson Collection. Purchased, 1889.

R. 1634. A middle dorsal vertebra, probably referable to this species; from Shornden. This specimen, in which the neural spine is wanting, agrees in size with the earlier dorsals of the preceding specimens. The costal facet is placed on the level of the neural platform, and thus resembles the figured dorsal of *I. dawsoni* (*suprà*, i. p. 197); but in addition to its much smaller dimensions the present specimen is distinguished by its much more compressed and carinated centrum. The length of the centrum is 0,110 (4 inches), its height 0,089 (3·5 inches), and its width 0,076 (3 inches).

Dawson Collection. Purchased, 1889.

Specifically Undetermined Specimens from the Wadhurst Clay near Hastings. All belong to the Dawson Collection. Purchased, 1889.

R. 1633. A left humerus, imperfect proximally; from Hollington.

R. 1633 a. The centrum of a dorsal vertebra, apparently associated with the preceding.

R. 1631. An imperfect ulna.

R. 1630. The two extremities of the tibia, and the imperfect astragalus of the left side. Although slightly the astragalus resembles the corresponding bone of No. R. 33 (see p. 262).

[1] In the text of the paper cited the specimen No. R. 8111, b (p. 263), is incorrectly given as the figured one.

R. 1628. A left fibula; from Shornden. Since this bone is larger than the fibula of *I. hollingtoniensis*, it may probably be referred to *I. dawsoni.*

R. 1628 a. The centrum of a late caudal vertebra, associated with the preceding.

R. 1628 b. The centrum of a later caudal vertebra, associated with the preceding.

The following specimen is from the Ashdown Sand, below the Wadhurst Clay.

R. 1637. A slightly imperfect cervical vertebra; from Ecclesbourne, Sussex. This specimen resembles the cervicals of *I. hollingtoniensis.* *Dawson Collection.*

Euiguanodont Group (suprà, i. p. 200).

The ilium is always shallow, with a reflected superior border.

Iguanodon bernissartensis (*suprà*, i. p. 201).

R. 1735. Cast of the left ilium. The original is part of the type skeleton of *I. seelyi*, and was obtained from the Wealden of Brook, Isle of Wight; it is figured by Hulke in the 'Quart. Journ. Geol. Soc.' vol. xxxviii. pl. iv. fig. 1 (*vide suprà*, i. p. 214, fig. 46). A comparison with Nos. 28685, 28686 (*suprà*, i. p. 213) shows that the latter ilium belongs to a somewhat larger individual, but (allowing for distortion) has the same general contour. The present specimen shows a similar longitudinal ridge on the outer aspect of the postacetabular portion, although, as stated in Part I., this is not represented in the figure.
Made in the Museum, 1889.

R. 1658. The original of No. 28679 (*suprà*, i. p. 217).
Presented by the Dorsetshire Museum, 1889.

R. 1657. The proximal phalangeal of the fourth digit of the right pes; from the Wealden of the Isle of Wight. The distal portion appears to have been crushed down.
Presented by the Dorsetshire Museum, 1889.

No. R. 1148 (*suprà*, i. p. 217) has been made the type of *I. hollingtoniensis.*

Iguanodon mantelli (*suprà*, i. p. 218).

The ilium No. R. 113 (i. p. 222) is figured by the writer in the 'Quart. Journ. Geol. Soc.' vol. xlvi. p. 37, fig. 1 D.

The calcanea Nos. R. 131 & 28671 (i. p. 225) belong to Sauropoda.

The following specimens are not specifically determined.

R. 1659. The terminal phalangeal of the middle digit of the pes of a large form; from the Wealden of the Isle of Wight. *Presented by the Dorsetshire Museum*, 1889.

[*Cetiosaurus brachyurus*, Owen [1].]

2109. The centrum and base of the neural arch of a vertebra apparently referable to the anterior part of the caudal region of a comparatively small *Iguanodon* (? *I. mantelli*); from the Wealden of Sussex. One of the types of *C. brachyurus*; noticed by Owen in the 'Rep. Brit. Assoc.' for 1841, p. 100. *Mantell Collection. Purchased*, 1838.

Generically Undetermined Specimen.

46770. An imperfect sacral vertebra; from the Wealden of the Isle of Wight. This specimen, which retains a portion of the rib of the left side, agrees precisely with the second sacral of the undetermined specimen No. R. 144 (Pt. I. p. 246). The sacrals, Nos. 46773 and 2127 (Pt. I. p. 230), are likewise of the same type. These vertebræ are much less compressed than those of the sacrum referred to *I. mantelli*, and in their flattened hæmal surface approximate to *I. hollingtoniensis*, although the centra are lower and narrower. If the above-mentioned sacrum be rightly referred to *I. mantelli*, the present type of sacrum may belong to *Sphenospondylus*. *Purchased*, 1875.

Order ICHTHYOPTERYGIA (*suprà*, ii. p. 1).

Family ICHTHYOSAURIDÆ (*suprà*, ii. p. 6).

Genus **OPHTHALMOSAURUS** (*suprà*, ii. p. 81).

The specimens in the collection of A. N. Leeds, Esq., show that in the type species at least the teeth were very small and few in number, and that the femur was about one half the size of the humerus.

[1] Rep. Brit. Assoc. for 1841, p. 100 (1842).

Ophthalmosaurus icenicus (*supra*, ii. p. 9).

In addition to the characters given in Pt. II. it may be observed that the postaxial facet of the humerus is narrow and terminates in a point, and that the radial and ulnar facets are also narrow, and the former with a pointed termination. The late cervical vertebræ have no channel on the hæmal surface.

No. R. 472 (*suprà*, ii. p. 12) is undoubtedly a femur of this species; while No. R. 472 *a* (*ibid* p. 29) is also a femur of this or an allied species, in which the third facet is scarcely marked. The humeri Nos. 47885, 4647 (*ibid.* p. 11) show all the characters of this species, and may therefore perhaps be from the Oxford Clay.

Fig. 62.

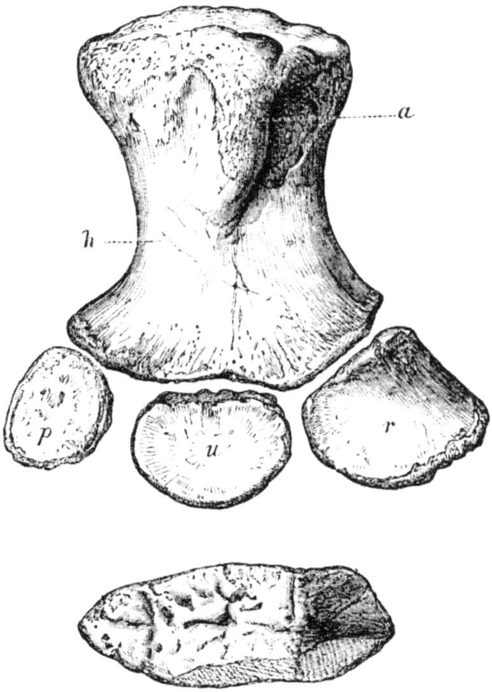

Ophthalmosaurus icenicus.—Dorsal aspect of part of the right pectoral limb, with a view of the distal extremity of the humerus. $\frac{1}{3}$. *h*, humerus; *a*, trochanteric ridge of do.; *r*, radius; *u*, ulna; *p*, pisiform.

R. 1667. The left scapula; from the Oxford Clay near Peterborough. Accords with the corresponding bone of the type.

Presented by A. N. Leeds, Esq., 1889.

R. 1668. (*Fig.*) The proximal portion of the right pectoral limb; from the Oxford Clay near Peterborough. These specimens comprise the humerus, radius, ulna, intermedium, and two other small bones. The humerus with the radius and ulna (fig. 62) are figured in the woodcut on the preceding page. *Presented by A. N. Leeds, Esq.*, 1889.

Ophthalmosaurus pleydelli, Lydekker (n. sp.).

Humerus shorter than in the preceding species, the postaxial facet being relatively larger and wider, with a blunt and rounded termination, and the radial and ulnar facets also wider and shorter, the latter having a rounded termination. Hindmost cervical vertebræ with a channel on the hæmal surface.

Fig. 63.

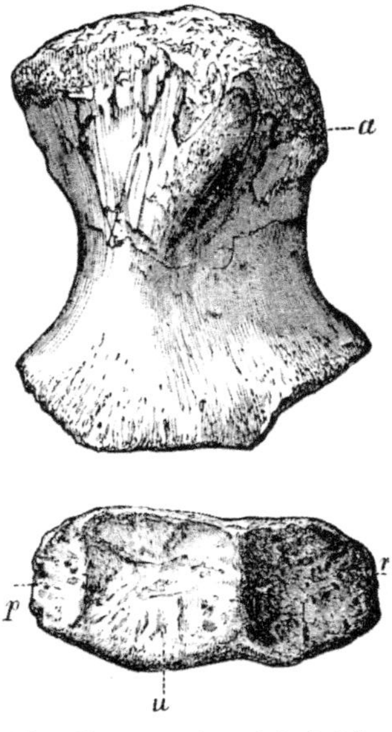

Ophthalmosaurus pleydelli.—Reversed view of the left humerus; from the Kimeridge Clay of Gillingham, Dorsetshire. ⅓ nat. size. *a*, trochanteric ridge; *r*, radial facet; *u*, ulnar do.; *p*, facet for pisiform.

There is no decisive evidence to prove that this provisional species is distinct from *Ichthyosaurus* (?) *dilatatus* (*suprà*, ii. p. 30), but if the specimens there provisionally referred to that species are correctly determined, and if (as is the case in the type species) large teeth are invariably wanting in *Ophthalmosaurus*, then the present form will be specifically distinct.

It is probable that the specimens Nos. 46491, 46473, 46407, 47885, and 46474 (*suprà*, ii. pp. 10, 11) are referable to this species.

Hab. Europe (England).

R. 1712. Cast of the left humerus. The original, which is the type, was obtained from the Kimeridge Clay of Gillingham, Dorsetshire, in association with other bones, and is preserved in the Museum of Dorset. It is figured in the woodcut on p. 268. A smaller propodial found with this bone is believed to be the humerus of an *Ichthyosaurus*, to which some large teeth also found at the same spot probably belong. *Made in the Museum*, 1889.

R. 1712 a. Cast of the centrum of one of the hindmost cervical vertebræ. The original (fig. 64), together with a number of other vertebræ, was found in association with the type

Fig. 64.

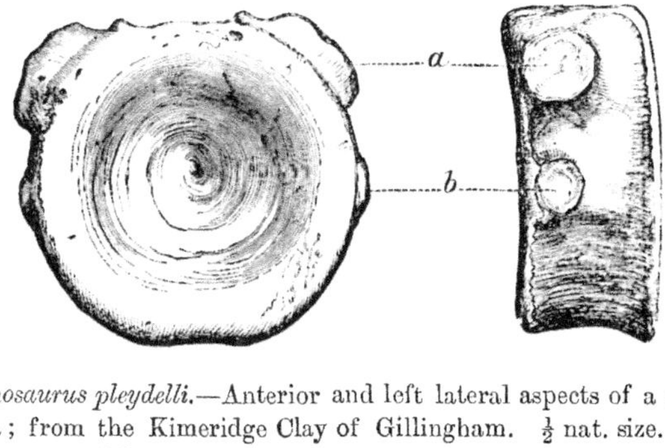

Ophthalmosaurus pleydelli.—Anterior and left lateral aspects of a cervical vertebra; from the Kimeridge Clay of Gillingham. ½ nat. size. *a*, upper; *b*, lower costal facet.

humerus, and is figured in the accompanying woodcut. The dimensions are, length 0,038 (1·15 inches), height 0,057 (2·25 inches), width (posterior face) 0,071 (2·8 inches). *Made in the Museum*, 1889.

Genus **ICHTHYOSAURUS** (*suprà*, ii. p. 12).

Ichthyosaurus (?) thyreospondylus (*suprà*, ii. p. 34).

R. 1684. The centrum of a caudal vertebra of the short type characteristic of this species; from the Portland Oolite of the Isle of Portland. This appears to be the only Ichthyopterygian specimen recorded from the English Portlandian.
Presented by G. Clifton, Esq., 1889.

Ichthyosaurus communis (*suprà*, ii. p. 41).

R. 1996. Slab showing the entire skeleton of a comparatively small individual; from the Lower Lias of Lyme-Regis. The dorsal aspect of the skull and limbs, and the left lateral aspect of the vertebræ are shown. The contour of the skull is better shown than in other specimens in the Museum. *Presented by F. Harford, Esq.*, 1889.

R. 1997. Slab showing the right lateral aspect of the skull and cervical vertebræ of an individual agreeing in size with the preceding, and probably referable to this species.
Presented by F. Harford, Esq., 1889.

Fig. 65.

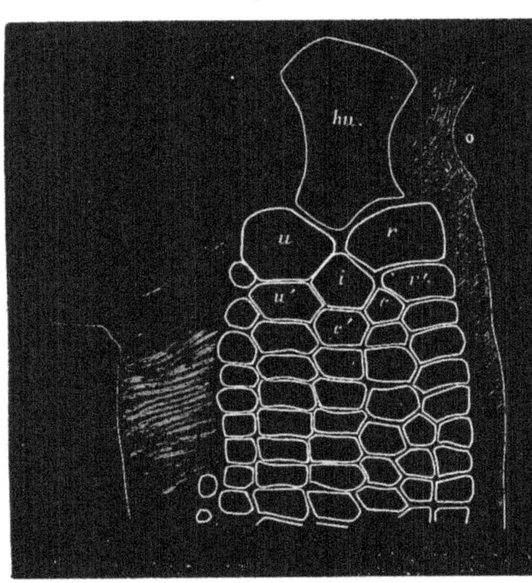

Ichthyosaurus intermedius.—Ventral aspect of part of the left pectoral limb from the Lower Lias of Barrow-on-Soar. *hu*, humerus; *r*, radius; *u*, ulna; *r'*, radiale; *i*, intermedium; *u'*, ulnare; *c*, centrale. ½ nat. size.

Ichthyosaurus intermedius (*suprà*, ii. p. 55).

R. 1664. (*Fig.*) One half of a split slab of rock, showing part of the thoracic region and the imperfect left pectoral limb of a small individual; from the Lower Lias of Barrow-on-Soar, Leicestershire. The other half of this specimen is preserved in the Museum at Leicester; the present half (fig. 65) is described and figured by the writer in the 'Geol. Mag.' dec. iii. vol. vi. p. 389. The contour of the soft parts of the paddle is clearly indicated on the matrix, which also exhibits transverse striæ which have been regarded as the impressions of muscular bundles in the postaxial flap of the paddle.

Presented by Montagu Browne, Esq., 1889.

Genus **TEMNODONTOSAURUS,** Lydekker[1].

Distinguished from *Ichthyosaurus* by its smooth and carinated teeth.

Temnodontosaurus platyodon (Conybeare).

Syn. *Ichthyosaurus platyodon*, Conybeare (*suprà*, ii. p. 94).

The type species.

Temnodontosaurus trigonodon (Theodori).

Syn. *Ichthyosaurus trigonodon*, Theodori (*suprà*, ii. p. 105).

Order SAUROPTERYGIA (*suprà*, ii. p. 118).

Since the publication of Part II. the writer has come to the conclusion that the elements of the pectoral girdle regarded by Mr. Hulke as the omosternum more probably represent the clavicles and interclavicle[2].

Family PLESIOSAURIDÆ (*suprà*, ii. p. 120).

Genus **PLIOSAURUS** (*suprà*, ii. p. 120).

Pliosaurus brachydirus (*suprà*, ii. p. 123).

Some or all of the following specimens may be referable to this species. The Portlandian specimens are the only recorded examples of the occurrence of this genus in that horizon.

[1] In Nicholson and Lydekker's Manual of Palæontology, 3rd ed. vol. ii. p. xi (1889).

[2] *Ibid.* p. 1069.

R. 1681. The crowns of three associated teeth; from the Kimeridge Clay of the Isle of Portland. One is of large size.
Presented by G. Clifton, Esq., 1889.

R. 1679. The head of an ischium: from the Portland Oolite of the Isle of Portland. *Presented by G. Clifton, Esq.*, 1889.

R. 1680. The distal portion of a propodial bone apparently referable to this genus; from the Portland Oolite of the Isle of Portland. *Presented by G. Clifton, Esq.*, 1889.

Pliosaurus ferox (*suprà*, ii. p. 145).

If the undermentioned be rightly referred to this species, it will be characterized by the absence of a central mammilla on the terminal faces of the cervical vertebræ, and by their prominent costal articulations; and also by the form of the epipodial bones, which are similar to those of *Peloneustes*, and thus more elongated than in the Kimeridgian species of *Pliosaurus*.

The originals of the following specimens, which belong to one individual, were obtained from the Oxford Clay near Peterborough, and are preserved in the collection of A. N. Leeds, Esq., of Eyebury, near that town. The casts were made in the Museum in 1889.

R. 1661. Cast of an anterior (? upper) tooth. The original is figured by the writer in the 'Quart. Journ. Geol. Soc.' vol. xlvi. pl. v. fig. 1. The ridges on the crown are placed more closely and the carinæ less defined than in typical examples.

R. 1661 a. Cast of the centrum and ribs of a late cervical vertebra. Original figured by the writer, *op. cit.* pl. v. fig. 2. The terminal faces are subcylindrical; the dimensions are, length 0,045 (1·75 inches), height 0,092 (3·6 inches), width 0,102 (4 inches).

R. 1661 b. Cast of the centrum of a later cervical vertebra. The original, which has been somewhat shortened by crushing, is noticed by the writer, *op. cit.* p. 51. The terminal faces are transversely ellipsoidal, and have a height of 0,095 (3·7 inches) and a width of 0,109 (4·3 inches). This vertebra has a strong general resemblance to the slightly smaller specimen, No. 47429 (*suprà*, ii. p. 146), but has more prominent costal articulations—a difference which may not improbably be explained partly by individual variation, and partly by the edges of the costal articulations of the latter having been broken. On the right side the two costal facets are separate, but on the left they have coalesced. This and the preceding specimen closely resemble the vertebræ from the Oxford Clay of Huntingdonshire, to which the name *P. pachydirus* (*vide suprà*, ii. p. 145) was applied.

Genus **PELONEUSTES** (*suprà*, ii. p. 151).

It appears that all the smaller forms hitherto included in *Pliosaurus* should be referred to this genus, in which the cervical centra are generally longer in proportion than in the latter, their length being usually greater than half the vertical diameter of the terminal faces. The vertebræ referred to *P. æqualis* (*suprà*, ii. p. 153) are, however, relatively shorter than in the Oxfordian species, unless this is due to the effects of pressure. It appears that the pelvic limb is larger than the pectoral, and that the humerus and femur have a nearly similar distal contour. In fig. 50, Pt. II. p. 154, the postaxial expansion of the distal extremity is broken away, the same being the case with the propodial of No. 47410 (*ibid.* p. 158). Both the latter specimen and the one represented in Pt. II. fig. 52, p. 155, are pelvic limbs.

Peloneustes evansi (Seeley).

Syn. *Pliosaurus evansi* (*suprà*, ii. p. 128).

A comparison of the undermentioned specimens with the cervical vertebræ of *P. philarchus* shows that they can only be distinguished by their larger dimensions. It is probable that the mandible and paddle in the Eyebury Collection, noticed in Pt. II. p. 154, belong to this form, since they agree in relative size with the vertebræ [1]

The originals of the following specimens belong to the type series obtained from the Oxford Clay near St. Neots, Huntingdonshire, and preserved in the Woodwardian Museum, Cambridge (*vide suprà*, ii. p. 129). *The casts were made in the Museum in* 1889.

R. 1713. Cast of the centrum of the fourth cervical vertebra. Original figured by Seeley in the 'Quart. Journ. Geol.

[1] See Quart. Journ. Geol. Soc. vol. xlvi. p. 52.

Soc.' vol. xxxiii. p. 718, and reproduced in Pt. II. p. 129 of this work [1], the dimensions being given on p. 128. Since the figure was drawn the anterior face of the specimen has been cleaned from matrix, and thus shows that the dotted extension of the upper part of the anterior face in the profile view is incorrect. Allowing for difference of serial position, this and the following specimen are undistinguishable in contour from the cervicals of *P. philarchus*, Nos. R. 1414, R. 1414 *a* (*suprà*, ii. p. 158).

R. 1713 a. Cast of the centrum of the fifteenth cervical vertebra.

Genus **CIMOLIOSAURUS** (*suprà*, ii. p. 180).

Cimoliosaurus valdensis (*suprà*, ii. p. 188).

R. 1612. The imperfect neural arch of a dorsal vertebra; from the Wadhurst Clay near St. Leonards. The left transverse process is wanting.

Dawson Collection. Purchased, 1889.

R. 1612 a. The centrum of a caudal vertebra, probably belonging to the same individual as the preceding specimen.

Dawson Collection.

Cimoliosaurus portlandicus (*suprà*, ii. p. 227).

The following specimens were obtained from the Portland Oolite of the Isle of Portland, and were presented by George Clifton, Esq., 1889.

R. 1674. The flattened centrum of a middle or posterior cervical vertebra. Of the type of No. 45904 (*suprà*, ii. p. 228), but probably larger.

R. 1674 a. One of the bones situated between the propodials and phalangeals.

R. 1674 b. A crushed phalangeal bone, associated with the two preceding specimens.

R. 1678. An associated series of 18 vertebral centra of a young individual. These comprise two anterior cervicals (*a*), 11 late cervicals and pectorals (*b*), 4 anterior dorsals (*c*), and

[1] The view of the terminal face is stated in Pt. II. to be from the anterior instead of the posterior aspect.

one later dorsal (*d*). The anterior dorsals are of the elongated type of No. 41238 *a* [1] (*suprà*, ii. p. 228), while the later cervicals are short like 45904 (*l. c.*). The anterior dorsals have elliptical terminal faces, while the faces of the later dorsal are subcylindrical.

R. 1676. Slab showing the dorsal aspect of the left coracoid, imperfect anteriorly. This specimen agrees in size with the scapula and coracoid, No. 41217 (*suprà*, i. p. 230); it is, however, much less imperfect than the latter.

R. 1677. A bone which is probably an ischium of an immature individual.

R. 1606. The slightly imperfect centrum of an anterior or middle cervical vertebra; from the Portland Oolite of the Isle of Portland. *No history.*

R. 1607. (*Fig.*) The slightly imperfect centra of two posterior cervical vertebræ; from the 'ash-bed' of the Purbeck of the Isle of Portland. Described, and one of the specimens figured by the donor in the 'Quart. Journ. Geol. Soc.' vol. xlvi. p. 48, fig. 5. The dimensions are—length 0,048 (1·9 inch), height 0,048 (1·9 inch), width 0,061 (2·4 inches). These are somewhat larger than those of the type cervical of *Plesiosaurus carinatus* figured by Phillips, and considerably larger than those of Nos. 41238 and 45904 (*suprà*, ii. p. 227). It was considered in Part II. that No. 41238 would agree in relative size with the type pelvic limb; but a subsequent comparison with the skeleton of *C. richardsoni* has shown that the present specimen presents nearer proportionate dimensions. The present specimen differs from Nos. 41238 and 45904 by the more prominent hæmal carina and the deeper depressions on either side; this being probably due to the waterworn condition and immaturity of the two former specimens. In this respect the present specimens agree with the one figured by Phillips. The undetermined propodial, No. 21974, mentioned in Part II. p. 227, may belong to this form, although indicating a small individual.

Presented by R. Lydekker, Esq., 1889.

[1] In Part II. pp. 228, 229, the number 41238 occurs three times, the second and third repetitions should respectively have the suffix *a* and *b*.

Cimoliosaurus richardsoni (*suprà*, ii. p. 240).

A figure of the proximal portion of the pectoral limb of the type specimen is given in fig. 66. An inspection of the type has shown

Fig. 66.

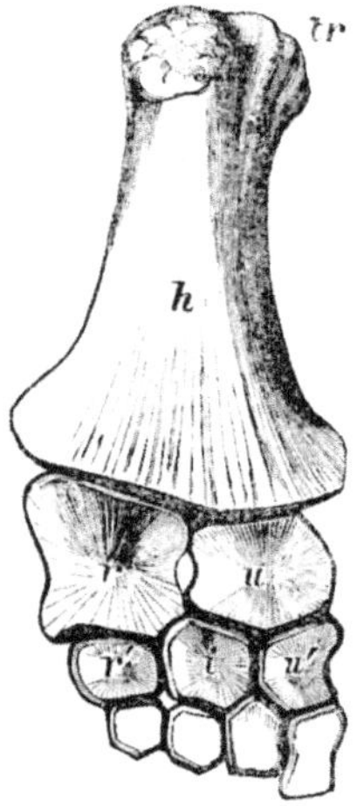

Cimoliosaurus richardsoni.—Ventral aspect of part of the right pectora limb; from the Oxford Clay of Weymouth. $\frac{1}{8}$. *h*, humerus; *tr*, trochanter of do.; *r*, radius; *u*, ulna; *r'*, radiale; *i*, intermedium; *u'*, ulnare. It is probable that *r'* and the subjacent bone are not the proper bones for the positions they occupy.

that the coracoids have a median bar connecting them with the scapulæ.

Specifically Undetermined Specimens.

R. 1683. The centrum of a caudal vertebra; from the Portland Oolite of the Isle of Portland. This specimen is characterized by its extreme shortness, in which respect it resembles the Kimeridgian *C. brachistospondylus*, although indicating a smaller individual than the types of the latter.
Presented by G. Clifton, Esq., 1889.

R. 1669. A left humerus; from the Oxford Clay near Peterborough. Except in being slightly longer, this bone is indistinguishable from the humerus of the type skeleton of *C. richardsoni* represented in woodcut, fig. 66, and indicates a form allied to that species or to *C. plicatus*.
Presented by A. N. Leeds, Esq., 1889.

R. 1670. An associated left radius and ulna; probably from the Kimeridge Clay; locality unknown. These bones have the general contour of those of *C. plicatus* and *C. richardsoni*, but the radius is more elongated; it is not improbable that they belong to *C. truncatus* or an allied form.
Presented by the Dorsetshire Museum, 1889.

R. 1682. A large humerus, wanting a portion of the shaft, probably referable to this genus; from the Kimeridge Clay of the Isle of Portland. This specimen closely resembles No. 31791 (*suprà*, ii. p. 148), and it is not improbable that both may be referable to *C. truncatus*.
Presented by G. Clifton, Esq., 1889.

Genus **PLESIOSAURUS** (*suprà*, ii. p. 252).

Plesiosaurus dolichodirus (*suprà*, ii. p. 255).

It may be observed that the name *Plesiosaurus priscus*, Parkinson[1], was applied to the vertebræ figured by Conybeare in his original description of the genus, but it cannot be determined to which of the forms subsequently named *P. dolichodirus* and *P. hawkinsi* these specimens are referable.

R. 1756. Slab showing a considerable part of the vertebræ and the pelvis, together with the paddles of the right side probably referable to the same individual; from the Lower Lias of Lyme-Regis. Sixteen cervical vertebræ now remain; these and the dorsals being seen from the right side, while the pelvis and anterior caudals have their dorsal surfaces exposed. The ilia are well displayed, and the paddles show all the features characteristic of this species. *Purchased*, 1889.

[1] Introduction to Oryctology, p. 294 (1822).

ALPHABETICAL INDEX

OF

GENERA AND SPECIES, INCLUDING SYNONYMS.

Actinodon, 183.
frossardi, 184.
latirostris, 184.
Ælurosaurus, 74.
curvimola, 77.
felinus, 74.
sp., 76.
Alligatorium, 234.
Andrias, 137.
scheuchzeri, 137.
tschudii, 139.
Anisopus, 221.
gracilis, 221.
Anomœpus, 221.
curvatus, 221.
gracillimus, 221.
intermedius, 221.
Anthodon, 112.
serrarius, 113.
sp., 114.
Anthracerpetum, 214.
crassosteum, 214.
Anthracosaurus, 158.
raniceps, 158.
russelli, 158.
Archegosaurus, 177.
decheni, 178.
latirostris, 184.
medius, 178.
minor, 178.
Arctosaurus, 250.
osborni, 251.

Batracocephalus, 212.
crassidens, 213.
validus, 214.
Bothriceps, 172.
australis, 172.
Bothriceps (*cont.*).
huxleyi, 172.
major, 174.
Bothriospondylus, 242.
elongatus, 237.
robustus, 236.
suffosus, 242.
Brachyops, 171.
laticeps, 171.
Branchiosaurus, 210.
amblystomus, 211.
gracilis, 211.
salamandroides, 211.
Brithopus, 98.
priscus, 98.
Brontozoum, 220.
giganteum, 220.
sillimani, 220.
validum, 220.
Bufo, 128.
melanostictus, 128.

Calamospondylus, 243.
foxi, 243.
Calochelys, 189.
lacertina, 189.
Camptosaurus, 257.
leedsi, 258.
prestwichi, 258.
valdensis, 257.
Capitosaurus, 147.
fronto, 149
nasutus, 147.
robustus, 149.
Cardiodon, 236.
rugulosus, 236.
Caulodon, 241.
precursor, 241.
Ceraterpetum, 197.
crassum, 198.
Ceraterpetum (*cont.*).
galvani, 197.
Ceratophrys, 127.
cornuta, 127.
Chælosaurus, 234.
Chelichnus, 218.
duncani, 218.
Chelyderpeton, 187.
vranii, 187.
Chelyosaurus, 187.
vranyi, 187.
Chirosaurus, 215.
barthi, 215.
herculis, 216.
stortonensis, 216.
Chirotherium, 215.
barthi, 215.
geinitzi, 215.
herculis, 216.
majus, 215.
minus, 217.
Cimoliosaurus, 274.
portlandicus, 274.
richardsoni, 276.
valdensis, 274.
Cirognathus, 49.
cordylus, 49.
Cistecephalus, 45.
arctatus, 48.
bathygnathus, 47.
chelydroides, 46.
leptorhinus, 47.
microrhinus, 46.
planiceps, 47.
Cochleosaurus, 186.
bohemicus, 186.
Cricotus, 176.
discophorus, 176.
heteroclitus, 176.
Crocodilæmus, 234.
jourdani, 234.

Crocodilus, 228.
spenceri, 228.
Cryptobranchus, 137.
alleghaniensis, 138.
diluvii-testis, 138.
maximus, 138.
primigenius, 137.
scheuchzeri, 137.
tschudii, 139.
Cryptodraco, 256.
eumerus, 256.
Cryptosaurus, 256.
eumerus, 256.
Cyamodus, 7.
laticeps, 8.
muensteri, 7.
rostratus, 7.
tarnowitzensis, 9.
Cyclotosaurus, 147.
robustus, 149.
Cynochampsa, 73.
laniaria, 73.
Cynodraco, 74.
major, 74.
serridens, 74.
Cynosuchus, 71.
suppostus, 71.
Cystignathus, 124.
labyrinthicus, 124.
ocellatus, 125.

Dacosaurus, 230.
maximus, 230.
Dawsonia, 213.
multidens, 213.
polydens, 213.
Dendrerpetum, 170, 223.
acadianum, 223.
deprivatum, 170.
oweni, 223.
pyriticum, 170.
Deuterosaurus, 99.
biarmicus, 99.
Diadectes, 104.
molaris, 104.
phaseolinus, 104.
sideropelicus, 104.
Dicynodon, 16.
baini, 24.
copei, 37.
curvatus, 23.
declivis, 34.
dubius, 21.
dunni, 36.
feliceps, 20.
lacerticeps, 18.
leoniceps, 19.
microtrema, 36.
murrayi, 37.
Dicynodon (*cont.*).
orientalis, 41.
pardiceps, 21.
parvidens, 26.
rectidens, 23
recurvidens, 19.
simocephalus, 22.
strigiceps, 30.
testudiceps, 24.
tigriceps, 24.
Dimetrodon, 106.
cruciger, 106.
Diplospondylus, 175.
Diplovertebron, 175.
Dolichosoma, 205.
angustatum, 206.
longissimum, 205.

Embolophorus, 108.
dollovorianus, 109.
fritillus, 108.
sp., 108.
Empedias, 101.
molaris, 102.
phaseolinus, 104.
Empedocles, 101.
molaris, 102.
Endothiodon, 64.
bathystoma, 65.
microps, 65.
uniseries, 65.
Eosaurus, 165.
acadianus, 166.
Epicordylus, 191.
Erpetocephalus, 168.
rugosus, 169.
Eryops, 191.
africanus, 193.
megacephalus, 191.
Eurosaurus, 98.
Eurycarpus, 53.
oweni, 50.
Euscelesaurus, 252.
bronni, 252.

Fritschia, 225
curtidentata, 225.

Galesaurus, 68.
planiceps, 68.
Garialis, 228.
dixoni, 228.
Gaudrya, 186.
latistoma, 187.
Gecsaurus, 230.
maximus, 230.
Glaridodon, 92.
Goniopholis, 229.
crassidens, 229.
minor, 229.
Gorgonops, 111.
tortus, 111.
Grallator, 222.
cuneatus, 222.

Heliarchon, 136.
furcillatus, 136.
Hoplosaurus, 243.
armatus, 243.
Hydrosalamandra, 137.
primigenia, 138.
prisca, 138.
Hylerpetum, 224.
dawsoni, 225.
longidentatum, 225.
Hylonomus, 201, 223.
aciedentatus, 224.
latidens, 224.
longicostatus, 201.
lyelli, 223.
wymani, 224.
Hyloplesion, 201.
longicostatum, 201.
Hyorhynchus, 80.
platyceps, 80.

Ichthyerpetum, 168.
bradleyæ, 169.
Ichthyosaurus, 270.
communis, 270.
intermedius, 271.
platyodon, 271.
trigonodon, 271.
Iguanodon, 259.
bernissartensis, 265.
dawsoni, 259.
fittoni, 260.
hollingtoniensis, 262.
mantelli, 266.
procursor, 241.
prestwichi, 258.

Keirognathus = **Cirognathus.**
Keraterpeton = **Ceraterpetum.**
Kistecephalus = **Cistecephalus.**

Labyrinthodon, 142.
brauni, 155.
diagnosticus, 153.
jægeri, 142.
salamandroides, 142.
schwarzenbergi, 160.
Latonia, 129.
gigantea, 129.
rugosa, 129.
seyfriedi, 129.

Lepidotosaurus, 214.
duffi, 214.
Lepterpetum, 199.
dobbsi, 199.
sp., 199.
Leptodactylus, 124.
ocellatus, 125.
pentadactylus, 124.
sp., 126.
Limnerpetum, 200.
elegans, 200.
laticeps, 200.
obtusatum, 200.
Loxomma, 163.
allmani, 164.
bavaricum, 164.
Lycosaurus, 78.
curvimola, 77.
pardalis, 78.
tigriceps, 78.

Macromerium, 160.
bayeri, 161.
schwarzenbergi, 160.
scoticum, 162.
simplex, 161.
Macropterna, 219.
divaricans, 219.
Macrosaurus, 163.
polyspondylus, 164.
Massospondylus, 248.
carinatus, 248.
hislopi, 248.
rawesi, 249.
Mastodonsaurus, 142.
giganteus, 142.
granulosus, 145.
indicus, 145.
jægeri, 142.
keuperinus, 145.
salamandroides, 142.
Megalobatrachus, 137.
scheuchzeri, 138.
Megalocephalus, 163.
macromma, 164.
Megalosaurus, 244.
dunkeri, 244.
oweni, 245.
Megalotriton, 135.
filholi, 135.
Melanerpetum, 209.
pulcherrimum, 209.
pusillum, 209.
Melosaurus, 186.
bohemicus, 186.
Metopias, 152.
diagnosticus, 153.
Metoposaurus, 152.
diagnosticus, 153.
Metriorhynchus, 233.
blainvillei, 134.
superciliosum, 234.
Microbrachis, 204.
mollis, 205.
pelicani, 204.
Microdon, 200.
laticeps, 200.
Micropholis, 174.
granulata, 174.
Molge, 136.
noachica, 136.
Morosaurus, 237.
becklesi, 237.
brevis, 237.

Naosaurus, 106.
claviger, 107.
cruciger, 106.
microdus, 107.
Neosodon, 239.
Nyrania, 166.
trachystoma, 166.
Nythosaurus, 68.
larvatus, 68.

Omosaurus, 251.
armatus, 252.
hastiger, 253.
Ophiderpetum, 206.
brownriggi, 206.
corvini, 208.
granulosum, 207.
nanum, 206.
pectinatum, 207.
vicinum, 207.
zieglerianum, 208.
Ophthalmosaurus, 266.
icenicus, 267.
pleydelli, 268.
Orinosaurus, 253.
capensis, 253.
Ornithoidichnites, 220.
giganteus, 220.
sillimani, 220.
Ornithopsis, 241, 243.
eucamerotus, 243.
hulkei, 243.
humerocristatus, 241.
leedsi, 242.
manseli, 241.
Orosaurus, 253
Orthocosta, 203.
microscopica, 203.
Orthopleurosaurus, 203.
microscopicus, 203.
Orthopus, 98.
primævus, 98.
Orthosaurus, 163.
pachycephalus, 164.
Oudenodon = **Udenodon.**
Oxyglossus, 121.
pusillus, 122.

Palæobatrachus, 130.
bohemicus, 132.
diluvianus, 130.
gigas, 131.
goldfussi, 130.
luedeckei, 132.
meyeri, 131.
sp., 133.
Palæosaurus, 246
platyodon, 246.
sternbergi, 189.
Pariasaurus, 114.
bombidens, 115.
serridens, 114.
Parioxys, 191.
Peloneustes, 273.
evansi, 273.
Pelorosaurus, 239.
armatus, 243.
becklesi, 237.
conybeari, 240.
humerocristatus, 241.
leedsi, 242.
manseli, 241.
Petrophryne, 174.
granulata, 174.
major, 174.
Phocosaurus, 82.
megischion, 83.
Pholidogaster, 195.
pisciformis, 195.
Placodus, 2.
andriani, 5.
bathygnathus, 2.
bombidens, 5.
gigas, 2.
hypsiceps, 2.
impressus, 6.
laticeps, 8.
muensteri, 7.
pachygnathus, 2.
quinimolaris, 6.
rostratus, 7.
Platyceps, 172.
wilkinsoni, 172.
Platyops, 182.
rickardi, 183.
stuckenbergi, 183.

Platyoposaurus, 182.
rickardi, 183.
stuckenbergi, 183.
Platypodosaurus, 62.
robustus, 62.
Plesiosaurus, 277.
dolichodirus, 277.
priscus, 277.
Pleurocœlus, 237.
nanus, 238.
valdensis, 237.
Pleurodus, 9.
bicolor, 9.
Pleuroneura, 210.
pellati, 210.
Pliosaurus, 271.
brachydirus, 271.
evansi, 273.
ferox, 271.
Probatrachus, 130.
Procolophon, 13.
griersoni, 13.
laticeps, 15.
minor, 14.
trigoniceps, 13.
Propappus, 118.
omocratus, 119.
Proteocordylus, 137.
diluvii, 137.
Protriton, 210.
petrolei, 210.
salamandroides, 211.
Ptychognathus, 33.
alfredi, 35.
boopis, 37.
declivis, 34.
depressus, 35.
latirostris, 35.
orientalis, 41.
pusillus, 49.
verticalis, 37.
Ptychosiagum, 33.
declive, 34.
latirostris, 35.
microtrema, 36.
murrayi, 37.
orientale, 41.

Rana, 122.
cornuta, 127.
diluviana, 130.
gigantea, 129.
labyrinthica, 124.
meriani, 122.

Rana (*cont*).
noeggerathi, 123.
ocellata, 125.
pentadactyla, 124.
pusilla, 122.
rugosa, 129.
Rhachitomus, 191.
valens, 191.
Rhamphorhynchus, 226.
gemmingi, 226.
jessoni, 226
Rhytidosteus, 194.
capensis, 194.
Ricnodon, 203.
trachylepis, 203.

Salamandra, 137.
gigantea, 137.
scheuchzeri, 137.
Salamandrella, 210.
petrolei, 210
Salamandroides, 142.
giganteus, 142.
Saurichnites, 219.
perlatus, 219.
rittlerianus, 219.
Scaloposaurus, 70.
constrictus, 71.
Scaphognathus, 226.
purdoni, 226.
Scincosaurus, 198.
crassus, 198.
Scelerocephalus, 185
hæuseri, 185.
Seeleya, 203
pusilla, 203.
Smilerpetum, 224.
aciedentatum, 229.
Sparagmites, 189.
lacertinus, 189.
Sparodus, 212
crassidens, 213.
validus, 212.
Sphenosaurus, 188.
sternbergi, 188
Stegosaurus, 251.
armatus, 252.
hastiger, 252.
Stelliosaurus, 201.
longicostatus, 201.
Steneosaurus, 234.
brevidens, 234.
dasycephalus, 232.

Stereorhachis, 105.
dominans, 105.
Suchodus, 231.
durobrivensis, 231.
Syngonosaurus, 254.
macrocercus, 254.

Tapinocephalus, 82.
atherstonei, 83.
Temnodontosaurus, 271
platyodon, 271.
trigonodon, 271.
Testudo, 218.
duncani, 218.
Thecodontosaurus, 246.
platyodon, 246.
Theriognathus, 64
microps, 65.
Tigrisuchus, 80
simus, 80.
Titanosuchus, 89.
ferox, 90.
Tomistoma, 228.
macrorhynchus, 228.
Trematosaurus, 154.
brauni, 155.
Tridentipes, 222.
elegantior, 222.
Trimerorhachis, 189.
insignis, 190.
Triton, 136.
noachicus, 136.
Tropidostoma, 36.
dunni, 36.

Udenodon, 28.
baini, 29.
brevirostris, 29.
greyi, 32.
magnus, 31.
megalops, 30.
prognathus, 31.
raniceps, 29.
rugosus, 29.
sp., 32.
strigiceps, 20.
Urocordylus, 196.
reticulatus, 196.
scalaris, 197.
wandesfordi, 196.

Weissia, 184.

ALPHABETICAL INDEX

OF

GENERA AND SPECIES TO THE ENTIRE WORK.

Acanthopholis, i. 183.
horridus, i. 183.
platypus, i 153.
Achelonia, iii 150.
formosa, iii 151.
Acichelys, iii. 150.
redenbacheri, iii. 151.
Acrosaurus, i. 293.
Actinodon, iv 183.
frossardi, iv 184.
latirostris, iv 184.
Adocus, iii 129
Ælurosaurus, iv 74.
curvimola, iv. 77.
felinus, iv. 74.
sp., iv. 76.
Æolodon, i. 101.
priscus, i. 101.
Aëtosaurus, i. 275.
Agomphus, iii 129.
Alligator, i. 44.
darwini, i. 46.
hantoniensis, i. 45.
latirostris, i. 45.
multiscutatus, i. 45.
punctatus, i. 45.
sp , i. 45.
Alligatorium, iv. 234.
Allopleuron, iii. 27.
hoffmanni, iii. 30.
Amphiemys, iii. 129.
Amphisaurus, i. 174.
Anchisaurus, i. 174.
Andrias, iv. 137.
scheuchzeri, iv. 137.
tschudii, iv. 139.
Anguisaurus, i. 293.
Anisopus, iv. 221.
gracilis, iv. 221.
Anomœpus, iv. 221.
curvatus, iv. 221.
gracillimus, iv. 221.
intermedius, iv. 221.
Anostira, iii. 143.
anglica, iii. 143.
ornata, iii. 143.
radiolina, iii. 144.
Anthodon, iv 112.
serrarius, iv. 113.
sp , iv. 114.
Anthracerpetum, iv. 214.
crassosteum, iv. 214
Anthracosaurus, iv. 158.
raniceps, iv. 158
russelli, iv. 158
Apatosaurus, i 145.
Aphelosaurus, i. 293.
lutevensis, i. 293.
Apholidemys, iii. 145.
Aplax, iii. 150.
oberndorferi, iii. 151.
Archæochelys, iii. 218.
valdensis, iii. 219.
Archegosaurus, iv. 177.
decheni, iv. 178.
latirostris, iv. 184.
medius, iv. 178.
minor, iv. 178.
Arctosaurus, iv. 250.
osborni, iv. 251
Ardeosaurus, i. 291.
brevipes, i. 291.
Argillochelys, iii. 40.
antiqua, iii. 41.
convexa, iii. 48.
cuneiceps, iii. 44.
sp., iii. 41, 44.
subcristata, iii. 47.
Aristosuchus, i. 157.
pusillus, i. 158.
Aspidonectes, iii. 7.
gergensi, iii. 11.
Atlantochelys, iii. 229.
Atlantosaurus, i. 144.
immanis, i. 145.
montanus. i. 145.
Atoposaurus, i. 302.
oberndorferi, i 302.
Aulacochelys, iii. 22.
circumsulcata, iii. 22.
Axestus, iii 7.

Baptanodon, ii. 60.
discus, ii. 60.
natans, ii. 60.
Baptemys, iii. 129, 143.
Batagur, iii. 120.
bakeri, iii. 124.
cautleyi, iii. 120.
dhongoka, iii. 125.
durandi. iii. 125.
falconeri, iii. 120.
kachuga, iii. 124.
lineata, iii. 124.
thurgi, iii. 120.
Batrachocephalus, iv. 212.
crassidens, iv. 213.
validus, iv. 214.
Batrachosaurus, i. 295.
Bellia, iii. 106.
crassicollis, iii. 107.
sivalensis, iii. 106.
theobaldi, iii 108.
Belodon, i. 124.
kapffi, i. 124.
plieningeri, i. 128.
Bothremys, iii. 174.
Bothriceps, iv. 172.
australis, iv. 172.
huxleyi, iv. 172.
major, iv. 174.

Bothriospondylus, i. 170, iv. 242.
elongatus, i. 142, iv. 237.
magnus, i. 142
robustus, i. 171, iv. 236.
suffosus, i. 170, iv. 242.
Brachydectes, i. 84.
major, i. 85.
minor, i. 85.
Brachyops, iv. 171.
laticeps, iv. 171.
Brachytrachelus, i. 27.
crassirostris, i. 28.
Branchiosaurus, iv. 210.
amblystomus, iv. 211.
gracilis, iv. 211.
salamandroides, iv. 211.
Brimosaurus, ii. 180.
grandis, ii. 247.
Brithopus, iv. 98.
priscus, iv. 98.
Brontosaurus, i. 144.
excelsus, i. 144.
Brontozoum, iv. 220.
giganteum, iv. 220.
sillimani, iv. 220.
validum, iv. 220.
Bufo, iv. 128.
melanostictus, iv. 128.

Cachuga, iii. 123.
dhongoka, iii. 125.
lineata, iii. 124.
sp, iii. 127.
tectum, iii. 127.
Caiman, i. 44.
Calamospondylus, iv. 243.
foxi, iv. 243.
Calochelys, iv. 189.
lacertina, iv. 189.
Camptonotus = **Camptosaurus.**
Camptosaurus, i. 192, iv. 257.
dispar, i. 192.
leedsi, iv. 258.
prestwichi, iv. 258.
valdensis, iv. 257.
Capitosaurus, iv. 147.
fronto, iv. 149.
nasutus, iv. 147.
robustus, iv. 149.
Cardiodon, iv. 236.
rugulosus, iv. 236.
Caulodon, iv. 241.
precursor, iv. 241.
Ceraterpetum, iv. 197.
crassum, iv. 197.
galvani, iv. 198.
Ceratochelys, iii. 159.
sthenurus, iii. 160, 166.
Ceratophrys, iv. 127.
cornuta, iv. 127.
Ceratosaurus, i. 157.
nasicornis, i. 157.
Cetiosaurus, i. 136.
brevis, i. 139, iv. 237.
glymptonensis, i. 138.
humerocristatus, i. 151.
longus, i 137.
medius, i. 137.
oxoniensis, i. 137.
Chælosaurus, iv 234.
Chaibassia, iii. 99.
theobaldi, iii 99.
tricarinata, iii. 99.
Chelichnus, iv 218.
duncani, iv 218.
Chelodina, iii. 168.
longicollis, iii. 168.
Chelone, iii. 27.
acuticeps, iii. 53.
antiqua, iii. 41.
belli, iii. 190.
benstedi, iii. 34.
breviceps, iii. 41, 64.
camperi, iii. 30.
convexa, iii. 44, 48.
costata, iii. 190.
crassicostata, iii. 60.
cuneiceps, iii. 44.
declivis, iii 48.
gigas, iii. 225.
girundica, iii. 30.
harvicensis, iii. 64.
hoffmanni, iii. 30.
jessoni, iii. 36.
latiscutata, iii. 60.
longiceps, iii. 57.
mantelli, iii. 190.
planiceps, iii. 203.
planimentum, iii. 64.
pulchriceps, iii. 176.
sp., iii. 30, 35.
subcarinata, iii. 47.
subcristata, iii. 47.
trigoniceps, iii. 53.
Chelonemys, iii. 184.
ovata, iii. 184.
plana, iii. 184.
Chelonia = **Chelone.**
Chelonides, iii. 197.
wittei, iii. 197.
Clelyderpeton, iv. 187.
vrani, iv. 187.
Chelydra, iii. 134.
decheni, iii. 136.
murchisoni, iii. 135.
œningensis, iii. 135.
rossignonii, iii. 135.
serpentina, iii. 135.
Chelydropsis, iii. 134.
Chelyosaurus, iv. 187.
vranyi, iv. 187.
Chelys, iii. 220.
blakei, iii. 220.
Chelytherium, iii. 222.
obscurum, iii 222.
Chirosaurus, iv. 215.
barthi, iv. 215.
herculis, iv. 216.
stortonensis, iv 216.
Chirotherium, iv. 215.
barthi, iv. 215.
geinitzi, iv. 215.
herculis, iv 216.
majus, iv. 215.
minus, iv 217.
stortonense, iv. 216.
Chitra, iii. 4.
indica, iii. 4.
Chlamydosaurus, i. 276.
bennetti, i. 276.
kingi, i. 276.
Chondrosteosaurus, i. 146.
gigas, i. 142, 146.
magnus, i. 146.
Chrysemys, iii. 118.
bicarinata, iii. 119.
testudiniformis, iii. 118.
Cimoliochelys, iii. 27.
benstedi, iii. 34.
Cimoliornis, i. 10.
diomedius, i. 13.
Cimoliosaurus, ii. 180, iv. 274.
australis, ii. 220.
bernardi, ii. 185.
biangulatus, ii. 227.
brachistospondylus, ii. 200.
brevior, ii. 243.
cantabrigiensis, ii. 183.
chilensis, ii. 222.
constrictus, ii. 212.
durobrivensis, ii. viii.
eurymerus, ii. 205.
grandis, 247.
haasti, ii. 215.
hoodi, ii. 245.
latispinus, ii. 222.
limnophilus, ii. 224.

Cimoliosaurus (*cont.*).
magnus, ii. 211.
neocomiensis, ii. 223.
oxoniensis, ii. 209.
planus, ii. 217.
plicatus, ii. 234.
portlandicus, ii. 227, iv. 274.
richardsoni, ii. 240, iv. 276.
smithi, ii. 215.
tenuis, ii. 188.
triangulatus, ii. 227.
trochanterius, ii. 190.
sp., ii. 210, 211, 227, 244, 445.
truncatus, ii. 230.
valdensis, ii. 188, iv. 274.
Circgnathus, iv. 49.
cordylus, iv. 49.
Cistecephalus, iv. 45.
arctatus, iv. 48.
bathygnathus, iv. 47.
chelydroides, iv. 46.
leptorhinus, iv. 47.
microrhinus, iv. 46.
planiceps, iv. 47.
Cistudo, iii. 103.
europæa, iii 103.
Claudius, iii 129.
Clemmys, iii. 105.
bicarinata, iii. 119.
hamiltoni, iii. 105.
hydaspica, iii. 106.
kargi, iii. 135.
palæindica, iii. 105.
punjabiensis, iii. 108.
sivalensis, iii. 106.
theobaldi, iii. 108.
watsoni, iii 120.
Clidastes, i. 272.
sp., i. 272.
Cochleosaurus, iv. 186
bohemicus, iv. 186.
Cœlurus, i. 156.
daviesi, i. 156.
fragilis, i. 156.
Coloborhynchus, i. 10.
cuvieri, i. 12.
sedgwicki, i. 15.
Colossochelys, iii. 71.
atlas, iii. 74.
Coluber, i. 250.
gervaisi, i. 252.
molurus, i. 253.
mucosus, i. 250.
oweni, i. 252.
papyraceus, i. 251.
sansaniensis, i. 251.
Coluber (*cont.*).
scalaris, i. 252.
Colymbosaurus, ii. 180.
dutertrei, ii. 248.
megadirus, ii. 190.
Compsemys, iii. 137.
Compsognathus, i. 156.
longipes, i. 156.
Conchiosaurus, ii. 295.
clavatus, ii. 295.
Coniasaurus, i. 289.
crassidens, i. 289.
Coryphodon, i. 250.
blumenbachi, i. 250.
Cretornis, i. 10.
hlavatschi, i. 14.
Cricosaurus, i. 95.
elegans, i. 98.
Cricotus, iv. 176.
discophorus, iv. 176.
heteroclitus, iv. 176.
Criorhynchus, i. 10.
simus, i. 16.
Crocodilæmus, i, 95, 98, iv. 234.
jordani, iv. 234.
Crocodilus, i. 53.
arduini, i. 60.
basifissus, i. 70.
biporcatus, i. 59.
bollensis, i. 109.
bombifrons, i. 54.
brauniorum, i. 46.
bruchi, i 46.
cadomensis, i. 118.
cantabrigiensis, i. 75.
champsoides, i. 60.
clavirostris, i. 70.
crassidens, i 71.
cultridens, i. 90.
ebertsi, i. 50
elaverensis, i 46.
gangeticus, i 65.
hastingsiæ, i. 45.
isorhynchus, i. 64.
leptodus, i. 68.
longirostris, i. 65.
macrorhynchus, i. 64.
medius, i. 46.
palæindicus, i. 58.
palustris, i. 54.
plenideus, i. 53.
porosus, i. 59.
priscus, i. 101.
rateli, i. 46.
rathi, i. 46.
rollinati, i. 74.
sauli, i. 77.
sivalensis, i. 55.
Crocodilus (*cont.*).
spenceri, i. 60.
superciliosus, i. 96.
temporalis, i. 106.
tenuirostris, i. 65.
toliapicus, i. 60.
Cryptobranchus, iv. 137.
alleghaniensis, iv. 138.
diluvii-testis, iv. 138.
maximus, iv. 138.
primigenius, iv. 137.
scheuchzeri, iv. 137.
tschudii, iv. 139.
Cryptodraco, iv. 256.
eumerus, iv. 256.
Cryptosaurus, iv. 256.
eumerus, iv. 256.
Cumnoria, i. 195, iv. 257.
prestwichi, i. 196, iv. 258.
Cyamodus, iv. 7.
laticeps, iv. 8.
muensteri, iv. 7.
rostratus, iv. 7.
tarnowitzensis, iv. 7.
Cyclotosaurus, iv. 147.
robustus, iv. 149.
Cycnorhamphus, i. 9.
suevicus, i. 10.
Cynochampsa, iv. 73.
laniaria, iv. 73.
Cynodraco, iv. 74.
major, iv. 74.
serridens, iv. 74.
Cynosuchus, iv. 71.
suppostus, iv. 71.
Cystignathus, iv. 124.
labyrinthicus, iv. 124.
ocellatus, iv. 125.

Dacochelys, iii. 173.
delabechei, iii. 173.
Dacosaurus, i. 92, iv. 230.
lissocephalus, i. 92.
manseli, i. 92.
maximus, i. 92, iv. 230.
primævus, i. 92.
Dakosaurus (*see* **Dacosaurus**).
Damonia, iii. 104.
hamiltoni, iii. 105.
Dawsonia, iv. 213.
multidens, iv. 213.
polydens, iv. 213.
Dendrerpetum, iv. 170.
deprivatum, iv. 170.
pyriticum, iv. 170.
Dermatemys, iii. 129.

Deuterosaurus, iv. 99.
biarmicus, iv. 99.
Diadectes, iv. 104.
molaris, iv. 102.
phaseolinus, iv. 104.
sideropelicus, iv. 104.
Dicynodon, iv. 16.
baini, iv. 24.
copei, iv. 37.
curvatus, iv. 23.
declivis, iv. 34.
dubius, iv. 21.
dunni, iv. 36.
feliceps, iv. 20.
lacerticeps, iv. 18.
leoniceps, iv. 19.
microtrema, iv. 36.
murrayi, iv 37.
orientalis, iv. 41.
pardiceps, iv. 21.
parvidens, iv 26.
rectidens, iv. 23.
recurvidens, iv. 19.
simocephalus, iv. 22.
strigiceps, iv. 30.
testudiceps, iv. 24.
tigriceps, iv. 24.
Digerrhum, iii. 205.
bullocki, iii. 206.
Dimetrodon, iv. 106.
cruciger, iv. 106.
Dimorphodon, i. 37.
banthensis, i. 37.
macronyx, i. 37.
Dinodocus, i 136.
mackesoni, i 136
Dinosaurus, i 171.
gresslyi, i. 172.
Diopecephalus, i. 4.
kochi, i. 6.
longicollum, i. 9.
rhamphastinus, i 8.
Diplocynodon, i. 45
darwini, i. 46.
gracilis, i. 50.
hantoniensis, i. 45.
plenidens, i. 50.
rateli, i. 46.
sp , i. 50.
Diplodocus, i. 132.
longus, i. 132
Diplospondylus, iv. 175.
Diplovertebron, iv. 175.
Discosaurus, ii. 180.
vetustus, ii. 211.
Dolichosaurus, i. 275.
longicollis, i. 275.
Dolichosoma, iv. 205.
angustatum, iv. 206.
longissimum, iv. 205.
Doratorhynchus, i. 10.
validus, i 26.
Dorygnathus, i. 37.
banthensis, i. 37.
Dracænosaurus, i. 288.
croizeti, i. 288
Dracosaurus, ii. 288.
brennii, ii. 288.
giganteus, ii. 293.
Dryptosaurus, i. 169
aquilunguis, i. 170.

Echinodon, i. 247.
becklesi, i. 247.
Edestosaurus, i 272
Elasmosaurus, ii 180
constrictus, ii 212.
orientalis, ii 247.
platyurus, ii. 215.
Embolophorus, iv. 108
dolloverianus, iv. 109.
fritillus, iv. 108.
sp., iv 108.
Empedias, iv 101.
molaris, iv. 102.
phaseolinus, iv. 104.
Empedocles, iv. 101.
molaris, iv. 102.
Emyda, iii. 22
ceylonensis, iii 23.
vittata, iii. 23.
Emydura, iii. 168.
macquariæ, iii. 169.
Emys, iii. 102
benstedi, iii 34.
bicarinata, iii 119
comptoni, iii. 93
conybeari, iii. 173.
crassus, iii. 110.
delabechei, iii. 173.
dhongoka, iii. 125.
europæa, iii. 103.
hamiltoni, iii 105.
hordwelliensis, iii. 110.
lævis, iii. 171.
lineata, iii. 174.
lutaria, iii. 103.
mantelli, iii. 190.
namadica, iii 127.
nicoleti, iii. 109.
orbicularis, iii. 103.
parkinsoni, iii. 64.
scutella, iii 91.
tecta, iii. 127
tentoria, iii. 127.
testudiniformis, iii. 118.
thurgi, iii. 120
turfa, iii. 103.
Enaliochelys, iii. 148.
chelonia, iii 148.
Endothiodon, iv. 64.
bathystoma, iv 65.
microps, iv 65
uniseries, iv. 65.
Engyommasaurus, i. 105.
brongniarti, i. 107.
Eosaurus, iv 165.
acadianus, iv. 166
Eosphargis, iii 225.
gigas, iii. 225.
Epicampodon, i. 174.
indicus, i. 174
Epicordylus, iv. 191.
Eretmosaurus, ii. 248.
rugosus, ii. 249.
sp., ii. 251.
Erpetocephalus, iv. 168.
rugosus, iv. 169.
Erquelinnesia, iii. 51.
gosseleti, iii 60.
Eryops, iv. 191.
africanus, iv. 193.
megacephalus, iv. 191.
Eucamerotus, i 146.
Euclastes, iii. 51.
gosseleti, iii. 60.
sp , iii 68
Eurosaurus, iv. 98.
Euryaspis, iii. 150.
radians, iii. 151.
Eurycarpus, iv. 50.
oweni, iv. 50.
Eurysternum, iii. 150.
crassipes, iii. 151.
wagleri, iii. 151.
Euscelesaurus, iv. 252.
browni, iv. 252.

Fritschia, iv. 225.
curtidentata, iv. 225.

Galesaurus, iv. 68.
planiceps, iv. 68.
Garialis, i. 65, iv. 228.
crassidens, i. 71.
dixoni, i. 69, iv. 228.
gangeticus, i. 65.
hysudricus, i. 67.
leptodus, i. 68.
macrorhynchus, i. 64.
neocesariensis, i. 70.
pachyrhynchus, i. 69.
priscus. i. 101.
Gaudrya, iv. 186.
latistoma, iv. 187.
Gavialis (*see* **Garialis**).
Gavialosuchus, i. 62.
eggenburgensis, i. 63.

Geoemyda, iii. 99.
tricarinata, iii. 99.
Geosaurus, i. 271, iv. 230.
giganteus, i. 271.
maximus, i. 92, iv. 230.
soemmerringi, i. 271.
Gigantosaurus, i. 146, iv. 239
megalonyx, i. 151.
Glaridodon, iv. 92
Glossochelys, iii. 51.
harvicensis, iii. 64.
Glyptosaurus, i. 278.
Gnathosaurus, i 95.
Goniopholis, i. 79, iv. 229.
crassidens, i. 79, iv. 229.
minor, i. 80, iv. 229.
pugnax, i. 80.
simus, i 83.
tenuidens, i. 83.
undidens, i. 80.
Gorgonops, iv. 111.
torvus, iv. 111.
Grallator, iv. 222.
cuneatus, iv 222.
Gresslysaurus, i 171.
ingens, i. 172
Gryphus, i. 12.

Hadrianus, iii. 72
Hadrosaurus, i. 243.
foulki, i. 244.
Halilimnosaurus, i. 271.
crocodiloides, i. 271.
Hardella, iii. 119.
thurgi, iii. 120.
Helemys, iii. 216.
serrata, iii. 216.
Heliarchon, iv. 136.
furcillatus, iv. 136.
Heterosuchus, i. 74
valdensis, i. 74.
Homœosaurus, i. 291.
brevipes, i. 291.
maximiliani, i. 291.
Homopus, iii 91.
comptoni, iii. 93.
scutella, iii. 91.
Hoplosaurus, iv. 243.
armatus, iv. 243.
Hydraspis, iii. 169.
leithi, iii. 170.
macquarii, iii. 169.
œningensis, iii. 135.
Hydrosalamandra, iv. 137.
primigenia, iv. 138.
prisca, iv. 138.

Hydrosaurus, i. 281.
Hylæochampsa, i. 76.
vectiana, i. 77.
Hylæochelys, iii. 185.
belli, iii 190.
emarginata, iii. 189.
lata, iii. 195.
latiscutata, iii. 186.
Hylæosaurus, i. 185.
oweni, i. 185.
Hylerpetum, iv. 224.
dawsoni, iv. 225.
longidentatum, iv. 225.
Hylonomus, iv. 201, 203.
aciedentatus, iv 224.
latidens, iv. 224.
longicostatus, iv. 201.
lyelli, iv. 223
wymani, iv. 224.
Hyloplesion, iv. 201.
longicostatum, iv 201.
Hyorhynchus, iv. 80.
platyceps, iv. 80.
Hyperodapedon, i. 297.
gordoni, i. 298.
huxleyi, i 299
Hyposaurus, i. 90.
derbianus, i 91.
Hypsilophodon, i. 193.
foxi, i. 193.

Ichthyerpetum, iv. 169.
bradleyæ, iv. 169.
Ichthyosaurus, ii. 12, iv. 270
acutirostris, ii. 73.
advenus, ii 114.
æqualis, ii. 113.
angustidens, ii 113.
atavus, ii. 113.
australis, ii 113.
bonneyi, ii 114.
brachyspondylus, ii. 34.
breviceps, ii. 52.
campylodon, ii 15.
carinatus, ii. 113.
ceramensis, ii 113.
chalorodirus, ii. 113
chiroligostinus, ii. 94.
chiroparamecostinus, ii 55.
chiropolyostinus, ii. 41.
chirostrongulostinus, ii 83.

Ichthyosaurus (*cont.*).
communis, ii. 41, iv. 270.
coniformis, ii. 113.
conybeari, ii. 53.
cornalianus, ii. 114.
crassicostatus, ii. 113.
crassimanus, ii. 105.
cuvieri, ii. 22.
dilatatus, ii. 30.
doughtyi, ii. 114.
entheciodon, ii. 32.
fortimanus, iii. xviii.
gaudensis, ii. 113.
giganteus, ii. 94.
grandipes, ii. 83.
hectori, ii. 113.
hexagonus, ii. 113.
hildesiensis, ii. 30.
hygrodirus, ii. 113.
indicus, ii. 22.
ingens, ii. 113.
integer, ii. 71.
intermedius, ii. 55, iv. 270.
laticeps, ii. 89.
latifrons, ii. 89.
latimanus, ii. 53.
leptospondylus, ii 31.
lonchiodon, ii. 92.
longifrons, ii. 76.
longimanus, iii. xviii.
longipinnis, ii. 73.
longirostris, ii. 89.
macrophthalmus, ii. 113.
marathonensis, ii. 113.
megalodirus, ii. 113.
microdon, ii. 73.
nordenskioeldi, ii. 113.
normanniæ, ii. 22.
ovalis, ii. 29.
planartus, ii. 114.
platymerus, ii. 114.
platyodon, ii. 94, iv. 271.
polaris, ii. 114.
polyptychodon, ii. 16.
posthumus, ii. 22.
quadriscissus, ii. 73.
rheticus, ii 114.
sinuatus, ii. 83.
strombecki, ii. 16.
tenuirostris, ii. 83.
thyreospondylus, ii. 34
trigonodon, ii. 105, iv. 271.
trigonus, ii. 22.
triscissus, ii 114.
walkeri, ii. 114.
zetlandicus, ii. 76.

Idiochelys, iii. 184.
fitzingeri, iii. 184.
wagneri, iii. 184.
wagnerorum, iii. 184.
Iguana, i. 277.
europæa, i. 277.
Iguanodon, i. 195, iv. 259.
bernissartensis, i. 201, iv. 265.
dawsoni, i. 196, iv. 259.
fittoni, iv. 260.
foxi, i. 193.
hollingtoniensis, iv. 262.
mantelli, i. 218, iv. 266.
precursor, iv. 241.
prestwichi, i. 196, iv. 258.
seelyi, i. 201.
sp., i. 226.
Ischyrodon, ii. 120.
meriani, ii. 131.
Ischyrosaurus, i. 146.
manseli, i. 152.

Jacare, i. 44.

Kachuga=**Cachuga.**
Keirognathus = **Cirognathus.**
Keraterpeton=**Ceraterpetum.**
Kistecephalus=**Cistecephalus.**

Labyrinthodon, iv. 142.
brauni, iv. 155.
diagnosticus, iv. 153.
jægeri, iv. 142.
salamandroides, iv. 142.
schwarzenbergi, iv. 160.
Lacerta, i. 287.
bifidentata, i. 287.
gangetica, i. 65.
gigantea, i. 271.
lamandini, i. 287.
Lælaps, i. 169.
aquilunguis, i. 170.
Lariosaurus, ii. 284.
balsami, ii. 284.
Latonia, iv. 129.
gigantea, iv. 129.
rugosa, iv. 129.
siefriedi, iv. 129.
Lepidotosaurus, iv. 214.
duffi, iv. 214.
Lepterpetum, iv. 199.
dobbsi, iv. 199.
sp., iv. 199.
Leptodactylus, iv. 124.
ocellatus, iv. 125.
pentadactylus, iv. 124.
sp., iv. 126.
Leptorhynchus, i. 65.
clifti, i. 65.
crassidens, i. 71.
gangeticus, i. 65.
Lestosasaurus, i. 269.
Limnerpetum, iv. 200.
elegans, iv. 200.
laticeps, iv. 200.
obtusatum, iv. 200.
Liodon, i. 264.
anceps, i. 265.
brumbyi, i. 267.
dyspelor, i. 267.
haumuriensis, i. 267.
lundgreni, i. 265.
perlatus, i. 267.
sp., i. 266.
Liopleurodon, i. 120.
bucklandi, i. 160.
ferox, i. 145.
grossouvrei, i. 130.
Loxomma, iv. 163.
allmani, iv. 164.
bavaricum, iv. 164.
Lutkesaurus, ii. 172.
sp., ii. 179.
Lutremys, iii. 102.
europæa, iii. 103.
Lycosaurus, iv. 78.
curvimola, iv. 77.
pardalis, iv. 78.
tigrinus, iv. 78.
Lytoloma, iii. 51.
cantabrigiense, iii. 68.
crassicostatum, iii. 60.
longiceps, iii. 57.
planimentum, iii. 64.
sp., iii. 68.
trigoniceps, iii. 53.

Macellodus, i. 289.
brodiei, i. 289.
Machimosaurus, i. 103.
hugii, i. 103.
mosæ, i. 104.
Macrochelys, iii. 224.
Macroclemmys, iii. 134.
temmincki, iii. 134.
Macromerium, iv. 160.
bayeri, iv. 161.
Macromerium (*cont.*).
schwarzenbergi, iv. 160.
scoticum, iv. 162.
simplex, iv. 161.
Macromerosaurus, ii. 286.
plinii, ii. 286.
Macropterna, iv. 219.
divaricata, iv. 219.
Macrorhynchus, i. 87.
meyeri, i. 88.
schaumburgensis, i. 87.
Macrosaurus, iv. 163.
polyspondylus, iv. 164.
Macrospondylus, i. 108.
bollensis, i. 109.
Macrotrachelus, i. 4.
longirostris, i. 5.
Macrurosaurus, i. 153.
semnus, i. 153.
Manouria, iii. 71.
emys, iii. 72.
Massospondylus, iv. 248.
carinatus, iv. 248.
hislopi, iv. 248.
rawesi, iv. 249.
Mastodonsaurus, iv. 142.
giganteus, iv. 142.
granulosus, iv. 145.
indicus, iv. 145.
jægeri, iv. 142.
keuperinus, iv. 145.
salamandroides, iv. 142.
Mauisaurus, ii. 180.
gardneri, ii. 212.
hasti, ii. 215.
latibrachialis, ii. 247.
Megalania, i. 281.
prisca, i. 284.
Megalobatrachus, iv. 137.
scheucherzi, iv. 138.
Megalocephalus, iv. 163.
macromma, iv. 164.
Megalochelys, iii. 74.
sivalensis, iii. 74.
Megalosaurus, i. 159, iv. 244.
bredai, i. 168.
bucklandi, i. 159.
dunkeri, i. 163, iv. 244.
insignis, i. 163.

Megalosaurus (*cont.*).
oweni, iv. 245.
sp., i. 163.
Megalotriton, iv. 135.
filholi, iv. 135.
Megasternum, iii. 205.
koenigi, iii. 206.
Melanerpetum, iv. 209.
pulcherrimum, iv. 209.
pusillum, iv. 209.
Melanochelys, iii. 99.
Melitosaurus, i. 62.
champsoides, i. 64.
Melosaurus, iv. 186.
bohemicus, iv. 186.
Mesosaurus, ii. 300.
tenuidens, ii. 300.
tumidus, ii. 302.
Metopias, iv. 152.
diagnosticus, iv. 153.
Metoposaurus, iv. 152.
diagnosticus, iv. 153.
Metriorhynchus, i. 95, iv. 233.
blainvillei, iv. 234.
elegans, i. 98.
moreli, i. 97.
sp., i. 98.
superciliosum, i. 96, iv. 234.
Microbrachis, iv. 204.
mollis, iv. 205.
pelicani, iv. 205.
Microdon, iv. 200.
laticeps, iv. 200.
Micropholis, iv. 174.
granulata, iv. 174.
Miolania, iii. 159.
minor, iii. 160.
oweni, iii. 166.
platyceps, iii. 160.
Mixosaurus, i. 114.
Molge, iv. 136.
noachica, iv. 136.
Monitor, i. 281.
dracæna, i. 282.
Morosaurus, i. 134, iv. 237.
becklesi, iv. 237.
brevis, iv. 237.
grandis, i. 133.
Mosasaurus, i. 261.
bavaricus, i. 271.
belgicus, i. 261.
camperi, i. 261.
dekayi, i. 264.
giganteus, i. 261.
gracilis, i. 273.
Mosasaurus (*cont.*).
hofmanni, i. 261.
maximus, i. 264.
stenodon, i. 265.
Mosellosaurus, i. 105.
rostro-minor, i. 106.
Muræenosaurus, ii. 180.
brachyspondylus, ii. 190.
leedsi, ii. 234.
manseli, ii. 170.
Mystriosaurus, i. 108.
bollensis, i. 109.
brevior, i. 111.
brongniarti, i. 107.
chapmani, 110.
egertoni, i. 109.
laurillardi, i. 110.
longipes, i. 109.
macrolepidotus, i. 110.
mandelslohi, i. 109.
minimus, i. 112.
muensteri, i. 109.
schmidti, i. 109.
speciosus, i. 110.
stukelyi, i. 109, 110.
tenuirostris, i 107.
tiedmanni, i. 109.

Nannosuchus, i. 84.
gracilidens, i. 84.
Naosaurus, iv. 106.
claviger, iv. 107.
cruciger, iv. 106.
microdus, iv. 107.
Nectoportheus, i. 264.
Neosodon, iv. 239.
Neusticosaurus, ii. 285.
edwardsi, ii 285.
pusillus, ii. 285.
Nicoria, iii. 99.
tricarinata, iii. 99.
Nothosaurus, ii. 288.
aduncidens, ii. 293.
andriani, ii. 299.
angustidens, ii. 299.
bergeri, ii. 299.
giganteus, ii. 293.
latifrons, iii. xviii.
mirabilis, ii. 288.
mougeoti, ii. 299.
muensteri, ii. 299.
schimperi, ii. 299.
venustus, ii. 295.
Notochelone, iii. 70.
costata, iii. 70.
Notochelys, iii. 70.
costata, iii. 70.
Nuthetes, i. 247.
destructor, i. 247.
Nyrania, iv. 166.
trachystoma, iv. 166.
Nythosaurus, iv. 66.
larvatus, iv. 66.

Ocadia, iii. 108.
crassa, iii. 110.
nicoleti, iii. 109.
oweni, iii. 115.
Oligosomus, ii. 180.
Omosaurus, i. 177, iv. 251.
armatus, i. 177, iv. 252.
hastiger, i. 179, iv. 253.
sp., i. 180.
Oölithes, i. 122.
bathonicæ, i. 122.
Ophiderpetum, iv. 206.
brownriggi, iv. 206.
corvini, iv. 208.
granulosum, iv. 207.
nanum, iv. 206.
pectinatum, iv. 207.
vicinum, iv. 207.
zieglerianum, iv. 208.
Ophisaurus, i. 278.
moguntinus, i. 278.
Ophthalmosaurus, ii. 8, iv. 266.
cantabrigiensis, ii. 9.
icenicus, ii. 9, iv. 267.
pleydelli, iv. 268.
sp., ii. 12.
Orinosaurus, iv. 253.
capensis, iv. 253.
Ornithocephalus, i. 4.
antiquus, i. 5.
banthensis, i. 37.
brevirostris, i. 4.
crassirostris, i 28.
gemmingi, i. 31.
grandis, i. 32.
longicaudatus, i. 29.
longirostris, i. 5.
meyeri, i. 4.
muensteri, i. 30.
rhamphastinus, i. 8.
Ornithochirus, i. 10.
clifti, i. 25.
compressirostris, i. 11.
curtus, i. 26.
cuvieri, i 12.
daviesi, i. 23.
diomedius, i. 13.
fittoni, i. 15.
giganteus, i. 12.

Ornithochirus (*cont.*).
hlavatschi, i. 14.
nobilis, i 24.
sedgwicki, i. 15.
simus, i. 16.
sp., i. 25.
validus, i. 26.
Ornithodesmus, i. 42.
cluniculus, i. 42.
Ornithoidichnites, iv. 220.
giganteus, iv. 222.
sillimani, iv. 221.
Ornithopsis, i. 146, iv. 241, 243.
eucamerotus, i. 146, iv 243.
hulkei, i. 146, iv. 243.
humerocristatus, i. 151, iv. 241.
leedsi, i. 151, iv. 242.
manseli, i. 152, iv 241
Orophosaurus, ii 182.
pauciporus, ii. 247
Orosaurus, iv. 253.
Orthocosta, iv. 203.
microscopica, iv. 203.
Orthomerus, i. 241.
dolloi, i. 241.
Orthopleurosaurus, iv. 203.
microscopicus, iv. 203.
Orthopus, iv. 98.
primævus, iv. 98.
Orthosaurus, iv. 163
pachycephalus, iv. 164.
Osteopygis, iii. 51.
Oudenodon = **Udenodon.**
Oweniasuchus, i. 84.
major, i. 85.
minor, i. 85.
Oxyglossus, iv. 121.
pusillus, iv. 122.

Pachypleura, ii. 285.
edwardsi, ii. 285.
Pachyrhamphus, i. 27.
crassirostris, i. 28.
Pachyrhynchus, iii. 51.
gosseleti, iii. 60.
longiceps, iii. 57.
planimentum, iii. 64.
trigoniceps, iii. 53.
Palæobatrachus, iv. 130.
bohemicus, iv. 132.
diluvianus, iv. 130.
Palæobatrachus (*cont.*)
gigas, iv 131.
goldfussi, iv. 130.
grandipes, iv. 132.
luedeckei, iv. 132.
meyeri, iv 131.
sp., iv. 133.
Palæochelys, iii. 101.
bussenensis, iii. 101.
Palæomedusa, iii. 150.
testa, iii. 151.
Palæophis, i. 257.
porcatus, i. 258.
sp , i. 260
toliapicus, i 257.
typhæus, i. 258.
Palæopython, i, 253
cadurcensis, i 254
filholi, i. 255.
Palæornis, i. 10.
clifti, i 25.
Palæosaurus, iv. 246.
platyodon, iv. 246.
strenbergi, iv. 188.
Palæovaranus, i. 279.
caylusi, i 279.
Paleryx, i. 253.
depressus, i. 255.
filholi, i. 255.
rhombifer, i. 254.
Pangshura, iii. 123.
flaviventris, iii. 127.
tecta, iii. 127.
tentoria, iii 127.
Parachelys, iii 195
eichstædtensis, iii. 195.
Parasuchus, i 130.
hislopi, i 130
Pariasaurus, iv. 114.
bombidens, iv. 115.
serridens, iv. 114.
Parioxys, iv 191.
Pelagosaurus, i. 105.
brongniarti, i. 107.
typus, i. 106.
Pelobatochelys, iii. 152.
blakei, iii. 153
Pelomedusa, iii 171.
galeata, iii. 171.
Peloneustes, ii 151, iv. 273.
æqualis, ii. 152.
evansi, iv. 273.
philarchus, ii 154.
Pelorosaurus, i. 145, iv. 239.
armatus, iv. 243.
becklesi, iv. 237.
Pelorosaurus (*cont.*).
conybeari, i 145, iv. 240.
humerocristatus, iv. 241.
leedsi, iv. 242.
manseli, iv. 241.
Peltocephalus, iii. 171.
lævis, iii 171.
Peltochelys, iii. 137.
duchastelli, iii. 138.
Peltosaurus, i. 278.
Periops, i. 252.
gervaisi, i 252.
Petrophryne, iv. 174.
granulata, iv. 174.
major, iv. 174.
Petrosuchus, i. 89.
lævidens, i 89
Phocosaurus, iv. 82.
megischion, iv 83
Pholidogaster, iv 195.
pisciformis, iv. 195.
Pholidosaurus, i. 87.
meyeri, i 88
schaumburgensis, i 87.
Phytosaurus, i. 124, iv. 235.
cubicodon, i. 124.
cylindricodon, i. 124.
plieningeri, i 128.
Pilemophis, i. 250.
sansaniensis, i. 251.
Piocormus, i. 292.
laticeps, i 292
Piptomerus, ii 180.
hexagonus, ii 247.
megaloporus, ii. 247.
microporus, ii 247.
Pistosaurus, ii 287.
longævus, ii. 287.
Placodus, iv. 2.
andriani, iv. 5.
bathygnathus, iv. 2.
bombidens, iv. 5.
gigas, iv 2.
hypsiceps, iv. 2.
impressus, iv. 6.
laticeps, iv. 8.
muensteri, iv. 7.
pachygnathus, iv. 2.
quinimolaris, iv. 6.
rostratus, iv. 7.
Placosaurus, i. 279.
margariticeps, i. 279.
Plastremys, iii. 185.
lata, iii. 195.
Platecarpus, i. 269.
curtirostris, i. 270.
mudgei, i. 271.
oweni, i. 270.
sp., i. 271.

Platemys, iii. 170.
bowerbanki, iii. 171.
bullocki, iii. 206.
concinna, iii. 206.
dixoni, iii. 190.
emarginata, iii. 189.
latiscutata, iii. 186.
mantelli, iii. 180.
ovata, iii. 206.
Platyceps, iv. 172
wilkinsoni, iv. 172.
Platychelys, iii. 216.
anglica, iii. 217.
oberndorferi, iii. 216.
Platycposaurus, iv. 182.
rickardi, iv. 183.
stuckenbergi, iv. 183.
Platyops, iv 182.
rickardi, iv. 183.
stuckenbergi iv. 183.
Platypodosaurus, iv. 62
robustus, iv. 62.
Plesiochelys, iii. 196.
brodiei, iii. 201.
hannoverana, iii 198.
rugosa, iii. 130
solodurensis, iii 197
sp., iii. 198, 199, 202
valdensis, iii. 199.
Plesicsaurus, ii. 251, iv. 277.
affinis, ii. 190.
arcuatus, ii 163.
australis, ii 220.
balticus, ii. 185.
bernardi, ii 185.
bitractensis, 283.
brachistospondylus, ii 200.
brachycephalus, ii. 266.
brachyspondylus, ii 139, 190
carinatus, ii. 168, 227.
chilensis, ii. 222.
cliduchus, ii. 255.
cœlospondylus, ii. 271.
constrictus, ii 212.
conybeari, ii. 269.
costatus, ii. 282.
cramptoni, ii. 161.
crassicostatus, ii. 220.
cycnodirus, ii. 246.
dædicomus, ii. 230.
degenhardti, ii. 224.
dewalquii, ii. 254.
dolichodirus, ii 255, iv 277.
eleutheraxon, ii. 259.
Plesiosaurus (*cont.*).
ellipsospondylus, ii. 172.
erraticus, 245.
etheridgei, ii 260.
eurymerus, ii. 205.
euryspondylus, ii. 246.
gardneri, ii. 212.
giganteus. ii. 123.
grandis, ii. 123.
gulo, ii. 247.
gurgitis, ii. 223.
hawkinsi, ii. 260.
helmerseni, ii. 212.
hexagonalis, ii. 234.
holmesi, ii. 247.
homalospondylus, ii. 253.
hoodi, ii. 245.
ichthyospondylus, ii 185.
indicus, ii 160.
infraplanus, ii. 234.
jenensis, ii. 299.
lævis, ii. 139
laticeps, ii. 269
latispinus, ii. 222, 269.
leedsi, ii 234
limnophilus, ii. 224
lockwoodi, ii 247
longirostris, ii 271
mackayi, ii. 247.
macrocephalus, ii 266.
macrodirus, ii 190
macrospondylus, ii. 247.
manseli, ii. 190.
megacephalus, ii 166.
megadirus, ii. 190.
neocomiensis, ii. 183, 223.
nordmanni, ii. 247.
ophiodirus. ii 246.
oxoniensis, ii. 209
pachyomus, ii. 185, 217.
pentagonus, ii 248.
philarchus, ii 154.
planus, ii. 217.
platydirus, ii. 246
plicatus, ii. 230, 234.
pœcilospondylus, ii. 246.
posidoniæ, ii. 284.
priscus, iv. 277
profundus, ii 299.
propinquus, ii 170.
recentior, ii. 139.
rostratus, ii. 271.
rugosus, ii. 249.
Plesiosaurus (*cont.*).
simpsoni, ii. 170.
smithi, ii. 215.
sp., ii. 264, 274.
speciosus, ii. 288.
sterrodirus, ii. 152.
subdepressus, ii. 230.
suevicus, ii. 248.
suprajurensis, ii. 248.
sutherlandi, ii. 247.
traversi, ii. 247.
trigonus, ii. 248.
triotarsostinus, ii. 260.
trochanterius, ii. 190.
truncatus, ii. 230.
validus, ii. 190.
winspitensis, ii. 227.
Plesiosuchus, i. 92.
manseli, i. 92.
Plestiodon, i. 281.
cadurcensis, i. 281
Pleurocœlus, i 134, iv. 237.
nanus, iv. 238.
valdensis, iv. 237.
Pleurodon, i. 53.
crocodiloides, i. 53.
Pleurodus, iv. 9.
bicolor, iv. 9.
Pleuroneura, iv. 210.
pellati, iv 210.
Pleurosaurus, i. 293.
goldfussi, i 293.
Pleurosternum, iii. 205.
bullocki, iii. 205.
concinnum, iii. 205
emarginatum, iii. 189, 205.
koeneni, iii 186.
latiscutatum, iii. 186.
ovatum, iii. 205.
oweni, iii. 207.
portlandicum, iii. 215.
sedgwicki, iii. 207.
typocardium, iii. 207.
vansittardi, iii. 207.
Plioplatecarpus, i. 261.
Pliosaurus, ii. 120, iv. 271.
æqualis, ii. 152.
archiachi, ii. 147.
brachydirus, ii. 123, iv. 271.
brachyspondylus, ii. 140.
evansi, ii. 128, iv 273.
ferox, ii. 145, iv. 271.
gamma, ii. 140.
giganteus, ii. 131.

Pliosaurus (*cont.*).
gigas, ii. 212.
grandis, ii. 131.
grossouvrei, ii. 130.
macromerus, ii. 131.
nitidus, ii. 140.
pachydirus, ii. 145.
planus, ii. 143.
portlandicus, ii. 127.
simplex, ii. 140.
suprajurensis, ii. 248.
trochanterius, ii. 190.
wosinskii, ii. 130.
Podocnemis, iii. 171.
bowerbanki, iii. 171.
delabechei, iii. 171.
indica, iii. 172.
lævis, iii. 171.
sextuberculata, iii. 172.
Pœkilopleuron, i. 159.
bucklandi, i. 159.
pusillus, i. 158.
Poikilopleuron = **Pœkilopleuron.**
Polacanthus, i. 189.
foxi, i. 189.
Polycotylus, ii. 180.
latipinnis, ii. 247.
suprajurensis, ii. 248.
tenuis, ii. 188.
Polyptychodon, ii. 172.
archiachi, ii. 147.
continuus, ii. 179.
ferox, ii. 145.
interruptus, ii. 173.
Pristichampsa, i. 74.
rollinati, i. 74.
Probatrachus, iv. 130.
Procolophon, iv. 13.
griersoni, iv. 13.
laticeps, iv. 15.
minor, iv. 14.
trigoniceps, iv. 13.
Proganochelys, iii. 222.
quenstedti, iii. 222.
Proiguana, i. 277.
europeana, i. 277.
Propappus, iv. 118.
omocratus, iv. 119.
Propleura, iii. 51.
Propseudopus, i. 278.
Proteocordylus, iv. 137.
diluvii, iv. 137.
Proteosaurus, ii. 12.
Proterosaurus, i. 301.
meyeri, i. 301.
speneri, i. 302.
Protochelys, iii. 220.
stricklandi, iii. 220.
Protosphargis, iii. 228.
veronensis, iii. 228.
Protostega, iii. 229.
anglica, iii. 229.
gigas, iii 229.
neptunia, iii. 229.
tuberosa, iii. 229.
Protriton, iv. 210.
petrolei, iv. 210
salamandroides, iv. 211.
Psammochelys, iii. 222.
keuperina, iii 222.
Psephoderma, iii. 231.
alpinum, iii. 231.
anglicum, iii 231.
Psephophorus, iii. 224.
scaldi, iii. 224.
sp., iii. 224.
Pseudopus, i 278.
moguntinus, i. 278.
Pseudotrionyx, iii. 145
delheidi, iii. 145.
Pterodactylus, i. 4.
aclandi, i. 36.
antiquus, i. 5.
bucklandi, i. 34.
clifti, i. 25.
compressirostris, i. 11.
conirostris i. 12.
crassirostris, i. 28.
curtus, i. 26.
cuvieri, i. 12.
daviesi, i. 23.
duncani, i. 34.
elegans, i. 7.
eurychirus, i. 10.
fittoni, i 15.
gemmingi, i. 31.
giganteus, i. 12.
grandis, i. 32.
kiddi, i. 34.
kochi, i. 6.
longicaudata, i 29.
longicollum, i. 9.
longirostris, i. 5.
macronyx, i 37.
macrurus, i. 26.
manseli, i. 40.
marderi, i. 37.
meyeri, i. 4.
muensteri, i. 30.
nobilis, i. 24.
pleydelli, i. 41.
pulchellus, i. 7.
rhamphastinus, i. 8.
scolopaciceps, i. 6.
Pterodactylus (*cont.*).
sinus, i. 16.
spectabilis, i. 7.
suevicus, i. 9.
validus, i. 26.
woodwardi, i. 16.
wurtembergicus, i. 10.
Ptyas, i. 250
mucosus, i. 250.
Ptychogaster, iii. 95.
cayluxensis, iii 98.
emydoides, iii. 95.
pomeli, iii. 97.
Ptychognathus, iv. 33.
alfredi, iv. 35.
boopis, iv 37.
declivis, iv 34
depressus, iv. 35.
latirostris, iv. 35.
orientalis, iv 41.
pusillus, iv. 40
verticalis, iv 37.
Ptychosiagum, iv. 33.
declive, iv. 34.
latirostris, iv. 35.
microtrema, iv. 36.
orientale, iv. 41.
Puppigerus, iii. 51.
breviceps, iii. 41, 64.
convexus, iii. 48.
crassicostatus, iii. 60.
latiscutatus, iii 60.
longiceps, iii. 57.
subcarinatus, iii. 47.
subcristatus, iii. 47.
Pylmophis = **Pilemophis.**
Python, i. 253.
cadurcensis, i 254.
molurus, i. 253.

Rana, iv. 122.
cornuta, iv. 127.
diluviana, iv. 130.
gigantea, iv. 129.
labyrinthica, iv. 124.
meriani, iv. 122.
nœggerathi, iv. 123.
ocellata, iv. 125.
pentadactyla, iv. 124.
pusilla, iv. 122.
rugosa, iv. 129.
Regnosaurus, i. 184.
northamptoni, i. 184.
Rhacheosaurus, i. 95.
Rhachitomus, iv. 191.
valens, iv. 191.

Rhamphocephalus, i. 33.
bucklandi, i. 34.
depressirostris, i. 36.
Rhamphorhynchus, i. 29, iv. 226.
bucklandi, i. 34.
crassirostris, i. 28.
depressirostris, i. 36.
gemmingi, i. 31. iv. 226.
grandis, i. 32.
jessoni, iv. 226.
longicauda, i. 29.
longicaudatus, i. 29.
longimanus, i. 31.
meyeri, i. 31.
muensteri, i. 30.
phylluris, i. 30.
Rhamphosaurus, i. 264.
Rhamphosuchus, i. 71.
crassidens, i 71.
Rhinochelys, iii. 175.
brachyrhina, iii. 179.
cantabrigiensis, iii. 176.
elegans, iii. 178.
jessoni, iii. 180.
macrorhina, iii. 177.
pulchriceps, iii. 176.
sp., iii. 180.
Rhinoclemmys, iii. 99.
Rhinosaurus, i. 264.
Rhomaleosaurus, ii. 158.
cramptoni, ii. 161.
Rhynchosaurus, i. 296.
articeps, i. 296.
Rhynchosuchus, i. 62.
Rhytidosteus, iv. 194.
capensis, iv. 194.
Ricnodon, iv. 203.
trachylepis, iv. 203.

Salamandra, iv. 137.
gigantea, iv. 137.
scheuchzeri, iv. 137.
Salamandrella, iv. 210.
petrolei, iv. 210.
Salamandroides, iv. 142.
giganteus, iv. 142.
Saniva, i. 278.
Sapheosaurus, i. 292.
laticeps, i. 292.
Sapheosaurus (*cont.*).
thiollierei, i. 292.
Sauranodon, ii. 60.
Saurichnites, iv. 219.
perlatus, iv. 219.
rittlerianus, iv. 219.
Saurillus, i. 289.
obtusus, i. 289.
Saurosternum, i. 295.
baini, i. 295.
Scaloposaurus, iv. 70.
constrictus, iv. 71.
Scaphognathus, i. 27, iv. 226.
crassirostris, i. 28.
purdoni, iv. 226.
Scelidosaurus, i. 181.
harrisoni, i. 181.
Scincosaurus, iv. 198.
crassus, iv 198.
Scincus, i. 288.
croizeti, i. 288.
Sclerocephalus, iv. 185.
hæuseri, iv. 185.
Seeleya, iv. 203.
pusilla, iv. 203.
Simosaurus, ii. 299.
guillardoti, ii. 299.
guillielmi, ii 299.
mougeoti, ii. 299.
pusillus, ii 285.
Smilerpetum, iv. 224.
aciedentatum, iv. 224.
Sparagmites, iv. 189.
lacertinus, iv. 189.
Sparodus, iv. 212.
crassidens, iv. 213.
validus, iv. 212.
Sphenosaurus, iv. 188.
sternbergi, iv. 188.
Sphenospondylus, i. 238.
gracilis, i. 238.
Spondylosaurus. ii. 120.
fahrenkohli, ii. 247.
fearsi, ii. 130.
Stagonolepis, i. 128.
robertsoni, i. 129.
Staurotypus, iii. 143.
Stegochelys, iii. 203, 233.
planiceps, iii. 233.
Stegosaurus, i. 177; iv. 251.
armatus, iv. 252.
hastiger, iv. 252.
stenops, i. 176.
Stelliosaurus, iv. 201.
longicostatus, iv. 201.
Steneosaurus, i. 108, iv. 234.
bollensis, i. 109.
brevidens, i. 115, iv. 234.
brevior, i. 111.
bronni, i. 106.
chapmani, i. 110.
dasycephalus, iv. 234.
edwardsi, i 117.
elegans, i. 98.
heberti, i. 109.
larteti, i. 114.
latifrons, i. 113.
longirostris, i. 116.
manseli, i 92.
megarhinus, i. 117.
megistorhynchus, i. 116.
minimus, i. 112.
rostro-major, i. 116.
sp., i. 117.
stephani, i. 114.
Stereorhachis, iv. 105.
dominans, iv. 105.
Stereosternum, ii. 300.
tumidum, ii. 302.
Streptospondylus, i. 201.
major, i. 201.
Stylemys, iii. 93, 196.
culbertsoni, iii. 94.
hannoverana, iii. 198.
nebrascensis, iii. 94.
sp., iii. 94.
Suchodus, iv. 231.
durobrivensis, iv. 231.
Suchosaurus, i. 89.
cultridens, i. 90.
lævidens, i. 90.
Syngonosaurus, iv. 254.
macrocercus, iv. 254.

Taniwhasaurus, i. 269.
oweni, i. 270.
Tanystrophæus, i. 155.
Tapinocephalus, iv. 82.
atherstonei, iv. 83.
Taphrosphys, iii. 174.
sp , iii. 174.
Teleidosaurus, i. 102.
calvadosi, i. 103.
joberti, i. 103.
Teleosaurus, i. 118.
bollensis, i. 109.
brevidens, i. 115.

Teleosaurus (*cont.*).
brevior, i. 111.
cadomensis, i. 118
chapmani, i. 110.
edwardsi, i. 117.
geoffroyi, i. 120.
gracilis, i. 101.
larteti, i. 114.
latifrons, i. 113.
megarhinus, i. 117.
megistorhynchus, i. 116.
minimus, i. 112.
mosæ, i. 104.
mosellana, i. 106.
oplites, i. 107.
priscus, i. 101.
subulidens, i. 121.
superciliosus, i. 96.
temporalis, i. 106.
Telerpeton, i. 295.
elginense, i. 295.
Temnodontosaurus, iv. 271.
platyodon, iv. 271.
trigonodon, iv. 271.
Teratosaurus, i. 171.
suevicus, i. 172.
Testudo, iii. 172.
atlas, iii. 74.
canetotiana, iii. 94.
cautleyi, iii. 96.
duncani, iv. 214.
emys, iii. 72.
escheri, iii. 89.
eurysternum, iii. 90.
frizaciana, iii. 94.
gigantea, iii. 90.
hemisphærica, iii. 94.
larteti, iii. 90.
lata, iii. 94.
leithi, iii. 170.
longicollis, iii. 168.
nebrascensis, iii. 94.
orbicularis, iii. 103.
oweni, iii. 94.
plana, iii. 60.
punjabiensis, iii. 87.
pygmæa, iii. 94.
robusta, iii. 75.
sloanei, iii. 89.
sp., 83, 84, 91.
spratti, iii. 74.
stricklandi, iii. 74.
Thalassemys, iii. 148.
hugii, iii. 148.
reutimeyeri, iii. 148.
Thalassochelys, iii. 49.
crassicostatus, iii. 60.
eocænica, iii. 50.
Thalassochelys (*cont.*).
planimentum, iii 64.
sp., iii. 50.
Thaumatosaurus, ii. 159.
arcuatus, ii. 163.
carinatus, ii. 168
cramptoni, ii. 161.
indicus, ii. 160.
megacephalus, ii. 166.
mosquensis, ii. 145.
oölithicus, ii. 160.
philarchus, ii. 154.
propinquus, ii. 170.
zetlandicus, ii. 167.
Thecodontosaurus, i. 174, iv. 246.
antiquus, i. 175.
platyodon, i. 174, iv. 246.
Thecospondylus, i. 152.
daviesi, i. 156.
horneri, i. 152.
Theriognathus, iv. 64.
microps, iv. 65.
Theriosuchus, i. 77.
pusillus, i. 78.
Thoracosaurus, i. 70.
grandis, i. 70.
macrorhynchus, i. 64; iv. 228.
neocesariensis, i. 70.
Tigrisuchus, iv. 80.
simus, iv. 80.
Titanosaurus, i. 134, 144.
indicus, i. 134.
sp. *a*, i. 135.
sp. *b*, i. 136.
Titanosuchus, iv. 89.
ferox, iv. 90.
Tomistoma, i 62, iv. 228.
champsoides, i. 64.
eggenburgense, i. 63.
macrorhynchus, i. 64, iv. 228.
Toxochelys, iii. 129.
Trachodon, i. 243.
cantabrigiensis, i. 244.
foulki, i. 244.
mirabilis, i. 244.
Trachyaspis, iii. 129.
ægyptiaca, iii. 131.
hantoniensis, iii. 131.
lardyi, iii. 131.
miocœna, iii. 130.
sanctæ-crucis, iii. 130.
Trachydermochelys, iii. 182.
phlyctænus, iii 182.
Trematosaurus, iv. 154.
brauni, iv. 155.
Tretosternum, iii. 137.
bakewelli, iii 138.
punctatum, iii. 141.
Tridentipes, iv. 222.
elegantior, iv. 222.
Trimerorhachis, iv. 189.
insignis, iv. 190.
Trinacromerum, ii. 180
bentonianum, ii. 247.
Trionyx, iii. 7
aquitanicus, iii. 10.
bakewelli, iii. 138.
barbaræ, iii. 13.
bowerbanki, iii. 19.
circumsulcatus, iii 22.
gangeticus, iii. 8.
gergensi, iii. 11.
henrici, iii 13.
hurum, iii 9.
incrassatus, iii 17.
marginatus, iii 13.
parisiensis, iii. 12.
phayrei, iii. 10.
planus, iii. 18.
rivosus, iii. 19.
vittatus, iii. 21.
Triton, iv. 136.
noachicus, iv. 136.
Tropidemys, iii. 155.
expansa, iii. 156.
gibba, iii. 156.
langi, iii. 156.
valanginensis, iii. 156.
Tropidonotus, i. 251.
atavus, i. 251.
Tropidostoma, iv. 36.
dunni, iv. 36.
Tupinambis, i 286.
bengalensis, i. 282.
nigropunctatus, i. 287.
teguixin, i. 286.
Tylosaurus, i. 264.

Udenodon, iv. 28.
baini, iv. 29.
brevirostris, iv. 29.
greyi, iv. 32.
magnus, iv. 31.
megalops, iv. 30.
prognathus, iv. 31.
raniceps, iv. 29.
rugosus, iv. 29.

Udenodon (*cont.*).
sp., iv. 321.
strigiceps, iv. 30.
Urocordylus, iv. 196.
reticulatus, iv. 196.
scalaris, iv. 197.
wandesfordi, iv. 196.
Uronautes, ii. 182.
cetiformis, ii. 247.

Varanus, i. 281.
bengalensis, i. 282.
dracæna, i. 282.
giganteus, i. 282.
margariticeps, i. 279.
priscus, i. 284.
sivalensis, i. 283.
sp., i. 282.

Weissia, iv. 184.

Zanclodon, i. 171.
lævis, i. 172.
sp. *a*, i. 172.
sp. *b*, i. 173.
suevicus, i. 172.

END OF PART IV.

PRINTED BY TAYLOR AND FRANCIS, RED LION COURT, FLEET STREET.

SELECTED LIST

OF THE

NATURAL HISTORY PUBLICATIONS

OF THE TRUSTEES OF THE

BRITISH MUSEUM.

REPORT on the Zoological Collections made in the Indo-Pacific Ocean during the voyage of H.M.S. 'Alert,' 1881–2. Edited by Dr. Günther. 54 Plates. 1884, 8vo. £1 10*s.*

MAMMALS.

Catalogue of Monkeys, Lemurs, and Fruit-eating Bats. 21 Woodcuts. By Dr. J. E. Gray. 1870, 8vo. 4*s.*

Catalogue of Carnivorous, Pachydermatous, and Edentate Mammalia. By Dr. J. E. Gray. 47 Woodcuts. 1869, 8vo. 6*s.* 6*d.*

Hand-List of Seals, Morses, Sea-Lions, and Sea-Bears. 30 Plates of Skulls. By Dr. J. E. Gray. 1874, 8vo. 12*s.* 6*d.*

Catalogue of Seals and Whales. By Dr. J. E. Gray. Second edition. 101 Woodcuts. 1866, 8vo. 8*s.*
Supplement. 11 Woodcuts. 1871, 8vo. 2*s.* 6*d.*

List of the Specimens of Cetacea in the Zoological Department. By Prof. Flower. 1885, 8vo. 1*s.* 6*d.*

Catalogue of Ruminant Mammalia (*Pecora*, Linnæus). By Dr. J. E. Gray. 4 Plates. 1872, 8vo. 3*s.* 6*d.*

Catalogue of the Marsupialia and Monotremata. By O. Thomas. 28 Plates (4 col.). 1888, 8vo. £1 8*s.*

BIRDS.

Catalogue of the Birds. By R. Bowdler Sharpe, H. Seebohm, H. Gadow, and P. L. Sclater. Vols. III.–XV., 1877–90, 8vo. Woodcuts and coloured Plates. 14*s.* to 28*s.* a volume. (Vols. I. and II. *out of print.*)

REPTILES AND BATRACHIANS.

Catalogue of the Chelonians, Rhynchocephalians, and Crocodiles. New edition. By G. A. Boulenger. 73 Woodcuts and 6 Plates. 1889, 8vo. 15*s.*

Gigantic Land Tortoises (living and extinct). By Dr. Günther. 54 Plates. 1877, 4to. £1 10*s.*

Catalogue of the Lizards. Second edition. By G. A. Boulenger. Plates. 3 vols. 1885–87, 8vo. Vols. I., II., 20*s.* each; III., 26*s.*

Catalogue of Colubrine Snakes. By Dr. Günther. 1858, 12mo. 4*s.*

Catalogue of the Batrachia Salientia. By Dr. Gunther. 12 Plates. 1858, 8vo. 6*s.*

Catalogue of the Batrachia Salientia s. Ecaudata. Second edition. By G. A. Boulenger. Woodcuts and 30 Plates. 1882, 8vo. £1 10*s.*

Catalogue of the Batrachia Gradientia s. Caudata, and Batrachia Apoda. Second edition. By G. A. Boulenger. 9 Plates. 1882, 8vo. 9*s.*

FISHES.

Catalogue of the Fishes. By Dr. Günther. Vols. II.–VIII. 1860–70, 8vo. Woodcuts. 7*s.* to 10*s.* 6*d.* a volume. (Vol. I. *out of print.*)

Catalogue of Lophobranchiate Fish. By Dr. J. J. Kaup. 4 Plates. 1856. 12mo. 2*s.*

MOLLUSCA.

Guide to the Systematic Distribution of Mollusca. Part I. By Dr. J. E. Gray. Woodcuts. 1857, 8vo. 5*s.*

Catalogue of Pulmonata. Part I. By Dr. L. Pfeiffer. Woodcuts. 1855, 12mo. 2*s.* 6*d.*

Catalogue of the Auriculidæ, Proserpinidæ, and Truncatellidæ. By Dr. L. Pfeiffer. Woodcuts. 1857, 12mo. 1*s.* 9*d.*

Catalogue of the Conchifera, or Bivalve Shells. By Mons. Deshayes. 2 Parts. 1853–54, 12mo. Part I., 3*s.*; II., 6*d.*

BRACHIOPODA.

Catalogue of Brachiopoda Ancylopoda, or Lamp Shells. By Dr. J. E. Gray. Woodcuts. 1853, 12mo. 3*s.*

POLYZOA.

Catalogue of Marine Polyzoa. Part III. Cyclostomata. By G. Busk. 38 Plates. 1875, 8vo. 5*s.*

CRUSTACEA.

Catalogue of Amphipodous Crustacea. By C. Spence Bate. 58 Plates. 1862, 8vo. £1 5*s.*

INSECTS.

Coleopterous Insects.

Catalogue of the Coleopterous Insects of Madeira. By T. V. Wollaston. 1 Plate. 1857, 8vo. 3*s.*

Catalogue of the Coleopterous Insects of the Canaries. By T. V. Wollaston. 1864, 8vo. 10*s.* 6*d.*

Illustrations of Typical Specimens of Coleoptera. Part I. Lycidæ. By C. O. Waterhouse. 18 Plates, coloured. 1879, 8vo. 16*s.*

Catalogue of Halticidæ. By Rev. H. Clark. 9 Plates. 1860, 8vo. 7*s.*

Catalogue of Hispidæ. By J. S. Baly. Part I. 9 Plates. 1858, 8vo. 6*s.*

Hymenopterous Insects.

Catalogue of Hymenopterous Insects. By F. Smith. Parts I.–VII. 1853–59, 12mo. 2*s.* to 6*s.* a Part.

Descriptions of New Species of Hymenoptera. By F. Smith. 1879, 8vo. 10*s.*

List of Hymenoptera, with descriptions and figures of the Typical Specimens. By W. F. Kirby. Vol. I. Tenthredinidæ and Siricidæ. 16 coloured Plates. 1882, 8vo. £1 18*s.*

Lepidopterous Insects.

Illustrations of Typical Specimens of Lepidoptera Heterocera. By A. G. Butler. Parts I.–VII. Coloured Plates. 1877–89, 4to. £2 to £2 10*s.* a Part. (Parts II. and IV. *out of print.*)

Catalogue of Diurnal Lepidoptera of the family Satyridæ. By A. G. Butler. 5 Plates. 1868, 8vo. 5*s.* 6*d.*

Catalogue of Diurnal Lepidoptera described by Fabricius. By A. G. Butler. 3 Plates. 1869, 8vo. 7*s.* 6*d.*

Specimen of a Catalogue of Lycænidæ. By W. C. Hewitson. 8 col. Plates. 1862, 4to. £1 1*s.*

Neuropterous Insects.

Catalogue of the Specimens of Neuropterous Insects. By Dr. H. Hagen. Part I. Termitina. 1858, 12mo. 6*d.*

Orthopterous Insects.

Catalogue of Orthopterous Insects. Part I. Phasmidæ. By J. O. Westwood. 48 Plates. 1859, 4to. £3.

Catalogue of the Blattariæ. By F. Walker. 1868, 8vo. 5*s.* 6*d.*

Catalogue of Dermaptera Saltatoria and Supplement to the Blattariæ. Gryllidæ, Blattariæ, Locustidæ. By F. Walker. 1869, 8vo. 5*s.*

———. Part II. Locustidæ (*continued*). 1869, 8vo. 4*s.* 6*d.*

———. Part III. Locustidæ (*continued*), Acrididæ. 1870, 8vo. 4*s.*

———. Part IV. Acrididæ (*continued*). 1870, 8vo. 6*s.*

———. Part V. Tettigidæ, and Supplements. 1870, 8vo. 6*s.*

Hemipterous Insects.

Catalogue of Heteropterous Hemiptera. By F. Walker. Parts I.-VIII. 1867-73, 8vo. 4*s*. to 6*s*. 6*d*. a Part.

VERMES.

Catalogue of Entozoa, or Intestinal Worms. By Dr. Baird. 2 Plates. 1853, 12mo. 2*s*.

ANTHOZOA.

Catalogue of Sea-pens or Pennatulariidæ. By Dr. J. E. Gray. Woodcuts. 1870, 8vo. 1*s*. 6*d*.

Catalogue of Lithophytes or Stony Corals. By Dr. J. E. Gray. Woodcuts. 1870, 8vo. 3*s*.

BRITISH ANIMALS.

Catalogue of British Fossorial Hymenoptera, Formicidæ, and Vespidæ. By F. Smith. 6 Plates. 1858, 12mo. 6*s*.

Catalogue of British Non-parasitical Worms. By Dr. G. Johnston. Woodcuts and 24 Plates. 1865, 8vo. 7*s*.

List of British Animals in the Collection of the British Museum :—

Part V. Lepidoptera. By J. F. Stephens. 2nd edition. 1856, 12mo. 1*s*. 9*d*.

Part XIII. Nomenclature of Hymenoptera. By F. Smith. 1853, 12mo. 1*s*. 4*d*.

Part XIV. Nomenclature of Neuroptera. By Adam White. 1853, 12mo. 6*d*.

Part XV. Nomenclature of Diptera, I. By Adam White. 1853, 12mo. 1*s*.

PLANTS.

List of British Diatomaceæ in the Collection of the British Museum. By Rev. W. Smith. 1859, 12mo. 1*s*.

FOSSILS.

Catalogue of the Fossil Mammalia. By R. Lydekker. Parts I.–V. Woodcuts. 1885–87, 8vo. *4s.* to *6s.* a volume.

Descriptive and Illustrated Catalogue of the Fossil Reptilia of South Africa. By Sir R. Owen. 70 Plates. 1876, 4to. £3 *3s.*

Catalogue of the Fossil Reptilia and Amphibia. By R. Lydekker. Parts I.–IV. Woodcuts. 1888–90, 8vo. *7s. 6d.* a volume.

Catalogue of the Fossil Fishes. Part I. By A. Smith Woodward. 13 Woodcuts and 17 Plates. 1889, 8vo. *21s.*

Catalogue of the Fossil Cephalopoda. Part I. By A. H. Foord. Woodcuts. 1888, 8vo. 10*s.* 6*d.*

A Catalogue of British Fossil Crustacea, with their Synonyms and the Range in Time of each Genus and Order. By Dr. H. Woodward. 1877, 8vo. *5s.*

Catalogue of the Blastoidea in the Geological Department. With an account of the morphology and systematic position of the group, and a revision of the genera and species. By R. Etheridge, jun., and P. H. Carpenter. 20 Plates, &c. 1886, 4to. 25*s.*

Catalogue of the Fossil Sponges in the Geological Department. With descriptions of new and little-known species. By Dr. G. J. Hinde. 38 Plates. 1883, 4to. £1 10*s.*

Catalogue of the Fossil Foraminifera. By T. Rupert Jones. 1882, 8vo. 5*s.*

Catalogue of the Palæozoic Plants in the Department of Geology and Palæontology. By R. Kidston. 1886, 8vo. 5*s.*

The above Publications can be purchased of *Messrs.* LONGMANS & Co., 39 *Paternoster Row ; Mr.* QUARITCH, 15 *Piccadilly ; Messrs.* ASHER & Co., 13 *Bedford Street, Covent Garden ; Messrs.* KEGAN PAUL, TRENCH, TRUBNER, & Co., 57 *Ludgate Hill*; and at the Natural History Museum, *Cromwell Road, London, S.W.* A more detailed list can be obtained on application to the Director of the Museum.

GUIDE-BOOKS.

A General Guide to the British Museum (Natural History). With 2 Plans and 2 Views. 8vo. *3d.*

Zoological Department.

Guide to the Galleries of Mammalia (Mammalian, Osteological, Cetacean) in the Department of Zoology. 57 Woodcuts and 2 Plans. Index. 8vo. *4d.*

Guide to the Gould Collection of Humming Birds. With Map showing the distribution of Humming Birds. 8vo. *2d.*

Guide to the Gallery of Reptilia in the Department of Zoology. 22 Woodcuts and 1 Plan. 8vo. *2d.*

Guide to the Galleries of Reptiles and Fishes in the Department of Zoology. 101 Woodcuts and 1 Plan. 8vo. *6d.*

Guide to the Shell and Star-fish Galleries in the Department of Zoology (Mollusca, Echinodermata, Vermes). 51 Woodcuts and 1 Plan. 8vo. *4d.*

[*Guides to other sections are in preparation.*]

Geological Department.

A Guide to the Exhibition Galleries of the Department of Geology and Palæontology :—

Part I. Fossil Mammals and Birds. 119 Woodcuts, table of Stratified Rocks, plan of Geological Galleries, and Index. 8vo. *6d.*

Part II. Fossil Reptiles, Fishes, and Invertebrates. 94 Woodcuts, table of Stratified Rocks, plan of Geological Galleries, and Index. 8vo. *6d.*

Guide to the Collection of Fossil Fishes in the Department of Geology and Palæontology. 81 Woodcuts. 8vo. *4d.*

Mineralogical Department.

A Guide to the Mineral Gallery. 8vo. 1*d.*

An Introduction to the Study of Minerals, with a Guide to the Mineral Gallery. Diagrams. Plan of the Mineral Gallery. Index. 8vo. 6*d.*

The Student's Index to the Collection of Minerals. 8vo. 2*d.*

An Introduction to the Study of Meteorites, with a List of the Meteorites represented in the Collection. Plan of the Mineral Gallery, and Index to the Meteorites represented in the Collection. 8vo. 3*d.*

The Guide-Books can only be obtained at the Museum.

PRINTED BY TAYLOR AND FRANCIS, RED LION COURT, FLEET STREET.

Printed in Dunstable, United Kingdom

91219491R00190